CATO

ON AGRICULTURE

VARRO

ON AGRICULTURE

LCL 283

MARCUS PORCIUS CATO

ON AGRICULTURE

MARCUS TERENTIUS VARRO

ON AGRICULTURE

WITH AN ENGLISH TRANSLATION BY

WILLIAM DAVIS HOOPER

REVISED BY HARRISON BOYD ASH

HARVARD UNIVERSITY PRESS
CAMBRIDGE, MASSACHUSETTS
LONDON, ENGLAND

First published 1934
Revised and Reprinted 1935
Reprinted 1954, 1960, 1967, 1979, 1993, 1999

ISBN 0-674-99313-6

Printed in Great Britain by St Edmundsbury Press Ltd,
Bury St Edmunds, Suffolk, on acid-free paper.
Bound by Hunter & Foulis Ltd, Edinburgh, Scotland.

CONTENTS

PREFACE

The translation of Cato and Varro *On Agriculture* was begun several years ago by Professor W. D. Hooper. The major part of the volume had reached the stage of page proof when reasons of health and pressure of other duties forced upon the translator the abandonment of the project. Desirous of carrying the work to completion, the editors were moved to call in further assistance in revising the printed proofs and seeing the work finally through the press. This assignment, a task made more difficult by the already fixed paging, has fallen to the writer of this Preface.

The reviser has ventured to alter the translation at a number of points, always mindful of the restrictions of pagination, and has inserted where space allowed a number of critical notes naming sources of accepted readings in the text. He has also emended or rewritten a large part of the commentary, and has added a Glossary to provide definitions of a few terms of weight, measure, etc., for which no room could be found in footnotes. Upon him, too, rests responsibility for the preparation of Introduction, Bibliography, and Index.

Grateful acknowledgment is made to Mr. R. J. H. De Loach, former Director of Agricultural Research of the Armour Company, Chicago, whose knowledge

of ancient farm machinery has benefited the translation through his collaboration with Professor Hooper in that technical aspect of the work. The writer is indebted to his colleague, Professor John C. Rolfe, and to Professor Edward Capps of Princeton, American editor of the series, for encouragement and valued suggestions freely given in the work of revision.

HARRISON BOYD ASH.

University of Pennsylvania,
July 1, 1933.

INTRODUCTION

Life and Works of Cato [1]

Marcus Porcius Cato (234–149 B.C.), known also as the Orator, the Censor, Cato Major, or the Elder, to distinguish him from his great-grandson Marcus Porcius Cato Uticensis, was born of an old plebeian family at Tusculum, an ancient town of Latium, within ten miles of Rome. His youth was spent on his father's farm near Reate, in the Sabine country. Here he acquired early in life those qualities of simplicity, frugality, strict honesty, austerity, and patriotism for which he was regarded by later generations as the embodiment of the old Roman virtues. His native ability and shrewdness, says Plutarch, gave him the surname Cato (" the shrewd ") replacing the earlier name of Priscus. Love of the soil, implanted in him in his youth, remained throughout his life; though not content with the agricultural limitations of a Sabine farmer he became in later years the owner of great plantations worked by slave labour.

[1] Ancient accounts of the life and character of Cato are found in Cicero's *Cato Maior* and in the " Lives " of Cornelius Nepos, Plutarch, and Aurelius Victor. See also Mommsen, *History of Rome*, Vol. III, Chaps. 11 and 14; Duff, *A Literary History of Rome* (to the close of the Golden Age), 250–52, 255–59, 262–64; Teuffel and Schwabe, *History of Roman Literature*, §§ 118–22.

Entering upon a military career at the age of seventeen, Cato served with distinction in the Second Punic War, and devoted the following twenty-six years of his life to military affairs. He accompanied Tiberius Sempronius as his lieutenant in his expedition into Thrace. He went with Manius Achilius Glabrio into Greece against Antiochus the Great in 191, in the capacity of tribune. As commander of the Roman army in Hither Spain during his consulship he won many victories and was a successful ruler, but noted for his cruelty to his defeated enemies. Always active in the affairs of the state, he showed himself an obstinate and vigorous opponent of the nobility, of luxurious living, and of the invasion into Italy of Greek culture; though it is said that he himself was taught Greek late in life by the poet Ennius.[1] Political offices came to him in due succession, a quaestorship in Sicily and Africa in 204, an aedileship in 199, a praetorship in Sardinia in 198, the consulship in 195 with Hither Spain as his province, and the censorship in 184. His innumerable speeches, political and judicial, delivered before the senate or popular assembly, were marked by eloquence, earnestness, and pungent wit, not without vainglory and narrowness of view. Always the champion of the common people, he stood out as the relentless foe of aristocratic factions. The vigour and severity with which he applied himself to the duties of the censorship, with his strict revision of the senatorial lists, gained for him the surname *Censorius*. Sent into Carthage on an official mission in 175, he conceived such hatred for the Cartha-

[1] Cf. Aurelius Victor, *de Viris Illustribus*, 47.

ginians, upon noting their obvious recovery from the effects of the Punic wars, that from that time on he closed every speech in the senate with the words *delenda est Carthago*, regardless of the occasion. The words of Cato became the policy of the senate and in 149, the year of Cato's death, the Third Punic War began.

Quintilian speaks [1] of the great versatility of Cato as general, philosopher, orator, historian, and outstanding expert in jurisprudence and agriculture. Cicero bears witness [2] to the breadth of his learning and the variety of his writings. He is praised by Pliny [3] as "the master of all good arts"; and is said by Columella [4] to have been the first to teach Rome to speak Latin. In the field of literary composition, for which he affected contempt, Cato was prolific. He was the first Roman to write out and publish on a large scale his own speeches, of which Cicero professed [5] to have known and read more than one hundred and fifty. He was the first Roman to leave to us prose writings of any consequence, and is regarded as the father of Latin prose [6] as Ennius is the father of Latin poetry. He rebelled against the prevailing annalistic treatment of history as written to flatter the vanity of the nobles, and omitted the names of all such from his account of the Second Punic War, preferring rather to sing the praises of a certain Surus, bravest elephant of the

[1] Quint., *Inst. Orat.*, XII, 11, 23.
[2] Cic., *de Orat.*, III, 135; *Brut.*, 61, 69, 294.
[3] Pliny, *N.H.*, XXV, 4.
[4] Colum., *de Re Rust.*, I, 1, 12.
[5] Cic., *Brut.*, 67.
[6] *Id.*, 16, 61.

Carthaginian army.[1] He paved the way in the field of encyclopaedic learning so dear to the Romans, and was the first of a group of Roman writers on husbandry. Yet of the great bulk of his writings comparatively little has been preserved to us in anything like completeness.[2] The speeches were familiar to Servius as late as the fourth century of our era,[3] but are known to us only through scattered references and occasional quotations. The same fate has overtaken the seven books of *Origines*, a work begun in his old age,[4] dealing with the ethnology, antiquities, and history of Italy from the founding of Rome down to the year 149, and deriving its title from its attempt to trace the origins of various Italian tribes. His greatest didactic work, an encyclopaedic handbook for his son, containing precepts on morals, sanitation, oratory, military science, agriculture, and other subjects, has perished, as has also a collection of aphorisms and witty sayings of others; what now passes under the name of *Catonis Disticha* is a collection of moral maxims in verse, perhaps spurious imitations of Cato, which circulated in the latter period of the Empire. The only work now surviving to represent Cato in its complete form is a miscellaneous collection of agricultural precepts which appear in the manuscripts under the name *De Agri Cultura*, and in the earlier printed editions as *De Re Rustica*.

[1] Pliny, *N.H.*, VIII, 11.

[2] The fragments in general are found in H. Jordan, *M. Catonis praeter librum de re rustica quae extant*, Leipzig, 1860.

[3] See Serv. *ad Aen.*, VII, 259, XI, 301.

[4] Nepos, *Cato*, 3, 3. Fragments of the *Origines* are included in H. Peter, *Historicorum Romanorum Fragmenta*, 1883.

INTRODUCTION

The *De Agri Cultura* constitutes our earliest extant specimen of connected, if often loosely connected, Latin prose. The work, with its notable lack of systematic arrangement, can hardly pass as literature. It resembles rather a farmer's notebook in which the author had jotted down in random fashion[1] all sorts of directions for the care of the farm, for his own private use or for the benefit of his friends and neighbours. Based on the writer's own first-hand experience and probably intended as a practical manual on the subject of husbandry, it contains all sorts of authoritative directions for the farm overseer. The work in its present form has lost in great measure its archaic diction, but the spirit of the stern old Roman remains. In its haphazard arrangement and abrupt Catonian style, a style best characterized by Aulus Gellius[2] as perhaps open to improvement in matters of clearness and fullness of expression, yet forceful and vigorous, it falls far short of the more finished work of Varro and the fluent, methodical treatise of Columella. So, too, it proved inadequate for the husbandmen of later generations; but Cato blazed the trail for his more eloquent successors in the field, and is often quoted by them as an authority. The work, despite its confused text, its difficulty of interpretation, and its problems still unsolved, is readable. Its greatest charm to-day lies in its severe simplicity, and its chief value in the picture which may be drawn from it of old Roman life in the best days of the Republic.

[1] Hörle (in his *Catos Hausbücher*, Paderborn, 1929) contends that the arrangement is not to be charged to Cato himself, but to some compiler of agricultural precepts contained in Cato's various works.

[2] Aul. Gell., *N.A.*, VI, 3, 17 f., 52 f.

Life and Works of Varro[1]

Marcus Terentius Varro (116–27 B.C.), sometimes called Varro Reatinus to distinguish him from his namesake Varro Atacinus, was born in the Sabine town of Reate, probably of a family of equestrian rank. Devoting himself early in life to the study of literature and antiquities, he came under the instruction of the learned antiquarian and philologist, Lucius Aelius Stilo, at Rome, and later studied at Athens under the Academic philosopher, Antiochus of Ascalon.

In public life Varro was attracted to the Pompeian party, and under the political banner of Pompey he held the offices of tribune, curule aedile, and praetor. He was commissioned by the First Triumvirate, in 59, as a member of the Board of Twenty delegated for the assignment of land grants to veterans in Campania. He served as pro-quaestor of Pompey in Spain, probably against Sertorius in 76. As lieutenant to Pompey in 67 he took part in the war which cleared the Mediterranean of pirates, winning the *corona navalis* for his personal exploits. He seems also to have served in the same capacity in the last war with Mithridates. In 49 he again commanded Pompeian forces in Spain, though soon forced to surrender to the Caesarians after the desertion of a part of his forces. Pardoned by

[1] Many facts of his life and works are given by Aulus Gellius in his *Attic Nights*. See also K. L. Roth, *Über das Leben des M. Terentius Varro*, Basel, 1857; G. Boissier, *La vie et les ouvrages de Varro*, Paris, 1861; Duff, 330–40; Teuffel and Schwabe, §§ 164–69.

Caesar, he rejoined Pompey in Greece, but played no active part in the remainder of the war. He returned to Rome after the battle of Pharsalus, to receive again the ready forgiveness of Caesar. Shortly thereafter he was commissioned by Caesar to superintend the collection and arrangement of a great library of Greek and Latin literature destined for public use.[1] In the same year he recovered, by Caesar's orders, some of the property which had been seized by Antony after the defeat of the Pompeians, including, it is said, his estate at Casinum. As a mark of gratitude to Caesar he dedicated to him the second part of his *Antiquities*. With the formation of the Second Triumvirate in 43, Varro again became the victim of Antony. His name appeared on the lists of the proscribed, and much of his property, including his library, was plundered. Barely escaping with his life, through the intervention of Octavianus, he spent the closing years of his long and active career in seclusion, devoting himself to study and writing.

In the field of learning Varro laboured with a diligence equalling, if not surpassing, that of the elder Pliny. The extent and variety of his erudition in nearly all branches of learning won for him the title *vir Romanorum eruditissimus*.[2] Retaining his keen mental vigour up to the time of his death, in his ninetieth year, he produced an enormous mass of literature. He himself claimed[3] at the beginning of his seventy-eighth year to have written "seventy times seven" books, many of which had been lost

[1] Suet., *Iul.*, 44.
[2] Quint., *Inst. Orat.*, X, 1, 95.
[3] Apud Aul. Gell., *N.A.*, III, 10, 17.

in the plundering of his library at the time of his proscription. From this statement, taken with Jerome's catalogue and other evidence, Ritschl estimates the number of his separate literary works to have been 74, and the grand total of books as close to 620. Out of this great bulk we possess, apart from some fragments, only six imperfect books on the Latin language and three books on agriculture.

Varro's writings may be grouped roughly under three heads: *belles lettres*, history and antiquities, and technical treatises on a great variety of subjects. To the first group belong the 150 books of *Menippean Satires*,[1] medleys in prose and verse in the manner of the dialogues of Menippus of Gadara, Cynic philosopher of the third century B.C.—a work of considerable importance in the history of literature. In the second group may be placed the 15 books of *Imagines*, consisting of 700 prose biographies of famous Greeks and Romans, each with a eulogy in verse, and a portrait; and the *Antiquities* in 41 books, dealing with "Things Human" and "Things Divine," a work dedicated in its latter part to Julius Caesar and frequently employed by St. Augustine in his *De Civitate Dei*. Mention is made, and some titles are given,[2] of treatises on the history of literature, to which we owe the canon of Plautine plays accepted as genuine; of a single collection of 76 books on historical, philosophical, and other subjects; and of less extensive works on biography and political and military history. A number of

[1] Fragments preserved chiefly by Nonius and edited by Oehler, 1844, and Buecheler, 1904.

[2] Aul. Gell., *passim*.

geographic, legal, and encyclopaedic works come under the third classification. Here, too, belong the 9 books of *Disciplines*, dealing with music, medicine, rhetoric, and grammar; the *Res Rusticae*; and the great work *On the Latin Language*.[1] This work, our first extant Roman treatise on grammar, consisted originally of 25 books in three divisions, dealing with etymology, inflection, and syntax. Only Books V–X survive, and these in somewhat mutilated condition, and with much absurd philology.

The *Res Rusticae* was begun in Varro's eightieth year, when he was warned that he "must pack his baggage for departure from this life."[2] The work as a whole is addressed to Varro's wife, Fundania, who had just purchased a farm, and was intended as a practical manual on husbandry. Its three books are devoted, respectively, to agriculture proper, domestic cattle, and the smaller stock of the farm, such as poultry, game birds, and bees. Each book, cast in the form of a dialogue, has its own appropriate setting and its own little drama. The names of the speakers are so chosen as to suggest the various topics under discussion, thus affording the author the opportunity for an occasional pun. Harsh and involved in style, its meaning often obscured by syntactical as well as technical difficulties, the treatise contains, nevertheless, a great wealth of information. Its rigidly systematic topical arrangement is lightened by introductory chapters of general interest and by occasional digressions from the technicalities of the subject. Again and again the author indulges his

[1] The latest edition of *De Lingua Latina* is that of Goetz and Schoell, 1910.

[2] *Res Rust.*, I, 1, 1.

bent for the derivations, in large part fanciful, of Latin words. Although inferior, on the whole, to the more lucid and voluminous work of Columella, who dealt with the same subject in the next century, the treatise was of immense practical value and was often quoted. Virgil undoubtedly derived from Varro much of the technical knowledge displayed in his *Georgics*; and many of the agricultural and veterinary precepts of the *Res Rusticae* reappear in Pliny's *Natural History*, in Columella's agricultural treatise and, through him, in the works of the fourth-century writers, Palladius and Vegetius.

Manuscripts and Important Editions

The manuscript text of Cato and Varro on agriculture has survived in very imperfect condition. The Cato text, in particular, is full of additions and repetitions, and its Latinity has been considerably modernized in the process of transmission. In the case of both authors textual corruptions have been multiplied by the tampering of unscientific Renaissance scholars.

Our knowledge of the manuscript tradition of these writers goes back only as far as a long-lost codex, the so-called Marcianus, once in the Library of St. Mark in Florence, last seen and used by Petrus Victorius (1499–1585) and described by him as "liber antiquissimus et fidelissimus." The readings of this ancient codex have been preserved by Angelo Politian (1454–94), who collated the manuscript in 1482 with a copy of the *editio princeps*, now in the National Library in Paris. The same codex

was used by Victorius in his edition of the Roman writers on agriculture which appeared in 1541; and many of its variant readings are discussed by him in his "Explicationes suarum in Catonem, Varronem, Columellam castigationum" of the following year. It has been demonstrated by the German scholars J. G. Schneider (1750–1822) and Heinrich Keil (1822–94) that all existing manuscripts of the agricultural works of Cato and Varro are directly or indirectly descended from the lost Marcian.

Of the small group of extant codices used by Keil and described in the Introduction to his edition, the two following are universally regarded as most trustworthy in preserving the readings of the lost archetype:

Cod. Parisinus 6842 A (= A), 12th/13th cents. The oldest of the extant manuscripts. It contains the agricultural writings of Cato and Varro in complete form, with corrections by different hands.

Cod. Laurentianus 51, 4 (= B), 15th cent. Contains the three books of Varro, mutilated at the end of the third book in the same way as the ancient codex collated by Politian and used by Victorius.

Other existing manuscripts, also employed and described by Keil, are:

Cod. Mediceus-Laurentianus 30, 10 (= m), 14th cent. It contains Cato, *de Agri Cultura*, and the three books of Varro appended to Vitruvius, *de Architectura*. It was used by Politian and Victorius, and often agrees with the Parisinus in emended passages where others differ.

Cod. Laurentianus 51, 1 (= f), 14th/15th cents. Contains Cato and Varro, with many corrections by a later hand.

Cod. Caesenas Malatestianus 24, 2 (= c), 15th cent. Three books of Varro written after the works of Cato and Columella, with same gap in the text as in the Marcian codex. The missing portion was added apparently at a later date.

Cod. Laurentianus 51, 2 (= b), 15th cent.

The efforts of scholars to emend the faulty manuscript text of Cato and Varro have received no little aid from the works of other writers of like interests, both Greek and Roman, from whom our authors borrowed or by whom they were quoted. In this way the text has been illuminated here and there by accounts of various agricultural operations contained in the works of Xenophon, Theophrastus, or Aristotle. The text and interpretation of Varro, especially, are clarified at times by the testimony of such successors as Virgil, Verrius Flaccus through Festus and Paulus, Columella, the elder Pliny, Nonius Marcellus, Palladius and Vegetius through Columella, Aulus Gellius, and Macrobius.

The agricultural works of Cato and Varro have usually appeared together, in printed editions as in the manuscripts, or in company with Columella and Palladius. The *editio princeps* was published at Venice in 1472, in the shop of Nicolaus Jenson and under the editorship of George Merula. The edition of Beroaldus at Bologna in 1494 improved a number of readings. The press of Aldus Manutius at Venice produced in 1514 the edition of Iucundus, the Veronese architect, which far surpassed its predecessors in learning and ingenuity of textual emendation. The edition of Victorius, in 1541, helped to preserve the readings of the lost Marcian codex and exerted a vast influence on later editors.

Gesner brought out in 1735 his great edition of *Scriptores Rei Rusticae*, with critical apparatus and valuable commentary, taking into account all earlier works of importance. This was reprinted by Ernesti in 1773. The most valuable of the older works and still the best of the annotated editions is the *Scriptores Rei Rusticae* of J. G. Schneider, published at Leipzig in 1794–96. The text was first put on a definite scientific basis by Heinrich Keil in his *editio maior* of 1884–94, with adequate critical apparatus and commentary. This was followed by his minor edition of Varro in 1889, and of Cato in 1895. Since the death of Keil, his text of both authors has undergone a complete and very careful revision at the hands of George Goetz, who has restored the text more nearly to that of the archetype. A second edition of Cato was published by Goetz, in the Teubner series, in 1922, and a second edition of Varro in 1929.

The difficulties of text and interpretation presented by Cato and Varro have attracted the attention of a large number of editors and commentators in addition to those already mentioned. Names of outstanding importance in this connection are, for the earlier scholars, those of Adrian Turnebus at the beginning of the sixteenth century, Joseph Scaliger in 1569, Fulvius Ursinus in 1587, J. F. Gronovius in the seventeenth century, and Julius Pontedera in 1740. Contributions to the study of one or both writers have been made in more recent or modern times by a large number of scholars, among whom may be named Theodor Mommsen, F. W. Ritschl, Hugo Reiter, Ernst Samter, O. Schöndorffer, H. Jordan, F. Zahlfeldt, W. Weise, Richard Krumbiegel,

Walter Kohlschmidt, V. Lundström, Theodor Birt, A. Kappelmacher, C. Howe, A. W. Van Buren, Josef Hörle, and Tenney Frank.

The text accompanying the present translation is based on that of the second Teubner editions of Goetz (Cato, *de Agri Cultura*, 1922; Varro, *Rerum Rusticarum libri tres*, 1929) as the latest and in every way the most trustworthy. Some slight changes have been made in punctuation and capitalization to conform more nearly with English and American usage. Occasional important divergences of reading are noted. The critical notes, drawn from the editions of Keil and Goetz and from independent investigation of other sources, aim to give, as far as the limited space permits, the source of the more important corrections together with the probable reading of the archetype.

Since the above was put in print, and while the volume was undergoing final revision, there has appeared *Cato the Censor on Farming*, with translation and commentary, by Ernest Brehaut (Columbia University Press, 1933). Dr. Brehaut's work, now inserted in the Bibliography of this volume, has been of service in the final reading of the proofs.

May 2, 1934.

BIBLIOGRAPHY

Principal Editions

Editio princeps. Venetiis apud Nicolaum Jensonum, 1472.

Iucundi Veronensis editio *De Re Rustica*, Venetiis apud Aldum, 1514.

Petri Victorii editio *De Re Rustica*, Lugduni apud Gryphium, 1541.

Gesner, J. M., *Scriptores Rei Rusticae*, Leipzig, 1735. Vol. I.

Schneider, J. G., *Scriptores Rei Rusticae*, Leipzig, 1794. Vol. I.

Keil, H., *M. Porci Catonis de Agri Cultura liber; M. Terenti Varronis Rerum Rusticarum libri tres*, Leipzig, 1884–94.

Goetz, G., *M. Porci Catonis de Agri Cultura liber*, Leipzig, 1922.

Goetz, G., *M. Terenti Varronis Rerum Rusticarum libri tres*, Leipzig, 1929.

Criticism

Hauler, E., *Zu Catos Schrift über das Landwesen*, Vienna, 1896.

Jordan, H., *Quaestionum Catonianarum capita duo* (Diss.), Berlin, 1856.

Keil, H., *Observationes criticae in Catonis et Varronis de Re Rustica libros*, Halle, 1849.

Keil, H., *Emendationes Varronianae*, Halle, 1883–84.

Krumbiegel, R., *De Varroniano scribendi genere quaestiones* (Diss.), Leipzig, 1892.

Reiter, H., *Quaestiones Varronianae grammaticae* (Diss.), Königsberg, 1892.

Ritschl, F. W., *Quaestiones Varronianae*, Bonn, 1845.

Samter, E., *Quaestiones Varronianae* (Diss.), Berlin, 1891.

Schöndorffer, O., *De genuina Catonis de Agri Cultura libri forma*, Pars I, De syntaxi Catonis (Strassburg Diss.), Königsberg, 1885.

Petri Victorii, *Explicationes suarum in Catonem, Varronem, Columellam castigationum*, Lugduni, 1542.

Weise, P., *Quaestionum Catonianarum capita V*, Gottingen, 1886.

Zahlfeldt, F., *Quaestiones criticae in Varronis Rerum Rusticarum libros tres*, Berlin, 1881.

Translations

Roman Farm Management: The Treatises of Cato and Varro done into English, with notes of modern instances. By a Virginia farmer ⟨Fairfax Harrison⟩, N. Y., 1913. (Partial translation of Cato, arranged by topics; complete translation of Varro.)

Brehaut, Ernest, *Cato the Censor on Farming*, New York, 1933.

Curcio, Gaetano, *La primitiva civiltà latina agricola e il libro dell' agricultura di M. Porcio Catone*, Florence, 1930.

Nisard, M., *Les Agronomes latins, Caton, Varron, Columelle, Palladius*, Paris, 1844.

Owen, Thomas, *Varro, Three Books concerning Agriculture*, Oxford, 1800.

Owen, Thomas, *M. Porcius Cato concerning Agriculture*, London, 1803.

Storr-Best, Lloyd, *Varro on Farming* (Bohn Library), London, 1912.

Miscellaneous

Dickson, Adam, *The Husbandry of the Ancients*, 2 Vols., Edinburgh, 1788.

Frank, Tenney, *An Economic History of Rome*, 2nd ed., rev., Baltimore, 1927.

Frank, Tenney, *An Economic Survey of Ancient Rome*, Vol. I., Baltimore, 1933.

Heitland, W. E., *Agricola*: A Study of Agriculture and Rustic Life in the Greco-Roman World from the Point of View of Labour, Cambridge, 1921.

Hörle, Josef, *Catos Hausbücher*: Analyse seiner Schrift De Agricultura nebst Wiederherstellung seines Kelterhauses und Gutshofes, Paderborn, 1929.

Keil, H., *De libris manuscriptis Catonis de Agri Cultura disputatio*, Halle, 1882.

Krumbiegel, R., *Index Verborum Catonis de Re Rustica*, Leipzig, 1897.

Krumbiegel, R., *Index Verborum in Varronis Rerum Rusticarum libris tribus*, Leipzig, 1902.

MARCUS CATO
ON AGRICULTURE

M. CATONIS

DE AGRI CULTURA

Est interdum praestare mercaturis rem quaerere, nisi tam periculosum sit, et item fenerari, si tam honestum sit. Maiores nostri sic habuerunt et ita in legibus posiverunt, furem dupli condemnari, feneratorem quadrupli. Quanto peiorem civem existimarint feneratorem quam furem, hinc licet existi-
2 mare. Et virum bonum quom laudabant, ita laudabant, bonum agricolam bonumque colonum. Amplissime laudari existimabatur qui ita laudabatur.
3 Mercatorem autem strenuum studiosumque rei quaerendae existimo, verum, ut supra dixi, peri-
4 culosum et calamitosum. At ex agricolis et viri fortissimi et milites strenuissimi gignuntur, maximeque pius quaestus stabilissimusque consequitur minimeque invidiosus, minimeque male cogitantes sunt qui in eo studio occupati sunt. Nunc ut ad rem redeam, quod promisi institutum [1] principium hoc erit.

I. Praedium quom parare cogitabis, sic in animo habeto, uti ne cupide emas neve opera [2] tua parcas

[1] *Institutum* is taken variously; as a genitive (Friedrich); a supine (Wünsch); as = *ad id institutum quod promisi* (Birt); as a pleonasm (Stangl). Birt is followed here.

[2] Pliny, *N.H.*, XVIII, 26, quotes this *operae*.

MARCUS CATO

ON AGRICULTURE

It is true that to obtain money by trade is sometimes more profitable, were it not so hazardous; and likewise money-lending, if it were as honourable. Our ancestors held this view and embodied it in their laws, which required that the thief be mulcted double and the usurer fourfold; how much less desirable a citizen they considered the usurer than the thief, one may judge from this. And when they would praise a worthy man their praise took this form: "good husbandman," "good farmer"; one so praised was thought to have received the greatest commendation. The trader I consider to be an energetic man, and one bent on making money; but, as I said above, it is a dangerous career and one subject to disaster. On the other hand, it is from the farming class that the bravest men and the sturdiest soldiers come, their calling is most highly respected, their livelihood is most assured and is looked on with the least hostility, and those who are engaged in that pursuit are least inclined to be disaffected. And now, to come back to my subject, the above will serve as an introduction to what I have undertaken.

I. When you are thinking of acquiring a farm, keep in mind these points: that you be not over-eager in buying nor spare your pains in examining,

visere et ne satis habeas semel circumire. Quotiens
2 ibis, totiens magis placebit quod bonum erit. Vicini
quo pacto niteant, id animum advertito: in bona
regione bene nitere oportebit. Et uti eo introeas
et circumspicias, uti inde exire possis. Uti bonum
caelum habeat, ne calamitosum siet, solo bono, sua
3 virtute valeat. Si poteris, sub radice montis siet,
in meridiem spectet, loco salubri, operariorum copia
siet, bonumque aquarium, oppidum validum prope
siet aut mare aut amnis, qua naves ambulant, aut
4 via bona celebrisque. Siet in his agris, qui non
saepe dominos mutant: qui in his agris praedia
vendiderint, eos pigeat vendidisse. Uti bene aedificatum
siet. Caveto alienam disciplinam temere
contemnas. De domino bono colono bonoque aedificatore
melius emetur. Ad villam cum venies, videto,
5 vasa torcula et dolia multane sient: ubi non erunt,
scito pro ratione fructum esse. Instrumenti ne
magni siet, loco bono siet. Videto, quam minimi
6 instrumenti sumptuosusque ager ne siet. Scito
idem agrum quod hominem, quamvis quaestuosus
7 siet, si sumptuosus erit, relinqui non multum. Praedium
quod primum siet, si me rogabis, sic dicam:
de omnibus agris optimoque loco iugera agri centum,
vinea est prima, si vino bono et [1] multo est, secundo

[1] *bono et* supplied by Leo from Varro, I, 7, 9.

---

[1] Others render, "Be careful not rashly to refuse to learn from others."

[2] A iugerum is approximately two-thirds of an acre. See Glossary, p. 531.

and that you consider it not sufficient to go over it once. However often you go, a good piece of land will please you more at each visit. Notice how the neighbours keep up their places; if the district is good, they should be well kept. Go in and keep your eyes open, so that you may be able to find your way out. It should have a good climate, not subject to storms; the soil should be good, and naturally strong. If possible, it should lie at the foot of a mountain and face south; the situation should be healthful, there should be a good supply of labourers, it should be well watered, and near it there should be a flourishing town, or the sea, or a navigable stream, or a good and much travelled road. It should lie among those farms which do not often change owners; where those who have sold farms are sorry to have done so. It should be well furnished with buildings. Do not be hasty in despising the methods of management adopted by others.[1] It will be better to purchase from an owner who is a good farmer and a good builder. When you reach the steading, observe whether there are numerous oil presses and wine vats; if there are not, you may infer that the amount of the yield is in proportion. The farm should be one of no great equipment, but should be well situated. See that it be equipped as economically as possible, and that the land be not extravagant. Remember that a farm is like a man—however great the income, if there is extravagance but little is left. If you ask me what is the best kind of farm, I should say: a hundred iugera[2] of land, comprising all sorts of soils, and in a good situation; a vineyard comes first if it produces bounti-

loco hortus inriguus, tertio salictum, quarto oletum, quinto pratum, sexto campus frumentarius, septimo silva caedua, octavo arbustum, nono glandaria silva.

II. Pater familias ubi ad villam venit, ubi larem familiarem salutavit, fundum eodem die, si potest, circumeat; si non eodem die, at postridie. Ubi cognovit, quo modo fundus cultus siet operaque quae facta infectaque sient, postridie eius diei vilicum vocet, roget, quid operis siet factum, quid restet, satisne temperi opera sient confecta, possitne quae reliqua sient conficere, et quid factum vini, 2 frumenti aliarumque rerum omnium. Ubi ea cognovit, rationem inire oportet operarum, dierum. Si ei opus non apparet, dicit vilicus sedulo se fecisse, servos non valuisse, tempestates malas fuisse, servos aufugisse, opus publicum effecisse, ubi eas aliasque causas multas dixit, ad rationem operum opera- 3 rumque vilicum revoca. Cum tempestates pluviae fuerint, quae opera per imbrem fieri potuerint, dolia lavari, picari, villam purgari, frumentum transferri,

---

[1] It is most significant that Cato places grain farming sixth in importance. The second Punic War had completely demoralized the Republic. The yeomanry had been conscripted and the fields desolated and burned. "Roman farmers torn from their homes for years and demoralized by the camps were unable or unwilling to settle down into the quiet routine of agricultural life. . . . Their farms passed into the hands of capitalists, and the rich lands of Italy fell back into pasture, and half-naked slaves tended herds of cattle." (Smith, *Rome and Carthage*, p. 230.) Grain farming was no longer profitable, and it had become the custom to import grain from Sicily and Africa. The new Rome that emerged from this horrible war centred around a nobility of wealth and was in a state of demoralization. Such a condition naturally caused the cultivation of grain to be less

fully wine of a good quality; second, a watered garden; third, an osier-bed; fourth, an oliveyard; fifth, a meadow; sixth, grain land;[1] seventh, a wood lot; eighth, an arbustum;[2] ninth, a mast grove.[3]

II. When the master arrives at the farmstead, after paying his respects to the god of the household, let him go over the whole farm, if possible, on the same day; if not, at least on the next. When he has learned the condition of the farm, what work has been accomplished and what remains to be done, let him call in his overseer the next day and inquire of him what part of the work has been completed, what has been left undone; whether what has been finished was done betimes, and whether it is possible to complete the rest; and what was the yield of wine, grain, and all other products. Having gone into this, he should make a calculation of the labourers and the time consumed. If the amount of work does not seem satisfactory, the overseer claims that he has done his best, but that the slaves have not been well, the weather has been bad, slaves have run away, he has had public work[4] to do; when he has given these and many other excuses, call the overseer back to your estimate of the work done and the hands employed. If it has been a rainy season, remind him of the work that could have been done on rainy days: scrubbing and pitching wine vats, cleaning the farmstead, shifting

important than that of the vine, the olive, domestic vegetables, or the rearing of cattle.

[2] The word is used of a plantation of trees, to which the vines were "wedded," or of an orchard. Columella gives a description, Book V, Chap. 6, but Cato seems not to use the word in the sense first given. [3] To furnish feed for live-stock.

[4] Possibly on the public roads, as in the French corvée.

stercus foras efferri, stercilinum fieri, semen purgari, funes sarciri, novos fieri; centones, cuculiones 4 familiam oportuisse sibi sarcire. Per ferias potuisse fossas veteres tergeri, viam publicam muniri, vepres recidi, hortum fodiri, pratum purgari, virgas vinciri, spinas runcari, expinsi far,[1] munditias fieri. Cum servi aegrotarint, cibaria tanta dari non oportuisse. 5 Ubi ea[2] cognita aequo animo sint, quae reliqua opera sint curare uti perficiantur: rationes putare argentariam, frumentariam, pabuli causa quae parata sunt; rationem vinariam, oleariam, quid venierit, quid exactum siet, quid reliquum siet, quid siet quod veneat: quae satis accipiunda sint, satis 6 accipiantur: reliqua quae sint uti conpareant. Siquid desit in annum, uti paretur: quae supersint, ut veneant: quae opus sint locato, locentur: quae opera fieri velit et quae locari velit, uti imperet et 7 ea scripta relinquat. Pecus consideret. Auctionem uti faciat: vendat oleum, si pretium habeat, vinum, frumentum quod supersit vendat; boves vetulos, armenta delicula, oves deliculas, lanam, pelles, plostrum vetus, ferramenta vetera, servum senem, servum morbosum, et siquid aliut supersit, vendat. Patrem familias vendacem, non emacem esse oportet.

III. Prima adulescentia patrem familiae agrum conserere studere oportet. Aedificare diu cogitare oportet, conserere cogitare non oportet, sed facere

[1] *far* Victorius: *lar* or *iar*.
[2] *ea* added by Keil: *haec* Iucundus.

---

[1] It was the regular custom among the Romans to let out certain work by contract in contrast with the work that was done by the farm organization under the management of the overseer.

grain, hauling out manure, making a manure pit, cleaning seed, mending old harness and making new; and that the hands ought to have mended their smocks and hoods. Remind him, also, that on feast days old ditches might have been cleaned, road work done, brambles cut, the garden spaded, a meadow cleared, faggots bundled, thorns rooted out, spelt ground, and general cleaning done. When the slaves were sick, such large rations should not have been issued. After this has been gone into calmly, give orders for the completion of what work remains; run over the cash accounts, grain accounts, and purchases of fodder; run over the wine accounts, the oil accounts—what has been sold, what collected, balance due, and what is left that is saleable; where security for an account should be taken, let it be taken; and let the supplies on hand be checked over. Give orders that whatever may be lacking for the current year be supplied; that what is superfluous be sold; that whatever work should be let out be let. Give directions as to what work you want done on the place, and what you want let out,[1] and leave the directions in writing. Look over the live stock and hold a sale. Sell your oil, if the price is satisfactory, and sell the surplus of your wine and grain. Sell worn-out oxen, blemished cattle, blemished sheep, wool, hides, an old wagon, old tools, an old slave, a sickly slave, and whatever else is superfluous. The master should have the selling habit, not the buying habit.

III. In his youth the owner should devote his attention to planting. He should think a long time about building, but planting is a thing not to be thought about but done. When you reach the age

oportet. Ubi aetas accessit ad annos XXXVI, tum aedificare oportet, si agrum consitum habeas. Ita aedifices, ne villa fundum quaerat neve fundus 2 villam.[1] Patrem familiae villam rusticam bene aedificatam habere expedit, cellam oleariam, vinariam, dolia multa, uti lubeat caritatem expectare: et rei et virtuti et gloriae erit. Torcularia bona habere oportet, ut opus bene effici possit. Olea ubi lecta siet, oleum fiat continuo, ne corrumpatur. Cogitato quotannis tempestates magnas venire et oleam 3 deicere solere. Si cito sustuleris et vasa parata erunt, damni nihil erit ex tempestate et oleum 4 viridius et melius fiet. Si in terra et tabulato olea nimium diu erit, putescet, oleum fetidum fiet. Ex quavis olea oleum viridius et bonum fieri potest, si 5 temperi facies. In iugera oleti CXX vasa bina esse oportet, si oletum bonum beneque frequens cultumque erit. Trapetos bonos privos inpares esse oportet, si orbes contriti sient, ut conmutare possis, funes loreos privos, vectes senos, fibulas duodenas, medipontos privos loreos. Trochileas Graecanicas binis 6 funibus sparteis ducunt: orbiculis superioribus octonis, inferioribus senis citius duces; si rotas voles facere, tardius ducetur, sed minore labore.

IV. Bubilia bona, bonas praesepis, faliscas clatratas, clatros interesse oportet pede. Si ita feceris, pabulum boves non eicient. Villam urbanam pro copia aedificato. In bono praedio si bene aedificaveris,

[1] *neve fundus villam* added by Iucundus from Col., I, 4, 8, and Plin., *N.H.*, XVIII, 32.

---

[1] The "planting" is, of course, of trees and vines.

[2] See Columella, I, 6; but Cato's villa had only two units, the *villa urbana*, or dwelling-house, and the *villa rustica*, for all other purposes.

[3] See Glossary

of thirty-six you should build, if you have your land planted.[1] In building, you should see that the steading does not lag behind the farm nor the farm behind the steading. It is well for the master to have a well-built barn [2] and storage room and plenty of vats for oil and wine, so that he may hold his products for good prices; it will redound to his wealth, his self-respect, and his reputation. He should have good presses, so that the work may be done thoroughly. Let the olives be pressed immediately after gathering, to prevent the oil from spoiling. Remember that high winds come every year and are apt to beat off the olives; if you gather them at once and the presses are ready, there will be no loss on account of the storm, and the oil will be greener and better. If the olives remain too long on the ground or the floor they will spoil, and the oil will be rancid. Any sort of olive will produce a good and greener oil if it is pressed betimes. For an oliveyard of 120 iugera there should be two pressing equipments, if the trees are vigorous, thickly planted, and well cultivated. The mills should be stout and of different sizes, so that if the stones become worn you may change. Each should have its own leather ropes, six sets of hand bars, six double sets of pins, and leather belts. Greek blocks [3] run on double ropes of Spanish broom; you can work more rapidly with eight pulleys above, and six below; if you wish to use wheels it will work more slowly but with less effort.

IV. Have good stalls, stout pens, and latticed feed-racks. The rack bars should be a foot apart; if you make them in this way the cattle will not scatter their feed. Build your dwelling-house [2] in accord-

bene posiveris, ruri si recte habitaveris, libentius et saepius venies; fundus melius erit, minus peccabitur, fructi plus capies; frons occipitio prior est. Vicinis bonus esto; familiam ne siveris peccare. Si te libenter vicinitas videbit, facilius tua vendes, opera facilius locabis, operarios facilius conduces; si aedificabis, operis, iumentis, materie adiuvabunt: siquid bona salute usus venerit, benigne defendent.

V. Haec erunt vilici officia. Disciplina bona utatur. Feriae serventur. Alieno manum abstineat, sua servet diligenter. Litibus familia supersedeat; siquis quid deliquerit, pro noxa bono modo vindicet. 2 Familiae male ne sit, ne algeat, ne esuriat; opere bene exerceat, facilius malo et alieno prohibebit. Vilicus si nolet male facere, non faciet. Si passus erit, dominus inpune ne sinat esse. Pro beneficio gratiam referat, ut aliis recte facere libeat. Vilicus ne sit ambulator, sobrius siet semper, ad cenam nequo eat. Familiam exerceat, consideret, quae dominus imperaverit fiant. Ne plus censeat sapere

---

[1] The content of this homely maxim appears in practically all the writers on agriculture, and has entered, in some form, into almost all proverbial wisdom. Perhaps its most popular modern form is taken from Poor Richard's Almanack: "The eye of a master will do more work than both his hands." Cf. Columella, I, 1, 18; Pliny, *N.H.*, XVIII, 31.

[2] See note 1, p. 8.

[3] *bona salute* is merely a formula to avoid the evil omen of mentioning misfortune.

ance with your means. If you build substantially on a good farm, placing the house in a good situation, so that you can live comfortably in the country, you will like to visit it, and will do so oftener; the farm will improve, there will be less wrongdoing, and you will receive greater returns; the forehead is better than the hindhead.[1] Be a good neighbour, and do not let your people commit offences. If you are popular in the neighbourhood it will be easier for you to sell your produce, easier to let out your work,[2] easier to secure extra hands. If you build, the neighbours will help you with their work, their teams, and their materials; if trouble comes upon you, which God forbid,[3] they will be glad to stand by you.

V. The following are the duties of the overseer:— He must show good management. The feast days must be observed. He must withhold his hands from another's goods and diligently preserve his own. He must settle disputes among the slaves; and if anyone commits an offence he must punish him properly in proportion to the fault. He must see that the servants are well provided for, and that they do not suffer from cold or hunger. Let him keep them busy with their work—he will more easily keep them from wrongdoing and meddling. If the overseer sets his face against wrongdoing, they will not do it; if he allows it, the master must not let him go unpunished. He must express his appreciation of good work, so that others may take pleasure in well-doing. The overseer must not be a gadabout, he must always be sober, and must not go out to dine. He must keep the servants busy, and see that the master's orders are carried out.

3 se quam dominum. Amicos domini, eos habeat sibi amicos. Cui iussus siet, auscultet. Rem divinam nisi Conpitalibus in conpito aut in foco ne faciat. Iniussu domini credat nemini: quod dominus crediderit, exigat. Satui semen, cibaria, far, vinum, oleum mutuum dederit nemini. Duas aut tres familias habeat, unde utenda roget et quibus det,
4 praeterea nemini. Rationem cum domino crebro putet. Operarium, mercennarium, politorem diutius eundem ne habeat die. Nequid emisse velit insciente domino, neu quid dominum celavisse velit. Parasitum nequem habeat. Haruspicem, augurem, hariolum, Chaldaeum nequem consuluisse velit. Segetem ne defrudet: nam id infelix est. Opus rusticum omne curet uti sciat facere, et id faciat
5 saepe, dum ne lassus fiat; si fecerit, scibit in mente familiae quid sit, et illi animo aequiore facient. Si hoc faciet, minus libebit ambulare et valebit rectius et dormibit libentius. Primus cubitu surgat, postremus cubitum eat. Prius villam videat clausa uti siet, et uti suo quisque loco cubet et uti iumenta pabulum habeant.

6 Boves maxima diligentia curatos habeto. Bubulcis opsequito partim, quo libentius boves curent.

---

[1] The festival held annually at the cross-roads, in honour of the Lares Compitales. It occurred soon after the Saturnalia, in December, on a day or days appointed by the praetor.

[2] Compare Horace's warning against "meddling with Babylonian calculations" (*Odes*, I, 11), and many others. Columella, I, 8, 6, emphasizes the warning.

He must not assume that he knows more than the master. He must consider the master's friends his own friends. He must pay heed to anyone to whom he has been bidden to listen. He must perform no religious rites, except on the occasion of the Compitalia [1] at the cross-roads, or before the hearth. He must extend credit to no one without orders from the master, and must collect the loans made by the master. He must lend to no one seed-grain, fodder, spelt, wine, or oil. He must have two or three households, no more, from whom he borrows and to whom he lends. He must make up accounts with the master often. He must not hire the same day-labourer or servant or caretaker for longer than a day. He must not want to make any purchases without the knowledge of the master, nor want to keep anything hidden from the master. He must have no hanger-on. He must not consult a fortune-teller, or prophet, or diviner, or astrologer.[2] He must not stint the seed for sowing, for that brings bad fortune. He must see to it that he knows how to perform all the operations of the farm, and actually does perform them often, but not to the extent of becoming exhausted; by so doing he will learn what is in his servants' minds, and they will perform their work more contentedly. Also, he will be less disposed to gad about, will be in better health, and will enjoy his sleep more. He must be the first out of bed, the last to go to bed. Before then he must see that the farmstead is closed, that each one is asleep in his proper place, and that the stock have fodder.

See that the draft oxen are looked after with the greatest care, and be somewhat indulgent to the

Aratra vomeresque facito uti bonos habeas. Terram cariosam cave ne ares, neve plostrum neve pecus inpellas. Si ita non caveris, quo inpuleris, trienni 7 fructum amittes. Pecori et bubus diligenter substernatur, ungulae curentur. Scabiem pecori et iumentis caveto; id ex fame et si inpluit fieri solet. Opera omnia mature conficias face. Nam res rustica sic est, si unam rem sero feceris, omnia opera sero facies. Stramenta si deerunt, frondem iligneam 8 legito, eam substernito ovibus bubusque. Sterculinum magnum stude ut habeas. Stercus sedulo conserva; cum exportabis, purgato et conminuito; per autumnum evehito. Circum oleas autumnitate ablaqueato et stercus addito. Frondem populneam, ulmeam, querneam caedito per tempus: eam condito non peraridam, pabulum ovibus. Item faenum cordum, sicilimenta de prato, ea arida condito. Post imbrem autumnum rapinam, pabulum lupinumque serito.

VI. Agrum quibus locis conseras, sic observari oportet. Ubi ager crassus et laetus est sine arboribus, eum agrum frumentarium esse oportet. Idem ager si nebulosus est, rapa, raphanos, milium, panicum, id maxime seri oportet. In agro crasso et caldo oleam conditivam, radium maiorem, Sallentinam, orcitem, poseam, Sergianam, Colminianam, albicerem, quam earum in iis locis optimam dicent

---

[1] The term is explained by Columella, II, 4, 5: "That is, when, after a long dry spell, a light rain wets the surface but does not sink in." The injunction is repeated there.

[2] For a description of the several varieties of olives, see Columella, V, 8. The Romans were experts in plant selection, and developed distinct varieties of all the leading horticultural and field crops.

teamsters to make them look after their stock with more pleasure. See that you keep your ploughs and ploughshares in good condition. Be careful not to plough land which is *cariosa* [1] or drive a cart over it, or turn cattle into it; if you are not careful about this, you will lose three years' crop of the land on which you have turned them. Litter the cattle and flocks carefully, and see that their hoofs are kept clean. Guard against the scab in flocks and herds; it is usually caused by under-feeding and exposure to wet weather. See that you carry out all farm operations betimes, for this is the way with farming: if you are late in doing one thing you will be late in doing everything. If bedding runs short, gather oak leaves and use them for bedding down sheep and cattle. See that you have a large dunghill; save the manure carefully, and when you carry it out, clean it of foreign matter and break it up. Autumn is the time to haul it out. During the autumn also dig trenches around the olive trees and manure them. Cut poplar, elm, and oak leaves betimes; store them before they are entirely dry, as fodder for sheep. Second-crop hay and aftermath should also be stored dry. Sow turnips, forage crops, and lupines after the autumn rains.

VI. This rule should be observed as to what you should plant in what places:—Grain should be sown in heavy, rich, treeless soil; and if this sort of soil is subject to fogs it should preferably be sown with rape, turnips, millet, and panic-grass. In heavy, warm soil plant olives [2]—those for pickling, the long variety, the Sallentine, the orcites, the posea, the Sergian, the Colminian, and the waxy-white; choose especially the varieties which are commonly agreed

esse, eam maxime serito. Hoc genus oleae in XXV
2 aut in XXX pedes conserito. Ager oleto conserundo, qui in ventum favonium spectabit et soli ostentus erit, alius bonus nullus erit. Qui ager frigidior et macrior erit, ibi oleam Licinianam seri oportet. Si in loco crasso aut calido severis, hostus nequam erit et ferundo arbor peribit et muscus
3 ruber molestus erit. Circum coronas et circum vias ulmos serito et partim populos, uti frondem ovibus et bubus habeas, et materies, siquo opus sit, parata erit. Sicubi in iis locis ripae aut locus umectus erit, ibi cacumina populorum serito et harundinetum. Id hoc modo serito: bipalio vortito, ibi oculos harundinis pedes ternos alium ab alio serito. Ibi
4 corrudam serito, unde asparagi fiant. Nam convenit harundinetum cum corruda, eo quia foditur et incenditur et umbram per tempus habet. Salicem Graecam circum harundinetum serito, uti siet qui vineam alliges.

Vineam quo in agro conseri oporteat, sic observato. Qui locus vino optimus dicetur esse et ostentus soli, Aminnium minusculum et geminum eugeneum, helvolum minusculum conserito. Qui locus crassus erit aut nebulosior, ibi Aminnium maius aut Murgentinum, Apicium, Lucanum serito. Ceterae vites, miscellae maxime, in quemvis agrum conveniunt.

VII. Fundum suburbanum arbustum maxime convenit habere; et ligna et virgae venire possunt, et

---

[1] *Corruda*: identified in the 5th (?) cent. herbal under the name of Pseudo-Apuleius (*Herb.*, 84) as "the wild asparagus which the Greeks call ὄρμινον or μυακανδον, and by other names."

[2] See Columella, III, 2 for a detailed discussion of varieties of grapes. In Chapter 9 he returns to the discussion of the Aminnian, and remarks that these were "almost the only varieties known to the ancients." [3] Cf. note 2, page 7.

to be the best for these districts. Plant this variety of olives at intervals of twenty-five or thirty feet. Land which is suitable for olive planting is that which faces the west and is exposed to the sun; no other will be good. Plant the Licinian olive in colder and thinner soil. If you plant it in heavy or warm soil the yield will be worthless, the tree will exhaust itself in bearing, and a reddish scale will injure it. Around the borders of the farm and along the roads plant elms and some poplars, so that you may have leaves for the sheep and cattle; and the timber will be available if you need it. Wherever there is a river bank or wet ground, plant poplar cuttings and a reed thicket. The method of planting is as follows:—turn the ground with the mattock and then plant the eyes of the reed three feet apart. Plant there also the wild asparagus,[1] so that it may produce asparagus; for a reed thicket goes well with the wild asparagus, because it is worked and burned over, and furnishes a shade when shade is needed. Plant Greek willows along the border of the thicket, so that you may have withes for tying up vines.

Choose soil for laying out a vineyard by the following rules:—In soil which is thought to be best adapted for grapes and which is exposed to the sun, plant the small Aminnian,[2] the double eugeneum, and the small parti-coloured; in soil that is heavy or more subject to fogs plant the large Aminnian, the Murgentian, the Apician, and the Lucanian. The other varieties, and especially the hybrids, grow well anywhere.

VII. It is especially desirable to have a plantation [3] on a suburban farm, so that firewood and faggots may

domino erit qui utatur. In eodem fundo suum quidquid conseri oportet; vitem compluria genera[1] Aminnium minusculum et maius et Apicium. Uvae
2 in olla in vinaceis conduntur; eadem in sapa, in musto, in lora recte conduntur. Quas suspendas duracinas Aminnias maiores, vel ad fabrum ferrarium
3 pro passis eae recte servantur. Poma, mala strutea, cotonea Scantiana, Quiriniana, item alia conditiva, mala mustea et Punica (eo lotium suillum aut stercus ad radicem addere oportet, uti pabulum malorum fiat),[2] pira volaema, Aniciana sementiva (haec condi-
4 tiva in sapa bona erunt), Tarentina, mustea, cucurbitiva, item alia genera quam plurima serito aut inserito. Oleas orcites, posias; eae optime conduntur vel virides in muria vel in lentisco contusae, vel orcites ubi nigrae erunt et siccae, sale confriato dies V; postea salem excutito, in sole ponito biduum, vel sine sale in defrutum condito. Sorba in sapa condere vel siccare; arida facias. Item pira facias.

VIII. Ficos mariscas in loco cretoso et aperto serito; Africanas et Herculaneas, Sacontinas, hibernas, Tellanas atras pediculo longo, eas in loco crassiore aut stercorato serito. Pratum si inrigivum habebis, si non erit siccum, ne faenum desiet, summittito.
2 Sub urbe hortum omne genus, coronamenta omne

[1] *compluria genera* Hauler: *compularia.*
[2] *pabulum malorum fiat* Gesner: *fabulim malorum fiant.*

---

[1] Cf. 143, 3, and Varro, I, 54, 2.
[2] A small or sharp wine made from the husks of grapes; cf. Varro, I, 54, 3.
[3] Cf. Varro, I, 59, 1, with note 2, page 294.
[4] Cf. Varro, I, 59, 3.
[5] This resin from the mastic-tree is used also to flavour a distilled liquor used in various countries, as Turkey, Greece, etc.

be sold, and also may be furnished for the master's use. On the same farm should be planted anything adapted to the soil, and several varieties of grapes, such as the small and large Aminnian and the Apician. Grapes are preserved in grape-pulp in jars;[1] also they keep well in boiled wine, or must, or after-wine.[2] You may hang up the hard-berried and the larger Aminnian and they will keep as well dried before the forge fire as when spread in the sun. Plant or ingraft all kinds of fruit—sparrow-apples, Scantian and Quirinian quinces,[3] also other varieties for preserving, must-apples and pomegranates (the urine or dung of swine should be applied around the roots of these to serve as food for the fruit); of pears, the volema, the Anician frost-pears (these are excellent when preserved in boiled wine),[4] the Tarentine, the must-pear, the gourd-pear, and as many other varieties as possible; of olives, the orcite and posea, which are excellent when preserved green in brine or bruised in mastic[5] oil. When the orcites are black and dry, powder them with salt for five days; then shake off the salt, and spread them in the sun for two days, or pack them in boiled must without salt. Preserve sorbs in boiled must; or you may dry them; make them quite free from moisture. Preserve pears in the same way.

VIII. Plant mariscan figs in chalky, open soil; the African, Herculanean, Saguntine, the winter variety, the black Tellanian with long pedicles, in soil which is richer or manured. Lay down a meadow, so that you may have a supply of hay—a water meadow if you have it, if not, a dry meadow. Near a town it is well to have a garden planted with all manner of

genus, bulbos Megaricos, murtum coniugulum et album et nigrum, loream Delphicam et Cypriam et silvaticam, nuces calvas, Abellanas, Praenestinas, Graecas, haec facito uti serantur. Fundum suburbanum, et qui eum fundum solum habebit, ita paret itaque conserat, uti quam sollertissimum habeat.

IX. Salicta locis aquosis, umectis, umbrosis, propter amnes ibi seri oportet; et id videto uti aut domino opus siet aut ut vendere possit. Prata inrigiva, si aquam habebis, id potissimum facito; si aquam non habebis, sicca quam plurima facito. Hoc est praedium quod ubi vis expedit facere.

X. Quo modo oletum agri iugera CCXL instruere oporteat. Vilicum, vilicam, operarios quinque, bubulcos III, asinarium I, subulcum I, opilionem I, summa homines XIII; boves trinos, asinos ornatos clitellarios qui stercus vectent tris, asinum molarium[1] I, 2 oves C; vasa olearia instructa iuga V, ahenum quod capiat Q. XXX, operculum aheni, uncos ferreos III, urceos aquarios III, infidibula II, ahenum quod capiat Q. V, operculum aheni, uncos III, labellum pollulum I, amphoras olearias II, urnam quinquagenariam unam, trullas tris, situlum aquarium I, pelvim I, matellionem I, trullium I, scutriscum I, matellam I, nassiternam I, trullam I, candelabrum I, sextarium I; plostra maiora III, aratra cum vomeribus VI, iuga cum loris ornata III, ornamenta bubus 3 VI; irpicem I, crates stercerarias IIII, sirpeas stercerarias III, semuncias III, instrata asinis III; ferramenta: ferreas VIII, sarcula VIII, palas IIII, rutra V, rastros quadridentes II, falces faenarias VIII, stra-

[1] *molarium* added by Meursius, from Varro, I, 19, 3.

---

[1] Pliny says (XV, 122) that the name is perhaps derived

vegetables, and all manner of flowers for garlands—Megarian bulbs, conjugulan myrtle,[1] white and black myrtle, Delphian, Cyprian, and wild laurel, smooth nuts, such as Abellan, Praenestine, and Greek filberts. The suburban farm, and especially if it be the only one, should be laid out and planted as ingeniously as possible.

IX. Osier-beds should be planted in damp, marshy, shady ground, near a stream. But be sure that the master will need them or that he can find a market for them. If you have a water supply, pay particular attention to water meadows; if not, have all the dry meadows possible. This is the sort of farm which it is profitable to make anywhere.

X. This is the proper equipment for an oliveyard of 240 iugera[2]: An overseer, a housekeeper, 5 labourers, 3 teamsters, 1 muleteer, 1 swineherd, 1 shepherd—a total of 13 persons; 3 yoke of oxen, 3 pack-asses to carry manure, 1 ass for the mill, and 100 sheep; 5 complete oil-pressing equipments, 1 copper vessel holding 30 quadrantals,[2] with copper cover, 3 iron hooks, 3 water-pots, 2 funnels, 1 copper vessel holding 5 quadrantals, with copper cover, 3 hooks, 1 small bowl, 2 oil jars, 1 jar holding 50 heminae (?),[2] 3 ladles, 1 water bucket, 1 basin, 1 small pot, 1 ewer, 1 platter, 1 chamber-vessel, 1 watering-pot, 1 ladle, 1 candlestick, 1 sextarius[2] measure; 3 large carts, 6 ploughs and ploughshares, 3 yokes fitted with straps, 6 sets of ox harness; 1 harrow, 4 manure hampers, 3 manure baskets, 3 pack-saddles, 3 pads for the asses; tools: 8 forks, 8 hoes, 4 spades, 5 shovels, 2 four-toothed rakes, 8 scythes, 5 straw-hooks,

from that for marriage (*coniugium*). The colours are those of the berries.

[2] See Glossary.

mentarias V, arborarias V, securis III, cuneos III, fistulam farrariam I, forpicis II, rutabulum I, focu-
4 los II; dolia olearia C, labra XII, dolia quo vinacios condat X, amurcaria X, vinaria X, frumentaria XX, labrum[1] lupinarium I, serias X, labrum eluacrum I, solium I, labra aquaria II, opercula doliis seriis priva; molas asinarias unas et trusatilis unas, Hispaniensis unas, molilia III, abacum I, orbes aheneos II, mensas II, scamna magna III, scamnum in cubiculo I,
5 scabilla III, sellas IIII, solia II, lectum in cubiculo I, lectos loris subtentos IIII et lectos III; pilam ligneam I, fullonicam I, telam togalem I, pilas II, pilum fabarium I, farrearium I, seminarium I, qui nucleos succernat I, modium I, semodium I; culcitas VIII, instragula VIII, pulvinos XVI, operimenta X, mappas III, centones pueris VI.

XI. Quo modo vineae iugera C instruere oporteat. Vilicum, vilicam, operarios X, bubulcum I, asinarium I, salictarium I, subulcum I, summa homines XVI; boves II, asinos plostrarios II, asinum molarium I; vasa torcula instructa III, dolia ubi quinque vindemiae esse possint culleum DCCC, dolia ubi vinaceos
2 condat XX, frumentaria XX, opercula doliorum et tectaria priva, urnas sparteas VI, amphoras sparteas IIII, infidibula II, cola vitilia III, cola qui florem demat III, urceos mustarios X; plostra II, aratra II, iugum plostrarium I, iugum vinarium I, iugum asinarium I, orbem aheneum I, molile I; ahenum quod capiat culleum I, operculum aheni I, uncos ferreos III, ahenum coculum quod capiat culleum I,
3 urceos aquarios II, nassiternam I, pelvim I, matellionem I, trulleum I, situlum aquarium I, scutriscum

[1] *labrum* added by Iucundus, from 11, 3.

---

[1] The watery residue left when the oil is drained from the

5 pruning-hooks, 3 axes, 3 wedges, 1 hand-mill, 2 tongs, 1 poker, 2 braziers; 100 oil jars, 12 pots, 10 jars for holding grape pulp, 10 for holding *amurca*,[1] 10 wine jars, 20 grain jars, 1 lupine vat, 10 large jars, 1 wash-tub, 1 bath-tub, 2 water-basins, several covers for jars and pots; 1 donkey-mill, 1 hand-mill, 1 Spanish mill, 3 collars and traces, 1 small table,[2] 2 copper disks, 2 tables, 3 large benches, 1 bedroom stool, 3 stools, 4 chairs, 2 arm-chairs, 1 bed in the bedroom, 4 beds on cords, and 3 common beds; 1 wooden mortar, 1 fuller's mortar, 1 loom, 2 mortars, 4 pestles—one for beans, one for grain, one for seed, one for cracking kernels; 1 modius[3] measure, 1 half-modius measure; 8 mattresses, 8 coverlets, 16 cushions, 10 table covers, 3 napkins, 6 servants' hoods.

XI. This is the proper equipment for a vineyard of 100 iugera: An overseer, a housekeeper, 10 labourers, 1 teamster, 1 muleteer, 1 willow-worker, 1 swineherd—a total of 16 persons; 2 oxen, 2 draft donkeys, 1 for the mill; 3 complete presses, vats for holding five vintages of 800 cullei,[3] 20 jars for holding grape pulp, 20 for grain, and the necessary covers and tops; 6 pots covered with Spanish broom, 4 amphorae[3] of the same kind, 2 funnels, 3 wicker strainers, 3 strainers for removing the flower, 10 vessels for juice; 2 carts, 2 ploughs, 1 wagon yoke, 1 *iugum vinarium*,[3] 1 donkey yoke; 1 copper disk, 1 mill harness, 1 copper vessel holding a culleus, 1 copper cover, 3 iron hooks, 1 copper boiler holding a culleus, 2 water pots, 1 watering-pot, 1 basin, 1 small pot, 1 wash-basin, 1 water-bucket, 1 platter, 1 ladle, 1

crushed olives. For uses, see Chapters 66–7, 69, 92–3, 95–101, 103, 128–130.

[2] Or kneading-trough.

[3] See Glossary.

I, trullam I, candelabrum I, matellam I, lectos IIII, scamnum I, mensas II, abacum I, arcam vestiariam I, armarium promptarium I, scamna longa VI, rotam aquariam I, modium praeferratum I, semodium I, labrum eluacrum I, solium I, labrum lupinarium I, 4 serias X; ornamenta bubus II, ornamenta asinis instrata III, semuncias III, sportas faecarias III, molas asinarias III, molas trusatilis unas; ferramenta: falces sirpiculas V, falces silvaticas VI, arborarias III, secures V, cuneos IIII, vomeres II, ferreas X, palas VI, rutra IIII, rastros quadridentes II, crates stercorarias IIII, sirpiam stercorariam I, falculas viniaticas XL, falculas rustarias X, foculos II, forpi-5 ces II, rutabulum I; corbulas Amerinas XX, quala sataria vel alveos XL, palas ligneas XL, luntris II, culcitas IIII, instragula IIII, pulvinos VI, operimenta VI, mappas III, centones pueris VI.

XII. In torcularium quae opus sunt. Vasis quinis prela temperata V, supervacanea III, suculas V, supervacaneam I, funes loreos V, subductarios V, melipontos V, troclias X, capistra V, assercula ubi prela sita sient V, serias III, vectes XL, fibulas XL constibilis ligneas, qui arbores conprimat, si dishiascent, et cuneos VI, trapetos V, cupas minusculas X, alveos X, palas ligneas X, rutra ferrea quinque.

XIII. In torcularium in usu quod opus est. Urceum I, ahenum quod capiat Q. V, uncos ferreos III, orbem aheneum I, molas [1], cribrum I, incerniculum I, securim I, scamnum I, seriam vinariam I, clavem torculari I, lectum stratum ubi duo custodes liberi cubent (tertius servus una cum factoribus uti cubet), fiscinas novas,[1] veteres,[1] epidromum I, pulvinum I, lucernas,[1] corium I, craticulas duas, carnarium I, scalas unas.

[1] The numeral has been lost.

candlestick, 1 chamber-vessel, 4 beds, 1 bench, 2 tables, 1 small table,[1] 1 clothes chest, 1 wardrobe, 6 long benches, 1 water-wheel, 1 iron-bound modius measure, 1 half-modius measure, 1 wash-tub, 1 bath-tub, 1 lupine vat, 10 large pots; 2 complete sets of ox-harness, 3 of donkey-harness, 3 pack-saddles, 3 baskets for wine-lees, 3 donkey-mills, 1 hand-mill; tools: 5 rush-hooks, 6 tree-hooks, 3 pruning-hooks, 5 axes and 4 wedges, 2 ploughs, 10 forks, 6 spades, 4 shovels, 2 four-toothed rakes, 4 manure-hampers, 1 manure-basket; 40 grape-knives, 10 broom-hooks, 2 braziers, 2 tongs, 1 poker; 20 Amerine baskets, 40 planting-baskets or troughs, 40 wooden scoops, 2 trays, 4 mattresses, 4 coverlets, 6 cushions, 6 table covers, 3 napkins, 6 servants' hoods.

XII. This is the necessary equipment for the pressing-room: For 5 vats, 5 mounted press-beams, with 3 spares; 5 windlasses, with 1 spare; 5 leather ropes, 5 hoisting ropes; 5 cables; 10 pulleys; 5 bands; 5 posts for the press-beams to rest on; 3 large jars; 40 levers; 40 stout wooden pins to brace the anchor-posts if they spread, and 6 wedges; 5 mills, 10 small casks, 10 troughs, 10 wooden spades, 5 iron shovels.

XIII. The following equipment is needed for the pressing-room at the time of pressing: A pitcher, 1 copper vessel holding 5 quadrantals, 3 iron hooks, 1 copper disk, — millstones, 1 strainer, 1 sieve, 1 axe, 1 bench, 1 large wine-jar, 1 key for the pressing-room, 1 complete bed for two free workmen who act as watchmen to sleep on (while the third, who is a slave, should sleep with the labourers), — new and — old baskets, 1 net-cord, 1 cushion, — lanterns, 1 hide, 2 gridirons, 1 meat-rack, 1 ladder.

[1] Or kneading-trough.

2 In cellam oleariam haec opus sunt. Dolia olearia, opercula, labra olearia XIIII, concas maioris II et minoris II, trullas aheneas tris, amphoras olearias II, 3 urceum aquarium I, urnam quinquagenariam I, sextarium olearium I, labellum I, infidibula II, spongeas II, urceos fictiles II, urnales II, trullas ligneas II, claves cum clostris in cellas II, trutinam I, centumpondium[1] I et pondera cetera.

XIV. Villam aedificandam si locabis novam ab solo, faber haec faciat oportet. Parietes omnes, uti iussitur, calce et caementis, pilas ex lapide angulari, tigna omnia, quae opus sunt, limina, postes, iugumenta, asseres, fulmentas, praesepis bubus hibernas 2 aestivas faliscas, equile, cellas familiae, carnaria III, orbem, ahenea II, haras X, focum, ianuam maximam et alteram quam volet dominus, fenestras, clatros in fenestras maioris bipedalis X, luminaria VI, scamna III, sellas V, telas togalis duas, paullulam pilam ubi triticum pinsat I, fulloniam I, antepagmenta, vasa 3 torcula II. Hae rei materiem et quae opus sunt dominus praebebit et ad opus dabit, serram I, lineam I (materiem dumtaxat succidet, dolabit, secabit facietque conductor), lapidem, calcem, harenam, aquam, paleas, terram unde lutum fiat. Si de caelo villa tacta siet, de ea re verba uti fiant.[2] Huic operi pretium ab domino bono, qui bene praebeat quae opus sunt et nummos fide bona solvat, in tegu- 4 las singulas II S. Tectum sic numerabitur: tegula integra quae erit, quae non erit, unde quarta pars aberit, duae pro una, conliciares quae erunt pro binis

[1] *centumpondium* Keil: *centumpondium incertum.*
[2] *verba uti fiant* Keil: *v(iri) b(oni) a(rbitratu) uti fiat* Brehaut.

The following equipment is needed for the oil cellar: Oil jars and covers, 14 oil vats, 2 large and 2 small oil flasks, 3 copper ladles, 2 oil amphorae, 1 water-jar, 1 jar holding fifty heminae (?),[1] 1 sextarius oil-measure, 1 pan, 2 funnels, 2 sponges, 2 earthenware pitchers, 2 half-amphora measures, 2 wooden ladles, 2 locks with bars for the cellar, 1 set of scales, 1 one-hundred-pound weight, and other weights.

XIV. If you are contracting for the building of a new steading from the ground up, the contractor should be responsible for the following:—All walls as specified, of quarry-stone set in mortar, pillars of solid masonry, all necessary beams, sills, uprights, lintels, door-framing, supports, winter stables and summer feed racks for cattle, a horse stall, quarters for servants, 3 meat-racks, a round table, 2 copper boilers, 10 coops, a fireplace, 1 main entrance and another at the option of the owner, windows, 10 two-foot lattices for the larger windows, 6 window-shutters, 3 benches, 5 stools, 2 looms, 1 small mortar for crushing wheat, 1 fuller's mortar, trimmings, and 2 presses. The owner will furnish the timber and necessary material for this and deliver it on the ground, and also 1 saw and 1 plumb-line (but the contractor will fell, hew, square, and finish the timber), stone, lime, sand, water, straw, and earth for making mortar. If the steading should be struck by lightning an expiatory prayer must be offered. The price of this work from an honest owner, who furnishes duly all necessary materials and pays conscientiously, one sesterce[1] per tile. The roof will be reckoned as follows: On the basis of a whole tile, one which is one-fourth broken is counted two

[1] See Glossary.

putabuntur; vallus quot erunt, in singulas quaternae numerabuntur.

Villa lapide calce. Fundamenta supra terram pede, ceteros parietes ex latere, iugumenta et ante-
5 pagmenta quae opus erunt indito. Cetera lex uti villae ex calce caementis. Pretium in tegulas singulas II S. Loco salubri bono domino haec quae supra pretia posita sunt: ex signo manipretium erit. Loco pestilenti, ubi aestate fieri non potest, bono domino pars quarta preti accedat.

XV. Macerias ex calce caementis silice. Uti dominus omnia ad opus praebeat, altam P. V et columen P. I, crassam P. I S, longam P. XIV, et uti sublinat locari oportet. Parietes villae si locet in P. C, id est P. X quoquo versum, libellis in ped. V et perticam I P. VIC. N. X. Sesquipedalem parietem dominus fundamenta faciat et ad opus praebeat calcis in P. singulos in longitudinem modium unum, harenae modios duos.

XVI. Calcem partiario coquendam qui dant, ita datur. Perficit et coquit et ex fornace calcem eximit calcarius et ligna conficit ad fornacem. Dominus lapidem, ligna ad fornacem, quod opus siet, praebet.

XVII. Robus materies, item ridica, ubi solstitium fuerit ad brumam semper tempestiva est. Cetera materies quae semen habet, cum semen maturum habet, tum tempestiva est. Quae materies semen

---

[1] This strange method of estimating the cost of a house has puzzled scholars, and various emendations have been proposed. The *vallus* is the semicylindrical tile overlapping the flat tiles at the line of juncture.

[2] This is really "adobe," clay mixed with straw and sundried.

[3] The rendering here given follows the Italian translation of Curcio.

for one; all gutter tiles are counted each as two; and all joint-tiles each as four.[1]

In a steading of stone and mortar groundwork, carry the foundation one foot above ground, the rest of the walls of brick[2]; add the necessary lintels and trimmings. The rest of the specifications as for the house of rough stone set in mortar. The cost per tile will be one sesterce. The above prices are for a good owner, in a healthful situation. The cost of workmanship will depend upon the count.[3] In an unwholesome situation, where summer work is impossible, the generous owner will add a fourth to the price.

XV. Construct the enclosure walls of mortar, rough stone, and rubble (the owner furnishing all the materials) five feet high, 1½ feet thick, with a one-foot coping, 14 feet long, and let out the plastering. If he lets the walls of the steading by the hundred feet, that is, ten feet on every side, 5 *libellae*[4] to the foot, and 10 *victoriati* for a strip one foot by ten.[5] The owner shall build the foundation 1½ feet thick, and will furnish one modius of lime and two modii of sand for each linear foot.

XVI. The following are proper terms of a contract for burning lime on shares: The burner prepares the kiln, burns the lime, takes it from the kiln, and cuts the wood for the kiln. The owner furnishes the necessary stone and wood for the kiln.

XVII. Oak wood and also wood for vine props is always ripe for cutting at the time of the winter solstice. Other species which bear seed are ripe when the seeds are mature, while those which are seedless are ripe when they shed bark. The pine,

[4] The *libella* was $\frac{1}{10}$ a *denarius*, the *victoriatus* ½.
[5] The *pertica* (= *decempeda*) being a 10-ft. measure.

non habet, cum glubebit, tum tempestiva est. Pinus[1] eo, quia[2] semen viride et maturum habet (id semen de cupresso, de pino quidvis anni legere possis), 2 item quidvis anni matura est et tempestiva. Ibidem sunt nuces bimae, inde semen excidet, et anniculae, eae ubi primum incipiunt hiascere, tum legi oportet; per sementim primum incipiunt maturae esse, postea usque adeo sunt plus menses VIII. Hornotinae nuces virides sunt. Ulmus, cum folia cadunt, tum iterum tempestiva est.

XVIII. Torcularium si aedificare voles quadrinis vasis, uti contra ora sient, ad hunc modum vasa conponito. Arbores crassas P. II, altas P. VIIII cum car- 2 dinibus, foramina longa P. III S exculpta digit. VI, ab solo foramen primum P. I S, inter arbores et parietes P. II, in II arbores P. I, arbores ad stipitem primum derectas P. XVI, stipites crassos P. II, altos[3] cum cardinibus P. X, suculam praeter cardines P. VIIII, prelum longum P. XXV, inibi lingulam P. II S, pavimentum binis vasis cum canalibus duabus P. XXX, IIII trapetibus locum dextra sinistra pavimentum 3 P. XX, inter binos stipites vectibus locum P. XXII, alteris vasis exadversum ab stipite extremo ad parietem qui pone arbores est P. XX: summa torculario vasis quadrinis latitudine P. LXVI, longitudine P. LII. Inter parietes, arbores ubi statues, fundamenta bona facito alta P. V, inibi lapides silices, totum forum longum P. V, latum P. II S, crassum 4 P. I S. Ibi foramen pedicinis duobus facito, ibi arbores pedicino in lapide statuito. Inter duas

[1] *Pinus* added by Keil.

[2] *eo quia* MSS. and ed. pr.: *ea quae* Iucundus, followed by many editors.

[3] *crassos . . . altos* Pontedera: *crassi . . . alti.*

because it has both green and ripe seed (such seed may be gathered from the cypress and the pine at any season) is ripe and ready at any season. The same tree has second-year cones from which the seed will fall, and first-year cones; when the latter are just beginning to open, they are ready for gathering. They begin to ripen at seed-time, and continue to ripen then for more than eight months. The first-year cones are green. The elm is fit for cutting a second time when the leaves fall.

XVIII. If you wish to build a pressing-room with four vats facing each other, lay off the vats as follows: Anchor-posts 2 feet thick, 9 feet high, including tenons; openings hollowed out 3½ feet long, 6 fingers wide, the bottom of the opening 1½ feet from the ground; 2 feet between anchor-post and wall; 1 foot between the two anchor-posts, and 16 feet straight to the first guide-posts; guide-posts 2 feet in diameter and 10 feet high, including the tenons; windlass 9 feet high, exclusive of mortice; press-beam 25 feet long, and the tongue on it 2½ feet long. Allow 30 feet of floor space for each pair of vats, with their conduits, and 20 feet for four mills, right and left. Allow 22 feet between the guide-posts of one press and those of the next for the levers. Allow 20 feet for the second set of vats facing them, from the last guide-post to the wall behind the anchor-posts. Total for the pressing-room with four vats, 66 feet by 52 feet. Between the walls, where you intend to mount the anchor-posts, make solid foundations 5 feet deep; cover the whole area 5 feet by 2½ feet with hard stones to a depth of 1½ feet; in this clear a place for two bolts, and fix the posts firmly in the stone with the

arbores quod loci supererit robore expleto, eo plumbum infundito. Superiorem partem arborum digitos VI altam facito siet, eo capitulum robustum indito, 5 uti siet stipites ubi stent. Fundamenta P. V facito, ibi silicem longum P. II S, latum P. II S, crassum P. I S planum statuito, ibi stipites statuito. Item alterum stipitem statuito. Insuper arbores stipitesque trabem planam inponito latam P. II, crassam P. I, longam P. XXXVII, vel duplices indito, si solidas non habebis. Sub eas trabes inter canalis et parietes extremos, ubi trapeti stent, trabeculam pedum XXIII S inponito sesquipedalem, aut binas 6 pro singulis eo supponito. In iis trabeculis trabes, quae insuper arbores stipites stant, conlocato; in iis tignis parietes extruito iungitoque materiae, uti oneris satis habeat. Aram ubi facies, pedes V fundamenta alta facito, lata P. VI, aram et canalem 7 rutundam facito latam P. IIII S, ceterum pavimentum totum fundamenta P. II facito. Fundamenta primum festucato, postea caementis minutis et calce harenato semipedem unum quodque corium struito. Pavimenta ad hunc modum facito: ubi libraveris, de glarea et calce harenato primum corium facito, id pilis subigito, idem alterum corium facito; eo calcem cibro subcretam indito alte digitos duo, ibi de testa arida pavimentum struito; ubi structum erit, pavito 8 fricatoque, uti pavimentum bonum siet. Arbores stipites robustas facito aut pineas. Si trabes minores facere voles, canalis extra columnam expolito. Si 9 ita feceris, trabes P. XXII longae opus erunt. Orbem olearium latum P. IIII Punicanis coagmentis facito, crassum digitos VI facito, subscudes iligneas adindito.

bolt. Fill the interval between the two anchor-posts with oak, and pour lead over it. Let the head of the anchor-posts project six fingers, and cap it with an oak head so as to make a place for the posts to stand. Make a 5-foot foundation and lay on it a flat stone, $2\frac{1}{2}$ by $2\frac{1}{2}$ by $1\frac{1}{2}$ feet, and set the posts on it. Mount the corresponding posts in the same way. Above the anchor-posts and the guide-posts lay a horizontal beam, 2 feet by 1, 37 feet long, or two beams if you have no solid ones of that size. Under these beams, between the conduits and the end walls, in the position of the mills, run a beam $1\frac{1}{2}$ feet square and $23\frac{1}{2}$ feet long, or two pieces. On these rest the beams which stand above the main posts, and on these timbers build a wall and join it to the timber to give it sufficient weight. Where you are to build a seat for the press make a foundation 5 feet deep, 6 feet across; the seat and circular conduit $4\frac{1}{2}$ feet in diameter. For the rest of the pavement make the foundation uniformly 2 feet deep. First pack down the bottom, and then spread successive half-foot layers of finely crushed stone and sanded lime. Construct the pavement as follows: After levelling, spread the first layer of gravel and sanded lime, and tamp it down; then spread a similar layer over it, sift lime with a sieve to the depth of two fingers, and then lay a pavement of dry pot-sherds. When completed, pack and rub down so as to have a smooth surface. All anchor-posts and guide-posts should be of oak or pine. If you wish to use shorter timbers, cut conduits on the outside; if this method is employed you will need 22-foot timbers. Make the disk 4 feet in diameter, 6 fingers thick, constructed in sections in the Punic style with

Eas ubi confixeris, clavis corneis occludito. In eum orbem tris catenas indito. Eas catenas cum orbi clavis ferreis corrigito. Orbem ex ulmo aut ex corylo facito: si utrumque habebis, alternas indito.

XIX. In vasa vinaria stipites arboresque binis pedibus altiores facito, supra foramina arborum, pedem quaeque uti absiet, unae fibulae locum facito. Semipedem quoquo versum in suculam sena foramina 2 indito. Foramen quod primum facies semipedem ab cardine facito, cetera dividito quam rectissime. Porculum in media sucula facito. Inter arbores medium quod erit, id ad mediam conlibrato, ubi porculum figere oportebit, uti in medio prelum recte situm siet. Lingulam cum facies, de medio prelo conlibrato, ut inter arbores bene conveniat, digitum pollicem laxamenti facito. Vectes longissimos P. XIIX, secundos P. XVI, tertios P. XV, remissarios P. XII, alteros P. X, tertios P. VIII.

XX. Trapetum quo modo concinnare oporteat. Columellam ferream, quae in miliario stat, eam rectam stare oportet in medio ad perpendiculum, cuneis salignis circumfigi oportet bene, eo plumbum effundere, caveat[1] ni labet columella. Si movebitur, eximito; denuo eodem modo facito, ne se moveat. 2 Modiolos in orbis oleagineos ex orcite olea facito, eos circumplumbato, caveto ne laxi sient. In cupam[2] eos indito. Cunicas[3] solidas latas digitum pollicem facito, labeam bifariam faciat habeant, quas figat clavis duplicibus, ne cadant.

XXI. Cupam facito P. X, tam crassam quam

[1] *caveat* MSS.: *caveto* Iucundus, followed by Schneider.
[2] The older editions read, *si autem labent in cupam.*
[3] *tunicas* in the older editions.

dovetailed oak. When you have fitted them together, fasten with pins of dogwood. Fit three crossbars to the disk, and fasten them with iron nails. Make the disk of elm or hazel; if you have both, lay them alternately.

XIX. For a wine press make the guide-posts and anchor-posts two feet higher, and above the holes in the anchor-posts, which should be one foot apart, make a place for one pin. Cut six openings, a half-foot square, in each of the windlass beams, placing the first a half-foot from the tenon, and the others at equal intervals. Set a hook in the middle of the windlass; the centre of the distance between the anchor-posts should correspond with the middle of the windlass, where the hook should be set, in order to have the press-beam exactly in the middle. When you set the tongue, measure from the centre of the press-beam so that it may be exactly midway between the anchor-posts; allow one thumb width play. The longest levers are 18 feet, the second size 16, the third 15; the hand-spikes are 12, 10, and 8 feet respectively.

XX. Method of mounting the mill. The iron pivot which stands on the post must stand straight upright in the centre; it should be fastened firmly on all sides with willow wedges, and lead should be poured over it to prevent it from shaking; if it moves, take it out and fasten it again in the same way, so that it will not move. Make the sockets for the stones of orcite olive wood, and fasten them with lead, being careful to keep them tight. Fix them on the axle. Make one piece bushings, a thumb wide, flanged at both ends and double-nailed to keep them from falling out.

XXI. Make a ten-foot bar as thick as the sockets

modioli postulabunt, media inter orbis quae conveniat. Crassam quam columella ferrea erit, eam mediam pertundito, uti columellam indere possis. Eo fistulam ferream indito, quae in columellam conveniat et in
2 cupam. Inter cupam dextra sinistra pertundito late digitos primoris IIII, alte digitos primoris III, sub cupa tabulam ferream, quam[1] lata cupa media erit, pertusam figito, quae in columellam conveniat. Dextra sinistra, foramina ubi feceris, lamnis circumplectito. Replicato in inferiorem partem cupae omnis quattuor lamminas; utrimque secus lamminas
3 sub lamminas pollulas minutas supponito, eas inter sese configito, ne foramina maiora fiant, quo cupulae minusculae indentur. Cupa qua fini in modiolos erit, utrimque secus imbricibus ferreis quattuor de suo sibi utrimque secus facito qui figas. Imbrices medias clavulis figito. Supra imbrices extrinsecus cupam pertundito, qua clavus eat, qui orbem cludat.
4 Insuper foramen librarium ferreum digitos sex latum indito, pertusum utrimque secus, qua clavus eat. Haec omnia eius rei causa fiunt, uti ne cupa in lapide conteratur. Armillas IIII facito, quas circum orbem indas, ne cupa et clavus conterantur intrinsecus.
5 Cupam materia ulmea aut faginea facito. Ferrum factum quod opus erit uti idem faber figat; HS LX opus sunt. Plumbum in cupam emito HS IIII. Cupam qui concinnet et modiolos qui indat et plumbet, operas fabri dumtaxat HS VIII; idem trapetum oportet accommodet. Summa sumpti HS LXXII praeter adiutores.

XXII. Trapetum hoc modo accommodare oportet. Librator uti statuatur pariter ab labris. Digitum minimum orbem abesse oportet ab solo mortari.

[1] *quam* added by Keil.

require, the mid-point to fit between the stones. Drill a hole in the middle as large as the iron pivot, so that the latter may be inserted in it. Insert here an iron casing to fit into the pivot and the bar. Make a hole in the bar, 4 finger-tips square and 3 finger-tips deep, and on the lower side of the bar fasten an iron plate of the breadth of the middle of the bar, perforated to fit over the pivot. After piercing the holes face them on both sides with metal plates, and bend back all four plates to the lower side of the bar; under these plates fasten thin metal strips on both sides, and fasten them together so that the holes in which the small handles are fitted may not spread. At the point where the bar enters the sockets be careful to face them on both sides with four trough-shaped iron plates and fasten them in the middle with nails. Above these plates pierce the bar on the outside for the bolt to fasten the stone. On top of the opening place a one-pound iron collar, 6 fingers wide, pierced on both sides to allow the bolt to enter. All this is for the purpose of preventing the bar from wearing on the stone. Make four rings to place around the stone to keep the bar and the bolt from wearing on the inside. Use elm or beech for the bar. The same smith should make and set the necessary iron work, at a cost of 60 sesterces; you can buy lead for the bar for 4 sesterces; wages of the workman who assembles and sets the sockets with lead, at least 8 sesterces, and the same man should adjust the mill. Total cost, 72 sesterces, exclusive of helpers.

XXII. The mill should be adjusted as follows: Level it so that the stones are set at equal distances from the rims and clearing the bottom of the mortar by a little finger's breadth; see that the stones do not

Orbes cavere oportet nequid mortarium terant. Inter orbem et miliarium unum digitum interesse oportet. Si plus intererit atque orbes nimium 2 aberunt, funi circumligato miliarium arte crebro, uti expleas quod nimium interest. Si orbes altiores erunt atque nimium mortarium deorsom teret, orbiculos ligneos pertusos in miliarium in columella supponito, eo altitudinem temperato. Eodem modo latitudinem orbiculis ligneis aut armillis ferreis temperato, usque dum recte temperabitur.

3 Trapetus emptus est in Suessano HS CCCC et olei P.L. Conposturae HS LX; vecturam boum, operas VI, homines VI cum bubulcis HS LXXII; cupam ornatam HS LXXII, pro oleo HS XXV; S. S. HS DCXXVIIII. Pompeis emptus ornatus HS CCCXXCIIII; vecturam HS CCXXC; domi melius concinnatur et accommodatur, eo sumpti opus est 4 HS LX: S. S. HS DCCXXIIII. Si orbes in veteres trapetos parabis, medios crassos P. I digitos III, altos P. I, foramen semipedem quoquo vorsum. Eos cum advexeris, ex trapeto temperato. Ii emuntur ad Rufri macerias HS CXXC, temperantur HS XXX. Tantidem Pompeis emitur.

XXIII. Fac ad vindemiam quae opus sunt ut parentur. Vasa laventur, corbulae sarciantur, picentur, dolia quae opus sunt picentur, quom pluet; quala parentur, sarciantur, far molatur, menae emantur, 2 oleae caducae salliantur. Uvas miscellas, vinum praeliganeum quod operarii bibant, ubi tempus erit,

---

[1] This disproportionate dimension suggests to Schneider a faulty text; cf. specifications for stones of the second size in 135, 6.

rub the basin at all. There should be a finger's breadth between the stone and the column; if the space is greater and the stones are too far distant, wind a cord around the column tightly several times so as to fill in the excessive space. If the stones are set too deep and rub the bottom of the basin too much, place perforated wooden disks over the pivot and on the column and thus regulate the height. In the same way adjust the spread with wooden disks or iron rings until the stones fit accurately.

A mill is bought near Suessa for 400 sesterces and fifty pounds of oil. The cost of assembling is 60 sesterces, and the charge for transportation by oxen, with six days' wages of six men, drivers included, is 72 sesterces. The bar complete costs 72 sesterces, and there is a charge of 25 sesterces for oil; the total cost is 629 sesterces. At Pompeii one is bought complete for 384 sesterces, freight 280 sesterces. It is better to assemble and adjust on the ground, and this will cost 60 sesterces, making a total cost of 724 sesterces. If you are fitting old mills with stones, they should be 1 foot 3 fingers thick at the centre and 1 foot in diameter,[1] with a half-foot square opening; alter them to fit the mill after they have been hauled. These can be bought at the yard of Rufrius for 180 sesterces, and fitted for 30 sesterces. The price is the same at Pompeii.

XXIII. Have everything that is needed ready for the vintage; let vats be cleaned, baskets mended and pitched, necessary jars be pitched on rainy days; let hampers be made ready and mended, spelt be ground, salt fish be bought, and windfall olives be salted. Gather the inferior grapes for the sharp wine for the hands to drink, when the time comes.

legito. Siccum puriter omnium dierum pariter in dolia dividito. Si opus erit, defrutum indito in mustum de musto lixivo coctum, partem quadragesimam addito defruti vel salis sesquilibram in 3 culleum. Marmor si indes, in culleum libram indito; id indito in urnam, misceto cum musto; id indito in doleum. Resinam si indes, in culleum musti P. III, bene conminuito, indito in fiscellam et facito uti in doleo musti pendeat; eam quassato crebro, uti 4 resina condeliquescat. Ubi[1] indideris defrutum aut marmor aut resinam, dies XX permisceto crebro, tribulato cotidie. Tortivum mustum circumcidaneum suo cuique dolio dividito additoque pariter.

XXIV. Vinum Graecum hoc modo fieri oportet. Uvas Apicias percoctas bene legito. Ubi delegeris, in eius musti culleum aquae marinae veteris Q. II indito[2] vel salis puri modium; eum in fiscella suspendito sinitoque cum musto distabescat. Si helvolum vinum facere voles, dimidium helvoli, dimidium Apicii vini indito, defruti veteris partem tricesimam addito. Quidquid vini defrutabis, partem tricesimam defruti addito.

XXV. Quom vinum coctum erit et quom legetur, facito uti servetur familiae primum suisque, facitoque studeas bene percoctum siccumque legere, ne vinum nomen perdat. Vinaceos cotidie recentis succernito lecto restibus subtento, vel cribrum illi rei parato. Eos conculcato in dolia picata vel in lacum vinarium picatum. Id bene iubeto oblini, quod des bubus per hiemem. Indidem, si voles, lavito paulatim. Erit lora familiae quod bibat.

[1] *Ubi* added by Keil. [2] *indito* added by Keil.

[1] Squeezed from the chopped up (*circumcidaneum*) mass of grape refuse after the ordinary pressing; cf. Varro, I, 54, 3, and Columella, XII, 36.

Divide the grapes gathered each day, after cleaning and drying, equally between the jars. If necessary, add to the new wine a fortieth part of must boiled down from untrod grapes, or a pound and a half of salt to the culleus. If you use marble dust, add one pound to the culleus; mix this with must in a vessel and then pour into the jar. If you use resin, pulverize it thoroughly, three pounds to the culleus of must, place it in a basket, and suspend it in the jar of must; shake the basket often so that the resin may dissolve. When you use boiled must or marble dust or resin, stir frequently for twenty days and press down daily. Divide the must of the second pressing[1] and add equally to each jar.

XXIV. Directions for making Greek wine: Gather carefully well-ripened Apician grapes, and add to the culleus of must two quadrantals of old sea-water, or a modius of pure salt. If the latter is used, suspend it in a basket and let it dissolve in the must. If you wish to make a straw-coloured wine, take equal parts of yellow and Apician wine and add a thirtieth of old boiled wine. Add a thirtieth part of concentrated must to any kind of blended wine.

XXV. When the grapes are ripe and gathered, let the first be kept for household use. See that they are not gathered until they are thoroughly ripe and dry, that the wine may not lose its reputation. Sift the fresh husks daily through a bed stretched on cords, or make a sieve for the purpose, and after treading place them in pitched jars or a pitched vat. Have this sealed tight, to feed to cattle through the winter; or if you wish you can soak some of it a while and you will have an after-wine for the hands to drink.

XXVI. Vindemia facta vasa torcula, corbulas, fiscinas, funis, patibula, fibulas iubeto suo quidquid loco condi. Dolia cum vino bis in die fac extergeantur, privasque scopulas in dolia facito habeas illi rei, qui labra doliorum circumfrices. Ubi erit lectum dies triginta, si bene deacinata erunt, dolia oblinito. Si voles de faece demere vinum, tum erit ei rei optimum tempus.

XXVII. Sementim facito, ocinum, viciam, faenum Graecum, fabam, ervum, pabulum bubus. Alteram et tertiam pabuli sationem facito. Deinde alias fruges serito. Scrobis in vervacto oleis, ulmis, vitibus, ficis; simul cum semine serito. Si erit locus siccus, tum oleas per sementim serito, et quae ante satae erunt, teneras tum supputato et arbores ablaqueato.

XXVIII. Oleas, ulmos, ficos, poma, vites, pinos, cupressos cum seres, bene cum radicibus eximito cum terra sua quam plurima circumligatoque, uti ferre possis; in alveo aut in corbula ferri iubeto. Caveto, quom ventus siet aut imber, effodias aut feras; nam 2 id maxime cavendum est. In scrobe quom pones, summam terram subdito; postea operito terra radicibus fini, deinde calcato pedibus bene, deinde festucis vectibusque calcato quam optime poteris; id erit ei rei primum. Arbores crassiores digitis V quae erunt, eas praecisas serito oblinitoque fimo summas et foliis alligato.

XXIX. Stercus dividito sic. Partem dimidiam in segetem, ubi pabulum seras, invehito, et si ibi olea

[1] Trigonella Faeno-Graecum, an annual leguminous plant.

[2] Columella, II, 8, 2, fixes the time "from the 24th of October to the time of the winter solstice." Varro, I, 34, gives the same dates.

[3] Compare Chapters 28, 32, 41, 42, 44, 45, 49, 51, 52; Varro, I, 40; Columella, V. 9.

XXVI. After the vintage is over order all the pressing utensils, hampers, baskets, ropes, props, and bars to be stored, each in its proper place. Have the jars containing wine wiped off twice a day, and see that you provide each jar with its own broom with which to wipe off the edges. Thirty days after the gathering, if the fermentation is complete, seal the jars. If you wish to draw off the wine from the lees, this will be the best time to do it.

XXVII. Sow clover, vetch, fenugreek,[1] beans, and bitter-vetch as forage for cattle. Make a second and a third sowing of forage; then plant the other crops. Dig trenches in fallow ground for olives, elms, vines, and figs, and plant at seed-time.[2] If the ground is dry, transplant[3] olives at seed-time, prune the young olives which had been planted before, and trench the trees.

XXVIII. In transplanting olives, elms, figs, fruit trees, vines, pines, and cypresses, dig them up carefully, roots and all, with as much of their own soil as possible, and tie them up so that you can transport them. Have them carried in a box or basket. Be careful not to dig them up or transport them when the wind is blowing or when it is raining, for this is especially to be avoided. When you place them in the trench, bed them in top soil, spread dirt over them to the ends of the roots, trample it thoroughly, and pack with rammers and bars as firmly as possible; this is the most important thing. Before transplanting, cut off the tops of trees which are more than five fingers in diameter and smear the scars with dung and wrap them in leaves.

XXIX. Divide your manure as follows: Haul one-half for the forage crops, and when you sow these, if

erit, simul ablaqueato stercusque addito: postea pabulum serito. Partem quartam circum oleas ablaqueatas, quom[1] maxime opus erit, addito terraque stercus operito. Alteram quartam partem in pratum reservato idque, quom maxime opus erit, ubi favonius flabit, evehito luna silenti.

XXX. Bubus frondem ulmeam, populneam, querneam, ficulneam, usque dum habebis, dato. Ovibus frondem viridem, usque dum habebis, praebeto; ubi sementim facturus eris, ibi oves delectato; et frondem usque ad pabula matura. Pabulum aridum quod condideris in hiemem quam maxime conservato, cogitatoque hiemis quam longa siet.

XXXI. Ad oleam cogendam quae opus erunt parentur. Vimina matura, salix per tempus legatur, uti sit unde corbulae fiant et veteres sarciantur. Fibulae unde fiant, aridae iligneae, ulmeae, nuceae, ficulneae, fac in stercus aut in aquam coniciantur; inde, ubi opus erit, fibulas facito. Vectes iligneos, acrufolios, laureos, ulmeos facito uti sient parati. Prelum ex 2 carpino atra potissimum facito. Ulmeam, pineam, nuceam, hanc atque aliam materiem omnem cum effodies, luna decrescente eximito post meridiem sine vento austro. Tum erit tempestiva, cum semen suum maturum erit, cavetoque per rorem trahas aut doles. Quae materies semen non habebit, cum glubebit,[2] tempestiva erit. Vento austro caveto nequam materiem neve vinum tractes nisi necessario.

XXXII. Vineas arboresque mature face incipias putare. Vites propages in[3] sulcos; susum vorsum,

[1] *quom* Keil: *quam*. [2] *glubebit* Keil: *glube*.
[3] *in* added by Iucundus.

---

[1] A harbinger of spring. Varro, I, 28, makes the date the 7th of February. Cf. Chapter 50.

this ground is planted with olives, trench and manure them at this time; then sow the forage crops. Add a fourth of the manure around the trenched olives when it is most needed, and cover this manure with soil. Save the last fourth for the meadows, and when most needed, as the west wind is blowing,[1] haul it in the dark of the moon.

XXX. Feed the cattle elm, poplar, oak, and fig leaves as long as these last; and keep the sheep supplied with green leaves as long as you have them. Fold sheep on land which you intend to plant, and feed them leaves there until the forage is full grown. Save as carefully as possible the dry fodder which you have stored against winter, and remember how long winter lasts.

XXXI. Let all necessary preparations be made for the olive harvest: Let ripe withes and willow branches be gathered betimes as material for making new baskets and mending old ones. Have dry oak, elm, nut, and fig sticks for making pins buried in the dunghill or in water, and make pins from them when needed. Have oak, ilex, laurel, and elm levers ready. Make the press-beam preferably of black hornbeam. Take out elm, pine, nut, and all other timber which you are felling, when the moon is on the wane, after noon, while there is no south wind. It is ready for cutting when the seed is ripe. Be careful not to haul or work it in the wet. Timber that has no seed is ready for cutting when the bark peels. Do not handle any timber or vine when the south wind is blowing, unless you are compelled to do so.

XXXII. See that you begin early to trim vines and trees. Layer vines into trenches, and, so far as

quod eius facere poteris, vitis facito uti ducas. Arbores hoc modo putentur, rami uti divaricentur, quos relinques, et uti recte caedantur et ne nimium crebri relin- 2 quantur. Vites bene nodentur; per omnes ramos diligenter caveto ne vitem praecipites et ne nimium praestringas. Arbores facito uti bene maritae sint vitesque uti satis multae adserantur et, sicubi opus erit, de arbore deiciantur, uti in terram deprimantur, et biennio post praecidito veteres.

XXXIII. Viniam sic facito uti curetur. Vitem bene nodatam deligato recte, flexuosa uti ne sit, susum vorsum semper ducito, quod eius poteris. Vinarios custodesque recte relinquito. Quam altissimam viniam facito alligatoque recte, dum ne nimium constringas. Hoc modo eam curato. Capita vitium 2 per sementim ablaqueato. Vineam putatam circumfodito, arare incipito, ultro citroque sulcos perpetuos ducito. Vites teneras quam primum propagato, sic occato; veteres quam minimum castrato, potius, si opus erit, deicito biennioque post praecidito. Vitem novellam resicari tum erit tempus, ubi valebit. 3 Si vinea a vite calva erit, sulcos interponito ibique viveradicem serito, umbram ab sulcis removeto crebroque fodito. In vinea vetere serito ocinum, si macra erit (quod granum capiat ne serito), et circum capita addito stercus, paleas, vinaceas, aliquid 4 horum, quo rectius valeat. Ubi vinea frondere coeperit, pampinato. Vineas novellas alligato crebro,

---

[1] The "knots" are the joints or nodes where the vine will make new growth; of importance for budding (Virg., *Georg.* II, 74–77) or for cleft grafting (Col., IV, 29, 8–9).

[2] *i.e.*, support the clinging vines. See note 2, page **7**.

[3] Defined by Columella (IV, 21, 3) as short cut stumps to sprout out and replace those which may die near them.

possible, train them to grow vertically. The trees should be trimmed as follows: The branches which you leave should spread out, should be cut straight up, and should not be left too thick. The vines should be well knotted[1]; and be especially careful not to bend them downward along any of the branches and not to tie them too tightly. See that the trees are well "wedded,"[2] and that a sufficient number of vines are planted for them; and wherever it is necessary let these be detached from the trees and buried in the ground, and two years later cut them off from the old stock.

XXXIII. Have the vineyard treated as follows: Tie a well-knotted vine straight up, keeping it from bending, and make it grow vertically, so far as you can. Leave fruit-bearing shoots and reserve stubs[3] at proper intervals. Train the vines as high as possible and tie them firmly, but without choking them. Cultivate as follows: At seed-time trench the soil around the crown of the vine, and after pruning cultivate around it. Begin ploughing, and run straight furrows back and forth. Set out young vines as early as possible, then harrow; prune the old ones very slightly, or rather, if you need cuttings, layer the branches and take off the cuttings two years later. The proper time for cutting back the young plant is when it is strong. If there are gaps in the rows, run furrows and plant rooted cuttings, keep the furrows clear of shade, and cultivate frequently. In an old vineyard sow clover if the soil is lean (do not sow anything that will form a head), and around the roots apply manure, straw, grape dregs, or anything of the sort, to make it stronger. When the vine begins to form leaves, thin them. Tie up the young vines at frequent

ne caules praefringantur, et quae iam in perticam ibit, eius pampinos teneros alligato leviter corrigitoque, uti recte spectent. Ubi uva varia fieri coeperit, vites subligato, pampinato uvasque expellito, circum capita sarito.

5 Salictum suo tempore caedito, glubito arteque alligato. Librum conservato, cum opus erit in vinea, ex eo in aquam coicito, alligato. Vimina, unde corbulae fiant, conservato.

XXXIV. Redeo ad sementim. Ubi quisque locus frigidissimus aquosissimusque erit, ibi primum serito. In caldissimis locis sementim postremum fieri oportet. 2 Terram cave cariosam tractes. Ager rubricosus et terra pulla, materina, rudecta, harenosa, item quae aquosa non erit, ibi lupinum bonum fiet. In creta et uligine et rubrica et ager qui aquosus erit, semen adoreum potissimum serito. Quae loca sicca et non herbosa erunt, aperta ab umbra, ibi triticum serito.

XXXV. Fabam in locis validis non calamitosis serito. Viciam et faenum Graecum quam minime herbosis locis serito. Siliginem, triticum in loco aperto celso, ubi sol quam diutissime siet, seri oportet. Lentim in rudecto et rubricoso loco, qui herbosus non siet, 2 serito. Hordeum, qui locus novus erit aut qui restibilis fieri poterit, serito. Trimestre, quo in loco sementim maturam facere non potueris et qui locus restibilis crassitudine fieri poterit, seri oportet. Rapinam et coles rapicii unde fiant et raphanum in loco stercorato bene aut in loco crasso serito.

---

[1] See note 1, page 16.

[2] A leguminous plant, used extensively for forage; see Columella, II, 10, 1.

[3] A variety of wheat, probably with thick husk; see Columella, II, 6.

intervals to keep the stems from breaking, and when they begin to climb the props tie the tender branches loosely, and turn them so that they will grow vertically. When the grapes begin to turn, tie up the vines, strip the leaves so as to expose the grapes, and dig around the stocks.

Cut willows at the proper time, strip the bark, and tie them in tight bundles. Save the bark, and when you need it for the vines, steep some of it in water to make tapes. Save the withes for making baskets.

XXXIV. I return to the matter of planting. Plant the coldest and most humid ground first, and then the rest of the ground in turn to the warmest, which should come last. Do not work ground which is *cariosa* [1] at all. Lupine [2] will do well in soil that is reddish, and also in ground that is dark, or hard, or poor, or sandy, or not wet. Sow spelt [3] preferably in soil that is chalky, or swampy, or red, or humid. Plant wheat in soil that is dry, free from weeds, and sunny.

XXXV. Plant beans in strong soil which is protected from storms; vetch and fenugreek [4] in places as clear of weeds as possible. Wheat and winter wheat should be sown on high, open ground, where the sun shines longest. Lentils should be planted in unfertile and reddish soil, free of weeds; barley in new ground, or ground which does not need to lie fallow. Spring wheat should be planted in ground in which you cannot ripen the regular variety, or in ground which, because of its strength, does not need to lie fallow. Plant turnips, kohlrabi seed, and radishes in land well manured or naturally strong.

[4] See note 1, page 44.

XXXVI. Quae segetem stercorent. Stercus columbinum spargere oportet in pratum vel in hortum vel in segetem. Caprinum, ovillum, bubulum, item ceterum stercus omne sedulo conservato. Amurcam spargas vel inriges ad arbores; circum capita maiora amphoras, ad minora urnas cum aquae dimidio addito, ablaqueato prius non alte.

XXXVII. Quae mala in segete sint. Si cariosam terram tractes. Cicer, quod vellitur et quod salsum est, eo malum est. Hordeum, faenum Graecum, ervum, haec omnia segetem exsugunt et omnia quae velluntur. Nucleos in segetem ne indideris.

2 Quae segetem stercorent fruges: lupinum, faba, vicia.

Stercus unde facias: stramenta, lupinum, paleas, fabalia, acus, frondem iligneam, querneam. Ex segeti vellito ebulum, cicutam et circum salicta herbam altam ulvamque;[1] eam substernito ovibus bubusque, frondem putidam. Partem de nucleis succernito et in lacum coicito, eo aquam addito, permisceto rutro bene; inde lutum circum oleas ablaqueatas addito, nucleos conbustos item addito.

3 Vitis si macra erit, sarmenta sua concidito minute et ibidem inarato aut infodito.

Per hiemem lucubratione haec facito: ridicas et palos, quos pridie in tecto posueris, siccos dolato, faculas facito, stercus egerito. Nisi intermestri

4 lunaque dimidiata tum ne tangas materiem. Quam effodies aut praecides abs terra, diebus VII proximis, quibus luna plena fuerit, optime eximetur. Omnino

[1] *ulvamque* Victorius: *uvamque.*

---

[1] See note 1, page 24.

XXXVI. Fertilizers for crops: Spread pigeon dung on meadow, garden, and field crops. Save carefully goat, sheep, cattle, and all other dung. Spread or pour amurca[1] around trees, an amphora to the larger, an urn to the smaller, diluted with half its volume of water, after running a shallow trench around them.

XXXVII. Things which are harmful to crops: If you work land which is *cariosa;* chick peas are harmful, because they are torn out by the roots and are salty; barley, fenugreek, bitter vetch, and all crops which are pulled out by the roots, exhaust the soil. Do not bury olive seeds in land intended for crops.

Crops which fertilize land: Lupines, beans, and vetch.

You may make compost of straw, lupines, chaff, bean stalks, husks, and ilex and oak leaves. Pull up the elder and hemlock bushes which grow in the grain fields, and the high grass and sedge around the willow bed; use them for bedding down sheep, and decayed leaves for cattle. Separate part of the olive seeds and throw them into a pit, add water, and mix them thoroughly with a shovel. Make trenches around the olive trees and apply this mixture, adding also burned seeds. If a vine is unhealthy, cut its shoots into small bits and plough or spade them in around it.

The following is evening work for winter: Work up into vine poles and stakes the wood which was brought under cover the day before to dry out; make faggots; and clear out manure. Do not touch timber except in the dark of the moon, or in its last phase. The best time to take out timber which you

caveto nequam materiem doles neu caedas neu tangas, si potes, nisi siccam neu gelidam neu roru-
5 lentam. Frumenta face bis sarias runcesque avenamque destringas. De vinea et arboribus putatis sarmenta degere et fascinam face et vitis et ligna in caminum ficulna et codicillos domino in acervum conpone.

XXXVIII. Fornacem calcariam pedes latam X facito, altam pedes XX, usque ad pedes tres summam latam redigito. Si uno praefurnio coques, lacunam intus magnam facito, uti satis siet ubi cinerem concipiat, ne foras sit educendus. Fornacemque bene struito; facito fortax totam fornacem infimam conplectatur.
2 Si duobus praefurniis coques, lacuna nihil opus erit. Cum cinere eruto opus erit, altero praefurnio eruito, in altero ignis erit. Ignem caveto ne intermittas quin semper siet, neve noctu neve ullo tempore intermittatur caveto. Lapidem bonum in fornacem quam candidissimum, quam minime varium indito.
3 Cum fornacem facies, fauces praecipites deorsum facito. Ubi satis foderis, tum fornaci locum facito, uti quam altissima et quam minime ventosa siet. Si parum altam fornacem habebis ubi facias, latere summam statuito aut caementis cum luto summam
4 extrinsecus oblinito. Cum ignem subdideris, siqua flamma exibit nisi per orbem summum, luto oblinito. Ventus ad praefurnium caveto ne accedat: inibi austrum caveto maxime. Hoc signi erit, ubi calx cocta erit, summos lapides coctos esse oportebit; item infimi lapides cocti cadent, et flamma minus fumosa exibit.

---

[1] See note 2, page 30.

dig up or fell is during the seven days following the full moon. Above all things, do not work, or fell, or, if you can avoid it, even touch timber which is wet, or frosted, or covered with dew. Hoe and weed grain twice, and strip the wild oats. Remove the twigs from the prunings of vines and trees, and make them into bundles; and heap the vine and fig sticks for the forge, and the split wood for the use of the master.

XXXVIII. Build the lime-kiln ten feet across, twenty feet from top to bottom, sloping the sides in to a width of three feet at the top. If you burn with only one door, make a pit inside large enough to hold the ashes, so that it will not be necessary to clear them out. Be careful in the construction of the kiln; see that the grate covers the entire bottom of the kiln. If you burn with two doors there will be no need of a pit; when it becomes necessary to take out the ashes, clear through one door while the fire is in the other. Be careful to keep the fire burning constantly, and do not let it die down at night or at any other time. Charge the kiln only with good stone, as white and uniform as possible. In building the kiln, let the throat run straight down. When you have dug deep enough, make a bed for the kiln so as to give it the greatest possible depth and the least exposure to the wind. If you lack a spot for building a kiln of sufficient depth, run up the top with brick,[1] or face the top on the outside with field stone set in mortar. When it is fired, if the flame comes out at any point but the circular top, stop the orifice with mortar. Keep the wind, and especially the south wind, from reaching the door. The calcining of the stones at the top will show that the whole has calcined; also, the calcined stones at the bottom will settle, and the flame will be less smoky when it comes out.

Si ligna et virgas non poteris vendere neque lapidem habebis, unde calcem coquas, de lignis carbones coquito, virgas et sarmenta, quae tibi usioni supererunt, in segete conburito. Ubi eas conbusseris, ibi papaver serito.

XXXIX. Ubi tempestates malae erunt, cum opus fieri non poterit, stercus in stercilinum egerito. Bubile, ovile, cohortem, villam bene purgato. Dolia plumbo vincito vel materie quernea vere sicca[1] alligato. Si bene sarseris aut bene alligaveris et in rimas medicamentum indideris beneque picaveris, quodvis dolium vinarium facere poteris. Medicamentum in dolium hoc modo facito: cerae P. I, resinae P. I, 2 sulpuris P. C' C'. Haec omnia in calicem novum indito, eo addito gypsum contritum, uti crassitudo fiat quasi emplastrum, eo dolia sarcito. Ubi sarseris, qui colorem eundem facias, cretae crudae partes duas, calcis tertiam conmisceto; inde laterculos facito, coquito in fornace, eum conterito idque inducito.

Per imbrem in villa quaerito quid fieri possit. Ne cessetur, munditias facito. Cogitato, si nihil fiet, nihilo minus sumptum futurum.

XL. Per ver haec fieri oportet. Sulcos et scrobes fieri, seminariis, vitiariis locum verti, vites propagari, in locis crassis et umectis ulmos, ficos, poma, oleas seri oportet. Ficos, oleas, mala, pira, vites inseri oportet luna silenti post meridiem sine vento austro. 2 Oleas, ficos, pira, mala hoc modo inserito. Quem

[1] *Vere sicca* Hauler, after Turnebus: *virisicca.*

If you cannot sell your firewood and faggots, and have no stone to burn for lime, make charcoal of the firewood, and burn in the field the faggots and brush you do not need. Where you have burned them plant poppies.

XXXIX. When the weather is bad and no other work can be done, clear out manure for the compost heap; clean thoroughly the ox stalls, sheep pens, barnyard, and farmstead; and mend wine-jars with lead, or hoop them with thoroughly dried oak wood. If you mend it carefully, or hoop it tightly, closing the cracks with cement and pitching it thoroughly, you can make any jar serve as a wine-jar. Make a cement for a wine-jar as follows: Take one pound of wax, one pound of resin, and two-thirds of a pound of sulphur, and mix in a new vessel. Add pulverized gypsum sufficient to make it of the consistency of a plaster, and mend the jar with it. To make the colour uniform after mending, mix two parts of crude chalk and one of lime, form into small bricks, bake in the oven, pulverize, and apply to the jar.

In rainy weather try to find something to do indoors. Clean up rather than be idle. Remember that even though work stops, expenses run on none the less.

XL. The following work should be done in the spring: Trenches and furrows should be made, ground should be turned for the olive and vine nurseries, vines should be set out; elms, figs, fruit trees, and olives should be planted in rich, humid ground. Figs, olives, apples, pears, and vines should be grafted in the dark of the moon, after noon, when the south wind is not blowing. The following is a good method of grafting olives, figs,

ramum insiturus eris, praecidito, inclinato aliquantum, ut aqua defluat; cum praecides, caveto ne librum convellas. Sumito tibi surculum durum, eum praeacuito, salicem Graecam discindito. Argillam vel cretam coaddito, harenae paululum et fimum bubulum, haec una bene condepsito, quam maxime uti lentum fiat. Capito tibi scissam salicem, ea stirpem 3 praecisum circumligato, ne liber frangatur. Ubi id feceris, surculum praeacutum inter librum et stirpem artito primoris digitos II. Postea capito tibi surculum, quod genus inserere voles, eum primorem praeacuito oblicum primoris digitos II. Surculum aridum, quem artiveris, eximito, eo artito surculum, quem inserere voles. Librum ad librum vorsum facito, artito usque adeo, quo praeacueris. Idem alterum surculum, tertium, quartum facito; quot 4 genera voles, tot indito. Salicem Graecam amplius circumligato, luto depsto stirpem oblinito digitos crassum tres. Insuper lingua bubula obtegito, si pluat, ne aqua in librum permanet. Eam linguam insuper libro alligato, ne cadat. Postea stramentis circumdato alligatoque, ne gelus noceat.

XLI. Vitis insitio una est per ver, altera est cum uva floret, ea optuma est. Pirorum ac malorum insitio per ver et per solstitium dies L et per vinde-2 miam. Oleae et ficorum insitio est per ver. Vitem sic inserito: praecidito quam inseres, eam mediam diffindito per medullam; eo surculos praeacutos

---

[1] Pliny, *Nat. Hist.*, XVII, 112, quoting this passage, merely comments that it is "a kind of plant." Ps. Apul. *Herb.*, 42 tit. says that it gets its name (the same as the Greek *buglossa*) "from the fact that it has rough leaves in the shape of the tongue of the ox." In the absence of a scientific description,

pears or apples: Cut the end of the branch you are going to graft, slope it a bit so that the water will run off, and in cutting be careful not to tear the bark. Get you a hard stick and sharpen the end, and split a Greek willow. Mix clay or chalk, a little sand, and cattle dung, and knead them thoroughly so as to make a very sticky mass. Take your split willow and tie it around the cut branch to keep the bark from splitting. When you have done this, drive the sharpened stick between the bark and the wood two finger-tips deep. Then take your shoot, whatever variety you wish to graft, and sharpen the end obliquely for a distance of two finger-tips; take out the dry stick which you have driven in and drive in the shoot you wish to graft. Fit bark to bark, and drive it in to the end of the slope. In the same way you may graft a second, a third, a fourth shoot, as many varieties as you please. Wrap the Greek willow thicker, smear the stock with the kneaded mixture three fingers deep, and cover the whole with ox-tongue,[1] so that if it rains the water will not soak into the bark; this ox-tongue must be tied with bark to keep it from falling off. Finally, wrap it in straw and bind tightly, to keep the cold from injuring it.

XLI. Vine grafting may be done in the spring or when the vine flowers, the former time being best. Pears and apples may be grafted during the spring, for fifty days at the time of the summer solstice, and during the vintage; olives and figs should be grafted during the spring. Graft the vine as follows: Cut off the stem you are grafting, and split

it is perhaps hazardous to identify it with our ox-tongue (Pieris Echioides, L.).

artito; quos inseres, medullam cum medulla conponito. Altera insitio est: si vitis vitem continget, utriusque vitem teneram praeacuito, obliquo inter 3 sese medullam cum medulla libro conligato. Tertia insitio est: terebra vitem quam inseres pertundito, eo duos surculos vitigineos, quod genus esse voles, insectos obliquos artito ad medullam; facito iis medullam cum medulla coniungas artitoque ea qua 4 terebraveris alterum ex altera parte. Eos surculos facito sint longi pedes binos, eos in terram demittito replicatoque ad vitis caput, medias vitis vinclis in terram defigito terraque operito. Haec omnia luto depsto oblinito, alligato integitoque ad eundem modum, tamquam oleas.

XLII. Ficos et oleas altero modo. Quod genus aut ficum aut oleam esse voles, inde librum scalpro eximito, alterum librum cum gemma de eo fico, quod genus esse voles, eximito, adponito in eum locum unde exicaveris in alterum genus facitoque uti conveniat. Librum longum facito digitos III S, latum digitos III. Ad eundem modum oblinito, integito, uti cetera.

XLIII. Sulcos, si locus aquosus erit, alveatos esse oportet, latos summos pedes tres, altos pedes quattuor, infimum latum P. I et palmum. Eos lapide consternito; si lapis non erit, perticis saligneis viridibus controversus conlatis consternito; si pertica non erit, sarmentis conligatis. Postea scrobes facito altos P. III S, latos P. IIII, et facito de scrobe aqua 2 in sulcum defluat: ita oleas serito. Vitibus sulcos

[1] A somewhat similar method of propagation is described by Columella nearly two centuries later, in a long chapter on vine-grafting (*De Re Rust.*, IV, 29, 7, 13–14; cf. *De Arb.*, 8, 3–4), as rarely practised in his day though common among the ancients.

the middle through the pith; in it insert the sharpened shoots you are grafting, fitting pith to pith. A second method is: If the vines touch each other, cut the ends of a young shoot of each obliquely, and tie pith to pith with bark. A third method is: With an awl bore a hole through the vine which you are grafting, and fit tightly to the pith two vine shoots of whatever variety you wish, cut obliquely. Join pith to pith, and fit them into the perforation, one on each side.[1] Have these shoots each two feet long; drop them to the ground and bend them back toward the vine stock, fastening the middle of the vine to the ground with forked sticks and covering with dirt. Smear all these with the kneaded mixture, tie them up and protect them in the way I have described for olives.

XLII. Another method of grafting figs and olives is: Remove with a knife the bark from any variety of fig or olive you wish, and take off a piece of bark containing a bud of any variety of fig you wish to graft. Apply it to the place you have cleared on the other variety, and make it fit. The bark should be three and a half fingers long and three fingers wide. Smear and protect as in the other operation.

XLIII. Ditches, if the ground is swampy, should be dug trough-shaped, three feet wide at the top, four feet deep, sloping to a width of one foot one palm at the bottom. Blind them with stones, or, lacking stones, with green willow sticks laid crosswise in layers; or, failing this, with bundles of brush. Then dig trenches three and a half feet deep, four feet wide, so placed that the water will run off from the trenches into the ditch; and so plant olives. Dig

et propagines ne minus P. II S quoquo versus facito. Si voles vinea cito crescat et olea, quam severis, semel in mense sulcos et circum capita oleaginea quot mensibus, usque donec trimae erunt, fodere oportet. Eodem modo ceteras arbores procurato.

XLIV. Olivetum diebus XV ante aequinoctium vernum incipito putare. Ex eo die dies XLV recte putabis. Id hoc modo putato. Qua locus recte ferax erit, quae arida erunt, et siquid ventus interfregerit, ea omnia eximito. Qua locus ferax non erit, id plus concidito aratoque. Bene enodato stirpesque levis facito.

XLV. Taleas oleagineas, quas in scrobe saturus eris tripedaneas decidito diligenterque tractato, ne liber laboret, cum dolabis aut secabis. Quas in seminario saturus eris, pedalis facito, eas sic inserito. Locus bipalio subactus siet beneque terra tenera siet bene-2 que glittus siet. Cum taleam demittes, pede taleam opprimito. Si parum descendet, malleo aut mateola adigito cavetoque ne librum scindas, cum adiges. Palo prius locum ne feceris, quo taleam demittas. 3 Si ita severis uti stet talea, melius vivet. Taleae ubi trimae sunt, tum denique maturae sunt, ubi liber sese vertet. Si in scrobibus aut in sulcis seres, ternas taleas ponito easque divaricato, supra terram ne plus IIII digitos transvorsos emineant; vel oculos serito.

---

[1] Columella, III, 13, gives an elaborate description of ditching and trenching.

[2] The trunk or thick branches were cut or sawed into truncheons, sticks from one to three feet long, which took root when planted.

[3] Cf. Chap. 49 and *Georgics*, II, 270. Both Cato and Virgil follow Theophrastus, *Hist. Plant.*, II, 5.

furrows and trenches for vines[1] not less than two and a half feet deep and the same distance wide. If you wish the vines and olives which you have planted to grow fast, spade the furrows once a month, and dig around the foot of the olives every month until they are three years old. Treat other trees in the same way.

XLIV. The trimming of the olive-yard should begin fifteen days before the vernal equinox; you can trim to advantage from this time for forty-five days. Follow this rule: If the land is very fertile, clear out all dead branches only and any broken by the wind; if it is not fertile, trim more closely and plough. Trim clean, and smooth the stems.

XLV. Cut olive slips[2] for planting in trenches three feet long, and when you chop or cut them off, handle them carefully so as not to bruise the bark. Those which you intend to plant in the nursery should be cut one foot long, and planted in the following way: The bed should be turned with the trenching spade until the soil is finely divided and soft. When you set the slip, press it in the ground with the foot; and if it does not go deep enough, drive it in with a mallet or maul, but be careful not to break the bark in so doing. Do not first make a hole with a stick, in which to set out the slip. It will thrive better if you plant it so that it stands as it did on the tree.[3] The slips are ready for transplanting at three years, when the bark turns. If you plant in trenches or furrows, plant in groups of three, and spread them apart. Do not let them project more than four finger-widths above the ground; or you may plant the eyes.

XLVI. Seminarium ad hunc modum facito. Locum quam optimum et apertissimum et stercorosissimum poteris et quam simillimum genus terrae eae, ubi semina positurus eris, et uti ne nimis longe semina ex seminario ferantur, eum locum bipalio vertito, delapidato circumque saepito bene et in ordine serito. In sesquipedem quoquo vorsum taleam
2 demittito opprimitoque pede. Si parum deprimere poteris, malleo aut mateola adigito. Digitum supra terram facito semina emineant fimoque bubulo summam taleam oblinito signumque aput taleam adponito crebroque sarito, si voles cito semina crescant. Ad eundem modum alia semina serito.

XLVII. Harundinem sic serito: ternos pedes oculos disponito.

Vitiarium eodem modo facito seritoque. Ubi vitis bima erit, resicato; ubi trima erit, eximito. Si pecus pascetur, ubi vitem serere voles, ter prius resicato, quam ad arborem ponas. Ubi V nodos veteres habebit, tum ad arborem ponito. Quotannis porrinam serito, quotannis habebis quod eximas.

XLVIII. Pomarium seminarium ad eundem modum atque oleagineum facito. Suum quidquid genus talearum serito.

Semen cupressi ubi seres, bipalio vertito. Vere
2 primo serito. Porcas pedes quinos latas facito, eo stercus minutum addito, consarito glebasque comminuito. Porcam planam facito, paulum concavam. Tum semen serito crebrum tamquam linum, eo

---

[1] Cf. Varro, I, 29, 3: "the ground between two furrows, the elevated earth, is called *porca,* because that presents (*porricit*) the grain."

XLVI. Make a nursery as follows: Choose the best, the most open, and the most highly fertilized land you have, with soil as nearly as possible like that into which you intend to transplant, and so situated that the slips will not have to be carried too far from the nursery. Turn this with a trench spade, clear of stones, build a stout enclosure, and plant in rows. Plant a slip every foot and a half in each direction, pressing into the ground with the foot; and if it does not go deep enough, drive it in with a mallet or maul. Let the slips project a finger above the ground, and smear the tops with cow dung, placing a mark by each; hoe often if you wish the slips to grow rapidly. Plant other slips in the same way.

XLVII. The reed bed should be planted as follows: Plant the eyes three feet apart.

Use the same method for making and planting the vine nursery. Cut back the vine when it is two years old and transplant when it is three. If the ground on which you wish to plant the vine is to be used for pasture, see that the vine has been cut back three times before it is tied up to the tree; it should not be trained on the tree until it has five old knots. Plant a leek-bed every year, and you will have something to take off every year.

XLVIII. In making the fruit nursery follow the method used in making the olive nursery. Plant separately each variety of slip.

Turn the ground with a trench spade where you are going to plant cypress seed, and plant at the opening of spring. Make ridges[1] five feet wide, add well-pulverized manure, hoe it in, and break the clods. Flatten the ridge, forming a shallow trough. Plant the seed as thickly as flax, sifting dirt a finger-

terram cribro incernito altam digitum transversum. Eam terram tabula aut pedibus conplanato, furcas circum offigito, eo perticas intendito, eo sarmenta aut cratis ficarias inponito, quae frigus defendant et solem. Uti subtus homo ambulare possit facito. Crebro runcato. Simul herbae inceperint nasci, eximito. Nam si herbam duram velles, cupressos simul evelles.

3 Ad eundem modum semen pirorum, malorum serito tegitoque. Nuces pineas ad eundem modum nisi tamquam alium serito.

XLIX. Vineam veterem si in alium locum transferre voles, dumtaxat brachium crassam licebit. 2 Primum deputato, binas gemmas ne amplius relinquito. Ex radicibus bene exfodito, usque radices persequito et caveto ne radices saucies. Ita uti fuerit, ponito in scrobe aut in sulco operitoque et bene occulcato, eodemque modo vineam statuito, alligato flexatoque, uti fuerit, crebroque fodito.

L. Prata primo vere stercerato luna silenti. Quae inrigiva non erunt, ubi favonius flare coeperit, cum prata defendes, depurgato herbasque malas omnis radicitus effodito.

2 Ubi vineam deputaveris, acervum lignorum virgarumque facito. Ficos interputato et in vinea ficos subradito alte, ne eas vitis scandat. Seminaria facito et vetera resarcito. Haec facito, antequam vineam fodere incipias.

Ubi daps profanata comestaque erit, verno arare incipito. Ea loca primum arato, quae siccissima erunt, et quae crassissima et aquosissima erunt, ea postremum arato, dum ne prius obdurescant.

---

[1] Screens made of fig branches. Cf. Columella, XII, 15, 1.
[2] *i.e.* if its branches are stout. [3] See Chapters 83 and 131.

breadth deep over it with a sieve. Level the ground with a board or the foot, and set forked stakes around the edges. Lay poles in the forks, and on these hang brush or fig-curtains,[1] to keep off cold and sun. Make the covering high enough for a person to walk under. Hoe often, and clear off the weeds as soon as they begin to grow; for if you pull up the growth when it is hard, you will pull up the cypress with it.

Plant and cover pear and apple seed in the same way. Use the same method for planting pine-nuts, but alter it slightly.

XLIX. You may transplant an old vine if you wish, up to the thickness of your arm.[2] First prune back so as to leave not more than two buds on each branch; clear the dirt thoroughly from the roots over their full length, and be careful not to injure them. Replace the vine just as it was, in a trench or furrow, cover with soil, and trample firmly. Plant, tie, and train it just as it was, and work it often.

L. Manure meadows at the opening of spring, in the dark of the moon. When the west wind begins to blow and you close the dry meadows to stock, clean them and dig up all noxious weeds by the roots.

After pruning vines, pile the wood and branches; prune fig trees moderately, and clear those in the vineyard to a good height, so that the vines will not climb them; make new nurseries and repair old ones. All this before you begin cultivating the vines.

As soon as the sacred feast has been offered and eaten,[3] begin the spring ploughing, working first the driest spots and last the heaviest and wettest, provided they do not get hard in the meantime.

LI. Propagatio pomorum, aliarum arborum. Ab arbore abs terra pulli qui nascentur, eos in terram deprimito extollitoque primorem partem, uti radicem capiat; inde biennio post effodito seritoque. Ficum, oleam, malum Punicum, cotoneum aliaque mala omnia, laurum, murtum, nuces Praenestinas, platanum, haec omnia a capite propagari eximique serique eodem modo oportet.

LII. Quae diligentius propagari voles, in aullas aut in qualos pertusos propagari oportet et cum iis in scrobem deferri oportet. In arboribus, uti radices capiant, calicem pertundito; per fundum aut qualum ramum, quem radicem capere voles, traicito; eum qualum aut calicem terra inpleto calcatoque bene, in arbore relinquito. Ubi bimum fuerit, ramum sub 2 qualo praecidito. Qualum incidito ex ima parte perpetuum, sive calix erit, conquassato. Cum eo qualo aut calice in scrobem ponito. Eodem modo vitem facito, eam anno post praecidito seritoque cum qualo. Hoc modo quod genus vis propagabis.

LIII. Faenum, ubi tempus erit, secato cavetoque ne sero seces. Priusquam semen maturum siet, secato, et quod optimum faenum erit, seorsum condito, per ver cum arabunt, antequam ocinum des, quod edint boves.

LIV. Bubus pabulum hoc modo parari darique oportet. Ubi sementim patraveris, glandem parari legique oportet et in aquam conici. Inde semodios singulis bubus in dies dari oportet, et si non laborabunt, pascantur satius erit, aut modium vinaceorum, quos in dolium condideris. Interdiu pascito, noctu

---

[1] Columella, *De Arb.*, 7, describes minutely three methods of layering, including this.

LI. Layering of fruit trees and other trees: Press into the earth the scions which spring from the ground around the trees, elevating the tip so that it will take root.[1] Then two years later dig up and transplant them. Fig, olive, pomegranate, quince, and all other fruit trees, laurel, myrtle, Praenestine nuts, and planes should all be layered, dug, and transplanted in the same way.

LII. When you wish to layer more carefully you should use pots or baskets with holes in them, and these should be planted with the scion in the trench. To make them take root while on the tree, make a hole in the bottom of the pot or basket and push the branch which you wish to root through it. Fill the pot or basket with dirt, trample thoroughly, and leave on the tree. When it is two years old, cut off the branch below the basket; cut the basket down the side and through the bottom, or, if it is a pot, break it, and plant the branch in the trench with the basket or pot. Use the same method with a vine, cutting it off the next year and planting it with the basket. You can layer any variety you wish in this way.

LIII. Cut hay in season, and be careful not to wait too long. Harvest before the seed ripens, and store the best hay by itself for the oxen to eat during the spring ploughing, before you feed clover.

LIV. Feed for cattle should be prepared and fed as follows: When the sowing is over, gather the acorns and soak them in water. A half-modius of this should be fed each ox per day, though if the oxen are not working it will be better to let them forage; or feed a modius of the grape husks which you have stored in jars. During the day let them forage, and

2 faeni P. XXV uni bovi dato. Si faenum non erit, frondem iligneam et hederaciam dato. Paleas triticeas et hordeaceas, acus fabaginum, de[1] vicia vel de lupino, item de ceteris frugibus, omnia condito. Cum stramenta condes, quae herbosissima erunt, in tecto condito et sale spargito, deinde ea pro faeno
3 dato. Ubi verno dare coeperis, modium glandis aut vinaceorum dato aut modium lupini macerati et faeni P. XV. Ubi ocinum tempestivum erit, dato primum. Manibus carpito, id renascetur:
4 quod falcula secueris, non renascetur. Usque ocinum dato, donec arescat: ita temperato. Postea viciam dato, postea panicum dato, secundum panicum frondem ulmeam dato. Si populneam habebis, admisceto, ut ulmeae satis siet. Ubi ulmeam non
5 habebis, querneam et ficulneam dato. Nihil est quod magis expediat, quam boves bene curare. Boves nisi per hiemem, cum non arabunt, pasci non oportet. Nam viride cum edunt, semper id expectant, et fiscellas habere oportet, ne herbam sectentur, cum arabunt.

LV. Ligna domino in tabulato condito, codicillos oleagineos, radices in acervo sub dio metas facito.

LVI. Familiae cibaria. Qui opus facient per hiemem tritici modios IIII, per aestatem modios IIII S, vilico, vilicae, epistatae, opilioni modios III, conpeditis per hiemem panis P. IIII, ubi vineam fodere coeperint, panis P. V, usque adeo dum ficos esse coeperint, deinde ad P. IIII redito.

LVII. Vinum familiae. Ubi vindemia facta erit,

[1] *de* added by Keil.

---

[1] Compare Columella, I, 8, 16. The field-hands, and especially the unruly, were chained together, and at night kept in an underground prison, the *ergastulum*.

at night feed 25 pounds of hay a head; if you have no hay, feed ilex and ivy leaves. Store wheat and barley straw, husks of beans, of vetch, of lupines, and of all other crops. In storing litter, bring under cover that which has most leaves, sprinkle it with salt, and feed it instead of hay. When you begin feeding in spring, feed a modius of mast, or grape husks, or soaked lupine, and 15 pounds of hay. When clover is in season feed it first; pull it by hand and it will grow again, for if you cut it with the hook it will not. Continue to feed clover until it dries out, after which feed it in limited quantities; then feed vetch, then panic grass, and after this elm leaves. If you have poplar leaves, mix them with the elm to make the latter hold out; and failing elm, feed oak and fig leaves. There is nothing more profitable than to take good care of cattle. They should not be pastured except in winter, when they are not ploughing; for when they once eat green food they are always expecting it; and so they have to be muzzled to keep them from biting at the grass while ploughing.

LV. Store firewood for the master's use on flooring, and cut olive sticks and roots and pile them out of doors.

LVI. Rations for the hands: Four modii of wheat in winter, and in summer four and a half for the field hands. The overseer, the housekeeper, the foreman, and the shepherd should receive three. The chain-gang[1] should have a ration of four pounds of bread through the winter, increasing to five when they begin to work the vines, and dropping back to four when the figs ripen.

LVII. Wine ration for the hands: For three

loram bibant menses tres; mense quarto heminas in dies, id est in mense congios II S: mense quinto, sexto, septimo, octavo in dies sextarios, id est in mense congios quinque; nono, decimo, undecimo, duodecimo[1] in dies heminas ternas, id est in mense[2] amphoram; hoc amplius Saturnalibus et Conpitalibus in singulos homines congios III S;[3] summa vini in homines singulos inter annum Q. VII. Conpeditis, uti quidquid operis facient, pro portione addito; eos non est nimium in annos singulos vini Q. X ebibere.

LVIII. Pulmentarium familiae. Oleae caducae quam plurimum condito. Postea oleas tempestivas, unde minimum olei fieri poterit, eas condito, parcito, uti quam diutissime durent. Ubi oleae comesae erunt, hallecem et acetum dato. Oleum dato in menses uni cuique S. I. Salis uni cuique in anno modium satis est.

LIX. Vestimenta familiae. Tunicam P. III S, saga alternis annis. Quotiens cuique tunicam aut sagum dabis, prius veterem accipito, unde centones fiant. Sculponias bonas alternis annis dare oportet.

LX. Bubus cibaria annua in iuga singula lupini modios centum viginti aut glandis modios CCXL, faeni pondo IↃXX, ocini . ,[4] fabae M̊. XX, viciae M̊. XXX. Praeterea granatui videto uti satis viciae seras. Pabulum cum seres, multas sationes facito.

LXI. Quid est agrum bene colere? bene arare. Quid secundum? arare. Quid tertium?[5] stercorare. Qui oletum saepissime et altissime miscebit, is

[1] *duodecimo* added by Iucundus.
[2] *in mense* added by Iucundus.
[3] III S added by Keil.
[4] The numeral has been lost.
[5] *quid tertium* Iucundus: *tertio.*

months following the vintage let them drink after-wine.[1] In the fourth month issue a hemina a day, that is, 2½ congii a month; in the fifth, sixth, seventh, and eighth months a sextarius a day, that is, 5 congii a month; in the ninth, tenth, eleventh, and twelfth months 3 heminae a day, that is, an amphora a month. In addition, issue 3½ congii per person for the Saturnalia and the Compitalia.[2] Total of wine for each person per year, 7 quadrantals; and an additional amount for the chain-gang proportioned to their work. Ten quadrantals of wine per person is not an excessive allowance for the year.

LVIII. Relish for the hands: Store all the windfall olives you can, and later the mature olives which will yield very little oil. Issue them sparingly and make them last as long as possible. When they are used up, issue fish-pickle and vinegar, and a pint of oil a month per person. A modius of salt a year per person is sufficient.

LIX. Clothing allowance for the hands: A tunic 3½ feet long and a blanket every other year. When you issue the tunic or the blanket, first take up the old one and have patchwork made of it. A stout pair of wooden shoes should be issued every other year.

LX. The following is a year's ration for a yoke of steers: 120 modii of lupines, or 240 of mast; 520 pounds of hay, and . . . of clover; 20 modii of beans; and 30 modii of vetch. See also that you sow enough vetch to allow some to go to seed. Make several sowings of forage crops.

LXI. What is good cultivation? Good ploughing. What next? Ploughing. What third? Manuring.

The planter who works his olives very often and

[1] See note 2, page 20. [2] See note 1, page 14.

tenuissimas radices exarabit. Si male arabit, radices susum abibunt, crassiores fient, et in radices vires oleae abibunt. Agrum frumentarium cum ares, 2 bene et tempestivo ares, sulco vario ne ares. Cetera cultura est multum sarire[1] et diligenter eximere semina et per tempus radices quam plurimas cum terra ferre; ubi radices bene operueris, calcare bene, ne aqua noceat. Siquis quaeret, quod tempus oleae serendae siet, agro sicco per sementim, agro laeto per ver.

LXII. Quot iuga boverum, mulorum, asinorum habebis, totidem plostra esse oportet.

LXIII. Funem torculum esse oportet extentum pedes LV, funem loreum in plostrum P. LX, lora retinacula longa P. XXVI, subiugia in plostrum P. XIIX, funiculum P. XV, in aratrum subiugia lora P. XVI, funiculum P. VIII.

LXIV. Olea ubi matura erit, quam primum cogi oportet, quam minimum in terra et in tabulato esse oportet. In terra et in tabulato putescit. Leguli volunt uti olea caduca quam plurima sit, quo plus legatur; factores, ut in tabulato diu sit, ut fracida sit, quo facilius efficiant. Nolito credere oleum in 2 tabulato posse crescere. Quam citissime conficies, tam maxime expediet, et totidem modiis collecta et[2]

[1] *sarire* Schneider: *serere.*
[2] *collecta et* Keil: *collectae.*

---

[1] Columella, II, 2, 24, advises deep ploughing, "especially in Italy, where land planted with fruit trees and olives needs to be broken and cut through deeply, so that the top roots of vines and olives may be cut off, for if they remain they impoverish the fruits; and so that the lower roots, when the soil is broken deeply, may not fail to take up the nourishment of the moisture." [2] Cf. Chap. 3, Secs. 2–4.

[3] See Columella, XII, 52, 18–19, for a discussion of the

very deep[1] will plough up the very slender roots; while bad ploughing will cause the roots to come to the surface and grow too large, and the strength of the tree will waste into the roots. When you plough grain land do it well and at the proper season, and do not plough with an irregular furrow. The rest of the cultivation consists in hoeing often, taking up shoots carefully, and transplanting, at the proper time, as many roots as possible, with their soil. When you have covered the roots well, trample them firmly so that the water will not harm them. If one should ask what is the proper time for planting olives, I should say, at seed-time in dry ground, and in spring in rich ground.

LXII. You should have as many carts as you have teams, either of oxen, mules, or donkeys.

LXIII. The press rope should be 55 feet long when stretched; there should be 60 feet of leather cordage for the cart, and 26 feet for reins; the yoke straps for the cart 18 feet, and the line 15; the yoke straps for the plough 16 feet and the line 8.

LXIV. When the olives are ripe they should be gathered as soon as possible, and allowed to remain on the ground or the floor as short a time as possible, as they spoil on the ground or the floor.[2] The gatherers want to have as many windfalls as possible, that there may be more of them to gather; and the pressers want them to lie on the floor a long time, so that they will soften and be easier to mill. Do not believe that the oil will be of greater quantity if they lie on the floor.[3] The more quickly you work them up the better the results will be, and you will get more

point, and the disbelief that the olives increase in size or the oil in amount.

plus olei efficiet et melius. Olea quae diu fuerit in terra aut in tabulato, inde olei minus fiet et deterius. Oleum, si poteris, bis in die depleto. Nam oleum quam diutissime in amurca et in fracibus erit, tam deterrimum erit.

LXV. Oleum viride sic facito. Oleam quam primum ex terra tollito. Si inquinata erit, lavito, a foliis et stercore purgato. Postridie aut post diem tertium, quam lecta erit, facito. Olea ubi nigra erit, stringito. Quam acerbissima olea oleum facies, tam oleum optimum erit. Domino de matura olea oleum fieri maxime expediet. Si gelicidia erunt, cum 2 oleam coges, triduum atque quatriduum post oleum facito. Eam oleam, si voles, sale spargito. Quam calidissimum torcularium et cellam habeto.

LXVI. Custodis et capulatoris officia. Servet diligenter cellam et torcularium. Caveat quam minimum in torcularium et in cellam introeatur. Quam mundissime purissimeque fiat. Vaso aheneo neque nucleis ad oleum ne utatur. Nam si utetur, oleum male sapiet. Cortinam plumbeam in lacum ponito, quo oleum fluat. Ubi factores vectibus prement, continuo capulator conca oleum, quam diligentissime poterit, tollat, ne cesset. Amurcam caveat ne tollat. 2 Oleum in labrum primum indito, inde in alterum dolium indito. De iis labris fraces amurcamque semper subtrahito. Cum oleum sustuleris de cortina, amurcam deorito.

---

[1] See note 1, page 24.

and better oil from a given quantity. Olives which have been long on the ground or the floor will yield less oil and of a poorer quality. If possible, draw off the oil twice a day, for the longer it remains on the amurca[1] and the dregs, the worse the quality will be.

LXV. Observe the following directions in making green oil: Pick the olives off the ground as soon as possible, and if they are dirty, wash them and clean off leaves and dung. Mill them a day or two days after they have been gathered. Pick olives after they have turned black; the more acid the olives the better the oil will be, but the master will find it most profitable to make oil only from ripe olives. If frost has fallen on the olives, mill them three or four days after gathering. You may sprinkle such olives with salt, if you wish; and keep a high temperature in the pressing-room and the storeroom.

LXVI. Duties of the watchman and the ladler: The watchman must keep a close watch on the storeroom and the pressing-room, and must see that there is as little passing in and out as possible. He must see that the work is done as neatly and cleanly as possible, that copper vessels are not used, and that no seeds are crushed for oil; otherwise it will have a bad flavour. Place a lead cauldron in the basin into which the oil flows. As soon as the workmen press down the levers, at once the ladler must take off the oil with a shell very carefully, and without stopping, being careful not to take off the amurca. Pour the oil into the first vessel, then into the second, each time removing the dregs and the amurca. When you take the oil from the cauldron, skim off the amurca.

LXVII. Item custodis officia. Qui in torculario erunt vasa pura habeant curentque uti olea bene perficiatur beneque siccetur. Ligna in torculario ne caedant. Oleum frequenter capiant. Factoribus det in singulos factus olei sextarios et in lucernam 2 quod opus siet. Fraces cotidie reiciat. Amurcam conmutet usque adeo, donec in lacum qui in cella est postremum pervenerit. Fiscinas spongia effingat. Cotidie oleo locum conmutet, donec in dolium pervenerit. In torculario et in cella caveat diligenter nequid olei subripiatur.

LXVIII. Ubi vindemia et oletas facta erit, prela extollito; funes torculos, melipontos, subductarios in carnario aut in prelo suspendito; orbes, fibulas, vectes, scutulas, fiscinas, corbulas, quala, scalas, patibula, omnia quis usus erit, in suo quidque loco reponito.

LXIX. Dolia olearia nova sic inbuito. Amurca inpleto dies VII, facito ut amurcam cotidie suppleas. 2 Postea amurcam eximito et arfacito. Ubi arebit, cummim pridie in aquam infundito, eam postridie diluito. Postea dolium calfacito minus, quam si picare velis, tepeat satis est; lenibus lignis facito calescat. Ubi temperate tepebit, tum cummim indito, postea linito. Si recte leveris, in dolium quinquagenarium cummim P. IIII satis erit.

LXX. Bubus medicamentum. Si morbum metues, sanis dato salis micas tres, folia laurea III, porri fibras III, ulpici spicas III, alii spicas III, turis

---

[1] Careless chopping of fire-wood might injure the presses; or the levers might be cut.

[2] So interpreted from *dolium quinquagenarium* in 112, 3.

LXVII. Further duties of the watchman: Those in the pressing-room must keep their vessels clean and see that the olives are thoroughly worked up and that they are well dried. They must not cut wood in the pressing-room.[1] They must skim the oil frequently. He must give the workmen a sextarius of oil for each pressing, and what they need for the lamp. He must throw out the lees every day and keep cleaning the amurca until the oil reaches the last vat in the room. He must wipe off the baskets with a sponge, and change the vessel daily until the oil reaches the jar. He must be careful to see that no oil is pilfered from the pressing-room or the cellar.

LXVIII. When the vintage and the olive harvest are over, raise up the press beams, and hang up the mill ropes, cables, and cords on the meat-rack or the beam. Put the stones, pins, levers, rollers, baskets, hampers, grass baskets, ladders, props, and everything which will be needed again, each in its proper place.

LXIX. To steep new oil jars: Fill them with amurca, maintaining a constant level, for seven days; then pour off the amurca and let the jars dry. When the drying is finished soak gum in water a day ahead, and the next day dilute it. Then heat the jar to a lower temperature than if you were to pitch it—it is sufficient for it to be warm, so heat it over a slow fire. When it is moderately warm, pour in the gum and rub it in. Four pounds of gum are enough for a jar holding 50 quadrantals,[2] if you apply it properly.

LXX. Remedy for oxen: If you have reason to fear sickness, give the oxen before they get sick the following remedy: 3 grains of salt, 3 laurel leaves, 3 leek leaves, 3 spikes of leek, 3 of garlic, 3 grains of

grana tria, herbae Sabinae plantas tres, rutae folia tria, vitis albae caules III, fabulos albos III, carbones vivos III, vini S. III. Haec omnia sublimiter legi teri darique oportet. Ieiunus siet qui dabit. Per triduum de ea potione uni cuique bovi dato. Ita dividito, cum ter uni cuique dederis, omnem absumas, bosque ipsus et qui dabit facito ut uterque sublimiter stent. Vaso ligneo dato.

LXXI. Bos si aegrotare coeperit, dato continuo ei unum ovum gallinaceum crudum; integrum facito devoret. Postridie caput ulpici conterito cum hemina vini facitoque ebibat. Sublimiter terat et vaso ligneo det, bosque ipsus et qui dabit sublimiter stet. Ieiunus ieiuno bovi dato.

LXXII. Boves ne pedes subterant, priusquam in viam quoquam ages, pice liquida cornua infima unguito.

LXXIII. Ubi uvae variae coeperint fieri, bubus medicamentum dato quotannis, uti valeant. Pellem anguinam ubi videris, tollito et condito, ne quaeras cum opus siet. Eam pellem et far et salem et serpullum, haec omnia una conterito cum vino, dato bubus bibant omnibus. Per aestatem boves aquam bonam et liquidam bibant semper curato; ut valeant refert.

LXXIV. Panem depsticium sic facito. Manus mortariumque bene lavato. Farinam in mortarium indito, aquae paulatim addito subigitoque pulchre. Ubi bene subegeris, defingito coquitoque sub testu.

---

[1] This seems to be part of the magic implied in the use of the number three. Some commentators interpret *sublimiter*, in this formula and the one following, as equivalent to *sub divo*, "under the open sky."

incense, 3 plants of Sabine herb, 3 leaves of rue, 3 stalks of bryony, 3 white beans, 3 live coals, and 3 pints of wine. You must gather, macerate, and administer all these while standing,[1] and he who administers the remedy must be fasting. Administer to each ox for three days, and divide it in such a way that when you have administered three doses to each you will have used it all. See that the ox and the one who administers are both standing, and use a wooden vessel.

LXXI. If an ox begins to sicken, administer at once one hen's egg raw, and make him swallow it whole. The next day macerate a head of leek with a hemina of wine, and make him drink it all. Macerate while standing, and administer in a wooden vessel. Both the ox and the one who administers must stand, and both be fasting.

LXXII. To keep oxen from wearing down their feet, smear the bottom of their hoofs with melted pitch before you drive them anywhere on a road.

LXXIII. Give the cattle medicine every year when the grapes begin to change colour, to keep them well. When you see a snake skin, pick it up and put it away, so that you will not have to hunt for one when you need it. Macerate this skin, spelt, salt, and thyme with wine, and give it to all the cattle to drink. See that the cattle always have good, clear water to drink in summer-time; it is important for their health.

LXXIV. Recipe for kneaded bread: Wash your hands and a bowl thoroughly. Pour meal into the bowl, add water gradually, and knead thoroughly. When it is well kneaded, roll out and bake under a crock.

LXXV. Libum hoc modo facito. Casei P. II bene disterat in mortario. Ubi bene distriverit, farinae siligineae libram aut, si voles tenerius esse, selibram similaginis eodem indito permiscetoque cum caseo bene. Ovum unum addito et una permisceto bene. Inde panem facito, folia subdito, in foco caldo sub testu coquito leniter.

LXXVI. Placentam sic facito. Farinae siligineae L. II, unde solum facias, in tracta farinae L. IIII et alicae primae L. II. Alicam in aquam infundito. Ubi bene mollis erit, in mortarium purum indito siccatoque bene. Deinde manibus depsito. Ubi bene subactum erit, farinae L. IIII paulatim addito. Id utrumque tracta facito. In qualo, ubi arescant, 2 conponito. Ubi arebunt, conponito puriter.[1] Cum facies singula tracta, ubi depsueris, panno oleo uncto tangito et circumtergeto unguitoque. Ubi tracta erunt, focum, ubi cocas, calfacito bene et testum. Postea farinae L. II conspargito condepsitoque. Inde facito solum tenue. Casei ovilli P. XIIII ne acidum et bene recens in aquam indito. Ibi macerato, aquam ter mutato. Inde eximito siccatoque bene paulatim manibus, siccum bene in mortarium 3 inponito. Ubi omne caseum bene siccaveris, in mortarium purum manibus condepsito conminuitoque quam maxime. Deinde cribrum farinarium purum

[1] *pariter* Turnebus.

---

[1] The *libum* and the *placenta*, especially, are important, as these cakes are employed in religious services. Horace, *Epistles*, I, 10, 10–11, compares himself to the runaway slave of the priest, who refuses *liba*, as he needs plain bread, which is better than *mellitis placentis*. Varro, I, 2, 28, is amused at the idea of treating such subjects in a work on agriculture. It should, of course, be remembered that honey took the

LXXV. Recipe for libum:[1] Bray 2 pounds of cheese thoroughly in a mortar; when it is thoroughly macerated, add 1 pound of wheat flour, or, if you wish the cake to be more dainty, ½ pound of fine flour, and mix thoroughly with the cheese. Add 1 egg, and work the whole well. Pat out a loaf, place on leaves, and bake slowly on a warm hearth under a crock.

LXXVI. Recipe for placenta:[1] Materials, 2 pounds of wheat flour for the crust, 4 pounds of flour and 2 pounds of prime groats for the *tracta*.[2] Soak the groats in water, and when it becomes quite soft pour into a clean bowl, drain well, and knead with the hand; when it is thoroughly kneaded, work in the 4 pounds of flour gradually. From this dough make the *tracta*, and spread them out in a basket where they can dry; and when they are dry arrange them evenly. Treat each *tractum* as follows: After kneading, brush them with an oiled cloth, wipe them all over and coat with oil. When the *tracta* are moulded, heat thoroughly the hearth where you are to bake, and the crock. Then moisten the 2 pounds of flour, knead, and make of it a thin lower crust. Soak 14 pounds of sheep's cheese (sweet and quite fresh) in water and macerate, changing the water three times. Take out a small quantity at a time, squeeze out the water thoroughly with the hands, and when it is quite dry place it in a bowl. When you have dried out the cheese completely, knead it in a clean bowl by hand, and make it as smooth as possible. Then take a clean

place of our sugar. But even so, these recipes cannot be considered alluring.

[2] Seemingly bits of pastry.

sumito caseumque per cribrum facito transeat in mortarium. Postea indito mellis boni P. IIII S. Id una bene conmisceto cum caseo. Postea in tabula pura, quae pateat P. I, ibi balteum ponito, folia laurea uncta supponito, placentam fingito. 4 Tracta singula in totum solum primum ponito, deinde de mortario tracta linito, tracta addito singulatim, item linito usque adeo, donec omne caseum cum melle abusus eris. In summum tracta singula indito, postea solum contrahito ornatoque focum de ve primo[1] temperatoque, tunc placentam inponito, testo caldo operito, pruna insuper et circum operito. Videto ut bene et otiose percoquas. Aperito, dum inspicias, bis aut ter. Ubi cocta erit, eximito et melle unguito. Haec erit placenta semodialis.

LXXVII. Spiram sic facito. Quantum voles pro ratione, ita uti placenta fit, eadem omnia facito, nisi alio modo fingito. In solo tracta cum melle oblinito bene. Inde tamquam restim tractes facito, ita inponito in solo, simplicibus conpleto bene arte. Cetera omnia, quasi placentam facias, facito coquitoque.

LXXVIII. Scriblitam sic facito. In balteo tractis caseo ad eundem modum facito, uti placentam, sine melle.

LXXIX. Globos sic facito. Caseum cum alica ad eundem modum misceto. Inde quantos voles facere facito. In ahenum caldum unguen indito. Singulos aut binos coquito versatoque crebro duabus rudibus, coctos eximito, eos melle unguito, papaver infriato, ita ponito.

[1] Corrupt. Hörle suggests *ornatoque*; *focum deverrito temperatoque*; *tunc*, etc., *Ph. W*. 48. 1389.

---

[1] The word *solum* seems to be used of the bottom crust,

flour sifter and force the cheese through it into the bowl. Add 4½ pounds of fine honey, and mix it thoroughly with the cheese. Spread the crust[1] on a clean board, one foot wide, on oiled bay leaves, and form the placenta as follows: Place a first layer of separate *tracta* over the whole crust,[1] cover it with the mixture from the bowl, add the *tracta* one by one, covering each layer until you have used up all the cheese and honey. On the top place single *tracta*, and then fold over the crust[1] and prepare the hearth . . . then place the placenta, cover with a hot crock, and heap coals on top and around. See that it bakes thoroughly and slowly, uncovering two or three times to examine it. When it is done, remove and spread with honey. This will make a half-modius cake.

LXXVII. Recipe for spira: For the quantity desired do everything in proportion just as for the placenta, except that you shape it differently. Cover the *tracta* on the crust thickly with honey; then draw out like a rope and so place it on the crust, filling in closely with plain *tracta*. Do everything else as in the case of the placenta, and so bake.

LXXVIII. Recipe for scriblita: Follow the same directions with respect to crust, *tracta*, and cheese, as for the placenta, but without honey.

LXXIX. Recipe for globi: Mix the cheese and spelt in the same way, sufficient to make the number desired. Pour lard into a hot copper vessel, and fry one or two at a time, turning them frequently with two rods, and remove when done. Spread with honey, sprinkle with poppy-seed, and serve.

while the *balteus* (belt) is used of the crust folded over; but the directions are vague.

LXXX. Encytum ad eundem modum facito, uti globos, nisi calicem pertusum cavum habeat. Ita in unguen caldum fundito. Honestum[1] quasi spiram facito idque duabus rudibus vorsato praestatoque. Item unguito coloratoque caldum ne nimium. Id cum melle aut cum mulso adponito.

LXXXI. Erneum sic facito tamquam placentam. Eadem omnia indito, quae in placentam. Id permisceto in alveo, id indito in irneam fictilem, eam demittito in aulam aheneam aquae calidae plenam. Ita coquito ad ignem. Ubi coctum erit, irneam confringito, ita ponito.

LXXXII. Spaeritam sic facito, ita uti spiram, nisi sic fingito. De tractis caseo melle spaeras pugnum altas facito. Eas in solo conponito densas, eodem modo conponito atque spiram itemque coquito.

LXXXIII. Votum pro bubus, uti valeant, sic facito. Marti Silvano in silva interdius in capita singula boum votum facito. Farris L. III et lardi P.[2] IIII S et pulpae P. IIII S, vini S.[3] III, id in unum vas liceto coicere, et vinum item in unum vas liceto coicere. Eam rem divinam vel servus vel liber licebit faciat. Ubi res divina facta erit, statim ibidem consumito. Mulier ad eam rem divinam ne adsit neve videat quo modo fiat. Hoc votum in annos singulos, si voles, licebit vovere.

[1] Corrupt, Keil.
[2] *pondo* Iucundus.
[3] *sestarios* Iucundus.

---

[1] The word is the Greek ἔγχυτον, literally "cast in a mould." Compare Athenaeus 644. C.
[2] A mixture of boiled must and honey.
[3] σφαιρίτης: a kind of round cake.

LXXX. Make the encytum[1] the same way as the globus, except that you use a vessel with a hole in the bottom; press it through this hole into boiling lard, and shape it like the spira, coiling and keeping it in place with two rods. Spread with honey and glaze while moderately warm. Serve with honey or with mulsum.[2]

LXXXI. The erneum is made in the same way as the placenta, and has the same ingredients. Mix it in a trough, pour into an earthenware jar, plunge into a copper pot full of hot water, and boil over the fire. When it is done, break the jar and serve.

LXXXII. The spaerita[3] is made in the same way as the spira, except that you shape it as follows: Mould balls as large as the fist, of *tracta*, cheese, and honey; arrange them on the crust as closely as in the spira, and bake in the same way.

LXXXIII. Perform the vow for the health of the cattle as follows: Make an offering to Mars Silvanus[4] in the forest during the daytime for each head of cattle: 3 pounds of meal, 4½ pounds of bacon, 4½ pounds of meat, and 3 pints of wine. You may place the viands in one vessel, and the wine likewise in one vessel. Either a slave or a free man may make this offering. After the ceremony is over, consume the offering on the spot at once. A woman may not take part in this offering or see how it is performed. You may vow the vow every year if you wish.

[4] W. Warde Fowler, *The Religious Experience of the Roman People*, identifies Silvanus with Mars (pp. 132–3), and is supported by Burriss, *Classical Journal*, XXI, 3, p. 221.

LXXXIV. Savillum hoc modo facito. Farinae selibram, casei P. II S una conmisceto quasi libum, mellis P. ≡ et ovum unum. Catinum fictile oleo unguito. Ubi omnia bene conmiscueris, in catinum indito, catinum testo operito. Videto ut bene percocas medium, ubi altissimum est. Ubi coctum erit, catinum eximito, melle unguito, papaver infriato, sub testum subde paulisper, postea eximito. Ita pone cum catillo et lingula.

LXXXV. Pultem Punicam sic coquito. Libram alicae in aquam indito, facito uti bene madeat. Id infundito in alveum purum, eo casei recentis P. III, mellis P. S, ovum unum, omnia una permisceto bene. Ita insipito in aulam novam.

LXXXVI. Graneam triticeam sic facito. Selibram tritici puri in mortarium purum indat, lavet bene corticemque deterat bene eluatque bene. Postea in aulam indat et aquam puram cocatque. Ubi coctum erit, lacte addat paulatim usque adeo, donec cremor crassus erit factus.

LXXXVII. Amulum sic facito. Siliginem purgato bene, postea in alveum indat, eo addat aquam bis in die. Die decimo aquam exsiccato, exurgeto bene, in alveo puro misceto bene, facito tamquam faex fiat. Id in linteum novum indito, exprimito cremorem in patinam novam aut in mortarium. Id omne ita facito et refricato denuo. Eam patinam in sole ponito, arescat. Ubi arebit, in aulam novam indito, inde facito cum lacte coquat.

LXXXVIII. Salem candidum sic facito. Am-

---

[1] *savillum*—some kind of sweet cake (*suavis*), "a cheese-cake."

LXXXIV. Recipe for the savillum:[1] Take ½ pound of flour, 2½ pounds of cheese, and mix together as for the libum; add ¼ pound of honey and 1 egg. Grease an earthenware dish with oil. When you have mixed thoroughly, pour into a dish and cover with a crock. See that you bake the centre thoroughly, for it is deepest there. When it is done, remove the dish, cover with honey, sprinkle with poppy-seed, place back under the crock for a while, then remove from the fire. Serve in the dish, with a spoon.

LXXXV. Recipe for Punic porridge: Soak a pound of groats in water until it is quite soft. Pour it into a clean bowl, add 3 pounds of fresh cheese, ½ pound of honey, and 1 egg, and mix the whole thoroughly; turn into a new pot.

LXXXVI. Recipe for wheat pap: Pour ½ pound of clean wheat into a clean bowl, wash well, remove the husk thoroughly, and clean well. Pour into a pot with pure water and boil. When done, add milk slowly until it makes a thick cream.

LXXXVII. Recipe for starch:[2] Clean hard wheat thoroughly, pour into a trough, and add water twice a day. On the tenth day drain off the water, squeeze thoroughly, mix well in a clean tray until it is of the consistency of wine-dregs. Place some of this in a new linen bag and squeeze out the creamy substance into a new pan or bowl. Treat the whole mass in the same way, and knead again. Place the pan in the sun and let it dry; then place in a new bowl and cook with milk.

LXXXVIII. Recipe for bleaching salt: Break off

[2] ἄμυλον (sc. ἄλευρον) is a fine meal or "starch," so called because it was not ground in a mill in the ordinary course.

phoram defracto collo puram inpleto aquae purae, in sole ponito. Ibi fiscellam cum sale populari suspendito et quassato suppletoque identidem. Id aliquotiens in die cotidie facito, usque adeo donec sal desiverit 2 tabescere biduum. Id signi erit: menam aridam vel ovum demittito; si natabit, ea muries erit, vel carnem vel caseos vel salsamenta quo condas. Eam muriam in labella vel in patinas in sole ponito. Usque adeo in sole habeto, donec concreverit. Inde flos salis fiet. Ubi nubilabitur et noctu sub tecto ponito; cotidie, cum sol erit, in sole ponito.

LXXXIX. Gallinas et anseres sic farcito. Gallinas teneras, quae primum parient, concludat. Polline vel farina hordeacia consparsa turundas faciat, eas in aquam intinguat, in os indat, paulatim cotidie addat; ex gula consideret, quod satis sit. Bis in die farciat et meridie bibere dato; ne plus aqua sita siet horam unam. Eodem modo anserem alito, nisi prius dato bibere et bis in die, bis escam.

XC. Palumbum recentem sic farcito. Ubi prensus erit, ei fabam coctam tostam primum dato, ex ore in eius os inflato, item aquam. Hoc dies VII facito. Postea fabam fresam puram et far purum facito et fabae tertia pars ut infervescat, tum far insipiat, puriter facito et coquito bene. Id ubi excluseris, depsito bene, oleo manum unguito, primum pusillum, postea magis depses, oleo tangito depsitoque, dum poterit facere turundas. Ex aqua dato, escam temperato.

the neck of a clean amphora, fill with clear water, and place in the sun. Suspend in it a basket filled with common salt and shake and renew from time to time. Do this daily several times a day until the salt ceases to dissolve for two days. You can find when it is saturated by this test: place a small dried fish or an egg in it, and if it floats you have a brine strong enough to pickle meat or cheese or salted fish. Place this brine in flat vessels or in pans and expose it to the sun. Keep it in the sun until it solidifies, and you will have a pure salt. In cloudy weather or at night put it under cover, but expose it to the sun every day when there is sunshine.

LXXXIX. To cram hens or geese: Shut up young hens which are beginning to lay; make pellets of moist flour or barley-meal, soak in water, and push into the mouth. Increase the amount daily, judging from the appetite the amount that is sufficient. Cram twice a day, and give water at noon, but do not place water before them for more than one hour. Feed a goose the same way, except that you let it drink first, and give water and food twice a day.

XC. To cram squabs: After catching the squab feed it first boiled and toasted beans, blowing them from your mouth into its mouth, and water the same way; do this for seven days. Then clean crushed beans and spelt; let one-third the quantity of beans come to a boil, then pour in the spelt, keeping it clean, and boil thoroughly. When you have turned it out of the pot, knead it thoroughly, after greasing the hand with oil—a small quantity first, then more—greasing and kneading until you can make pellets. Feed the food in moderate quantities, after soaking it.

XCI. Aream sic facito. Locum ubi facies confodito. Postea amurca conspargito bene sinitoque conbibat. Postea conminuito glebas bene. Deinde coaequato et paviculis verberato. Postea denuo amurca conspargito sinitoque arescat. Si ita feceris, neque formicae nocebunt neque herbae nascentur.

XCII. Frumento ne noceat curculio neu mures tangant. Lutum de amurca facito, palearum paulum addito, sinito macerescant bene et subigito bene; eo granarium totum oblinito crasso luto. Postea conspargito amurca omne quod lutaveris. Ubi aruerit, eo frumentum refrigeratum condito; curculio non nocebit.

XCIII. Olea si fructum non feret, ablaqueato. Postea stramenta circumponito. Postea amurcam cum aqua conmisceto aequas partes. Deinde ad oleam circumfundito. Ad arborem maxumam urnam conmixti sat est; ad minores arbores pro ratione indito. Et idem hoc si facies ad arbores feraces, eae quoque meliores fient. Ad eas stramenta ne addideris.

XCIV. Fici uti grossos teneant, facito omnia quo modo oleae, et hoc amplius, cum ver adpetet, terram adaggerato bene. Si ita feceris, et grossi non cadent et fici scabrae non fient et multo feraciores erunt.

XCV. Convolvolus in vinia ne siet. Amurcam condito, puram bene facito, in vas aheneum indito congios II. Postea igni leni coquito, rudicula agitato crebro usque adeo, dum fiat tam crassum quam mel. Postea sumito bituminis tertiarium et sulpuris quar-

[1] See note 1, page 24.

XCI. To make a threshing-floor: Turn the soil for the floor and pour amurca[1] over it thickly, letting it soak in. Then break up the clods carefully, level the ground, and pack it with rammers; then cover again with amurca and let it dry. If you build in this way the ants will not injure it, and weeds will not grow.

XCII. To keep weevils and mice from injuring grain, make a slime of amurca with a little chaff added, leaving it quite thin and working thoroughly.[2] Cover the whole granary with the thick slime, and then add a coat of amurca over the whole. After it has dried, store cooled grain there, and the weevils will not injure it.

XCIII. If an olive tree is sterile, trench it and wrap it with straw. Make a mixture of equal parts of amurca and water and pour it around the tree; an urna[3] is sufficient for a large tree, and a proportionate quantity for the smaller trees. If you do the same thing for bearing trees they will be even more productive; do not wrap these with straw.

XCIV. To make fig trees retain their fruit, do everything as for the olive, and in addition bank them deep in early spring. If you do this the fruit will not drop prematurely, the trees will not be scaly, and they will be much more productive.

XCV. To keep caterpillars off the vines: Strain stored amurca and pour 2 congii into a copper vessel; heat over a gentle fire, stirring constantly with a stick until it reaches the consistency of honey. Take one-third sextarius of bitumen, and

[2] A similar slime is recommended in Chapter 128; but there the verb used is *fracescat*. *Macrescant* or *macerescant* would have about the same meaning. [3] See Glossary.

2 tarium. Conterito in mortario seorsum utrumque. Postea infriato quam minutissime in amurcam caldam et simul rudicula misceto et denuo coquito sub dio caelo. Nam si in tecto coquas, cum bitumen et sulpur additum est, excandescet. Ubi erit tam crassum quam viscum, sinito frigescat. Hoc vitem circum caput et sub brachia unguito; convolvolus non nascetur.

XCVI. Oves ne scabrae fiant. Amurcam condito, puram bene facito, aquam in qua[1] lupinus deferverit et faecem de vino bono, inter se omnia conmisceto pariter. Postea cum detonderis, unguito totas,
2 sinito biduum aut triduum consudent. Deinde lavito in mari; si aquam marinam non habebis, facito aquam salsam, ea lavito. Si haec sic feceris, neque scabrae fient et lanae plus et meliorem habebunt, et ricini non erunt molesti. Eodem in omnes quadripedes utito, si scabrae erunt.

XCVII. Amurca decocta axem unguito et lora et calciamenta et coria; omnia meliora facies.

XCVIII. Vestimenta ne tiniae tangant. Amurcam decoquito ad dimidium, ea unguito fundum arcae et extrinsecus et pedes et angulos. Ubi ea adaruerit, vestimenta condito. Si ita feceris, tiniae non noce-
2 bunt. Et item ligneam supellectilem omnem si ungues, non putescet, et cum ea terseris, splendidior fiet. Item ahenea omnia unguito, sed prius extergeto bene. Postea cum unxeris, cum uti voles, extergeto. Splendidior erit, et aerugo non erit molesta.

[1] *in qua* added by Keil.

one-fourth sextarius of sulphur, pulverize each in a mortar separately, and add in very small quantities to the warm amurca, at the same time stirring with a stick, and let it boil again in the open; for if you boil it under cover it will blaze up when the mixture of bitumen and sulphur is added. When it reaches the consistency of glue let it cool. Apply this around the trunk and under the branches, and caterpillars will not appear.

XCVI. To keep scab from sheep; Take equal parts of old strained amurca, water in which lupines have been boiled, and dregs of good wine, and mix all together. After shearing, smear the whole body with this, and let them sweat two or three days. Then wash them in the sea, or, if you have no sea-water, make a brine and wash them in it. If you do this as directed, they will not have the scab, will bear more wool and of better quality, and ticks will not bother them. Use the same remedy for all quadrupeds if they have the scab.

XCVII. Grease the axle, belts, shoes, and hides with boiled amurca; you will make them all better.

XCVIII. To protect clothing from moths: Boil amurca down to one-half its volume and rub it over the bottom, the outside, the feet, and the corners of the chest. After it is dry, store the clothing and the moths will not attack it. Also, if you rub it over the whole surface of wooden furniture it will prevent decay, and the article when rubbed will have a higher polish. You may also use it as a polish for any kind of copper vessel, after cleaning the article thoroughly. After applying the amurca, rub the vessel when it is to be used; it will have a lustre, and will be protected from rust.

XCIX. Fici aridae si voles uti integrae sint, in vas fictile condito. Id amurca decocta unguito.

C. Oleum si in metretam novam inditurus eris, amurca, ita uti est cruda, prius conluito agitatoque diu, ut bene conbibat. Id si feceris, metreta oleum non bibet, et oleum melius faciet, et ipsa metreta firmior erit.

CI. Virgas murteas si voles cum bacis servare et item aliut genus quod vis, et si ramulos ficulneos voles cum foliis, inter se alligato, fasciculos facito, eos in amurcam demittito, supra stet amurca facito. Sed ea quae demissurus eris sumito paulo acerbiora. Vas, quo condideris, oblinito plane.

CII. Si bovem aut aliam quamvis quadrupedem serpens momorderit, melanthi acetabulum, quod medici vocant zmurnaeum, conterito in vini veteris hemina. Id per nares indito et ad ipsum morsum stercus suillum adponito. Et idem hoc, si usus venerit, homini facito.

CIII. Boves uti valeant et curati bene sint, et qui fastidient cibum, uti magis cupide adpetant, pabulum quod dabis amurca spargito; primo paululum, dum consuescant, postea magis, et dato rarenter bibere conmixtam cum aqua aequabiliter. Quarto quinto quoque die hoc sic facies. Ita boves et corpore curatiores erunt, et morbus aberit.

CIV. Vinum familiae per hiemem qui utatur. Musti Q. X in dolium indito, aceti acris Q. II eodem 2 infundito, sapae Q. II, aquae dulcis Q. L. Haec

---

[1] See Glossary.

[2] The word *melanthium* seems to mean either "cultivated fennel," or "camomile"; and *smyrnaeum* is derived from *smyrna*, myrrh.

[3] Cf. Columella, VI, 4, 4.

XCIX. If you wish to keep dried figs from spoiling, place them in an earthenware vessel and coat this with boiled amurca.

C. If you intend to store oil in a new jar, first wash down the jar with crude amurca, shaking for a long time so that it may soak up the amurca thoroughly. If you do this, the jar will not soak up the oil, it will make the oil better, and the jar itself will be stronger.

CI. To preserve myrtle or any other twigs with the berries, or fig branches with the leaves, tie them together into bundles and plunge them into amurca until they are covered. But the fruit to be preserved should be picked a little before it is ripe, and the vessel in which it is stored should be sealed tight.

CII. When a serpent has bitten an ox or any other quadruped, macerate an acetabulum[1] of fennel flower, which the physicians call smyrnaeum,[2] in a hemina of old wine. Administer through the nostrils, and apply swine's dung to the wound itself. Treat a person in the same way if occasion arises.

CIII. To keep cattle well and strong, and to increase the appetite of those which are off their feed, sprinkle the feed which you give with amurca. Feed in small quantities at first to let them grow accustomed to it, and then increase. Give them less often a draught of equal parts of amurca and water. Do this every fourth or fifth day. This treatment will keep them in better condition, disease will stay away from them.[3]

CIV. Wine for the hands to drink through the winter: Pour into a jar 10 quadrantals of must, 2 quadrantals of sharp vinegar, 2 quadrantals of boiled must, 50 quadrantals of fresh water. Stir

rude misceto ter in die dies quinque continuos. Eo addito aquae marinae veteris sextarios LXIIII et operculum in dolium inponito et oblinito post dies X. Hoc vinum durabit tibi usque ad solstitium. Siquid superfuerit post solstitium, acetum acerrimum et pulcherrimum erit.

CV. Qui ager longe a mari aberit, ibi vinum Graecum sic facito. Musti Q. XX in aheneum aut plumbeum infundito, ignem subdito. Ubi bullabit vinum, ignem subducito. Ubi id vinum refrixerit, in dolium quadragenarium infundito. Seorsum in vas aquae dulcis Q. I infundito, salis M̊ I, sinito 2 muriam fieri. Ubi muria facta erit, eodem in dolium infundito. Schoenum et calamum in pila contundito, quod satis siet, sextarium unum eodem in dolium infundito, ut odoratum siet. Post dies XXX dolium oblinito. Ad ver diffundito in amphoras. Biennium in sole sinito positum esse. Deinde in tectum conferto. Hoc vinum deterius non erit quam Coum.

CVI. Aquae marinae concinnatio. Aquae marinae Q. I ex alto sumito, quo aqua dulcis non accedit. Sesquilibram salis frigito, eodem indito et rude misceto usque adeo, donec ovum gallinaceum coctum natabit, desinito miscere. Eodem vini veteris vel Aminnii vel miscelli albi congios II infundito, misceto probe. Postea in vas picatum confundito et oblinito. Siquid plus voles aquae marinae concinnare, pro portione ea omnia facito.

CVII. Quo labra doliorum circumlinas, ut bene odorata sint et nequid viti in vinum accedat. Sapae congios VI quam optimae infundito in aheneum aut in plumbeum et iris aridae contusae heminam et

[1] The making of Coan wine is described in Chapter 112.

with a stick thrice a day for five consecutive days. Then add 64 sextarii of old sea-water, cover the jar, and seal ten days later. This wine will last you until the summer solstice; whatever is left over after the solstice will be a very sharp and excellent vinegar.

CV. If your place is far from the sea, you may use this recipe for Greek wine: Pour 20 quadrantals of must into a copper or lead boiler and heat. As soon as the wine boils, remove the fire; and when the wine has cooled, pour into a jar holding 40 quadrantals. Pour 1 modius of salt and 1 quadrantal of fresh water into a separate vessel, and let a brine be made; and when the brine is made pour it into the jar. Pound rush and calamus in a mortar to make a sufficient quantity, and pour 1 sextarius into the jar to give it an odour. Thirty days later seal the jar, and rack off into amphorae in the spring. Let it stand for two years in the sun, then bring it under cover. This wine will not be inferior to the Coan.[1]

CVI. Preparation of sea-water: Take 1 quadrantal of water from the deep sea where no fresh water comes; parch 1½ pounds of salt, add it, and stir with a rod until a boiled hen's egg will float; then stop the stirring. Add 2 congii of old wine, either Aminnian or ordinary white, and after mixing thoroughly pour into a pitched jar and seal. If you wish to make a larger quantity of sea-water, use a proportionate amount of the same materials.

CVII. To coat the brim of wine jars, so as to give a good odour and to keep any blemish from the wine: Put 6 congii of the best boiled must in a copper or lead vessel; take a hemina of dry crushed

sertam Campanicam P. V bene odoratam una cum iri contundas quam minutissime, per cribrum cernas et una cum sapa coquas sarmentis et levi flamma. 2 Conmoveto, videto ne aduras. Usque coquito, dum dimidium excoquas. Ubi refrixerit, confundito in vas picatum bene odoratum et oblinito et utito in labra doliorum.

CVIII. Vinum si voles experiri duraturum sit necne, polentam grandem dimidium acetabuli in caliculum novum indito et vini sextarium de eo vino quod voles experiri eodem infundito et inponito in carbones; facito bis aut ter inferveat. Tum id percolato, 2 polentam abicito. Vinum ponito sub dio. Postridie mane gustato. Si id sapiet, quod in dolio est, scito duraturum; si subacidum erit, non durabit.

CIX. Vinum asperum quod erit lene et suave si voles facere, sic facito. De ervo farinam facito libras IIII et vini cyatos IIII conspargito sapa. Postea facito laterculos. Sinito conbibant noctem et diem. Postea conmisceto cum eo vino in dolio et oblinito post[1] dies LX. Id vinum erit lene et suave et bono colore et bene odoratum.

CX. Odorem deteriorem demere vino. Testam de tegula crassam puram calfacito in igni bene. Ubi calebit, eam picato, resticula alligato, testam demittito in dolium infimum leniter, sinito biduum oblitum dolium. Si demptus erit odor deterior, id optime; si non, saepius facito, usque dum odorem malum dempseris.

CXI. Si voles scire, in vinum aqua addita sit necne, vasculum facito de materia hederacia. Vinum id, quod putabis aquam habere, eo demittito. Si

[1] *post* added by Gronovius.

iris and 5 pounds of fragrant Campanian melilot, grind very fine with the iris, and pass through a sieve into the must. Boil the whole over a slow fire of faggots, stirring constantly to prevent scorching; continue the boiling, until you have boiled off a half. When it has cooled, pour into a sweet smelling jar covered with pitch, seal, and use for the brims of wine jars.

CVIII. If you wish to determine whether wine will keep or not, place in a new vessel half an acetabulum of large pearl barley and a sextarius of the wine you wish to test; place it on the coals and bring it to a boil two or three times; then strain, throw away the barley, and place the wine in the open. Taste it the next morning. If it is sweet, you may know that the wine in the jar will keep; but if it is slightly acid it will not.

CIX. To make sharp wine mild and sweet: Make 4 pounds of flour from vetch, and mix 4 cyathi of wine with boiled must; make into small bricks and let them soak for a night and a day; then dissolve with the wine in the jar, and seal sixty days later. The wine will be mild and sweet, of good colour and of good odour.

CX. To remove a bad odour from wine: Heat a thick clean piece of roofing-tile thoroughly in the fire. When it is hot coat it with pitch, attach a string, lower it gently to the bottom of the jar, and leave the jar sealed for two days. If the bad odour is removed the first time, that will be best; if not, repeat until the bad odour is removed.

CXI. If you wish to determine whether wine has been watered or not: Make a vessel of ivy wood and put in it some of the wine you think has water

habebit aquam, vinum effluet, aqua manebit. Nam non continet vinum vas hederaceum.

CXII. Vinum Coum si voles facere, aquam ex alto marinam sumito mari tranquillo, cum ventus non erit, dies LXX ante vindemiam, quo aqua dulcis non perveniet. Ubi hauseris de mari, in dolium infundito, nolito inplere, quadrantalibus quinque minus sit quam plenum. Operculum inponito, relin-
2 quito qua interspiret. Ubi dies XXX praeterierint, transfundito in alterum dolium puriter et leniter, relinquito in imo quod desiderit. Post dies XX in alterum dolium item transfundito; ita relinquito usque ad vindemiam. Unde vinum Coum facere voles, uvas relinquito in vinea, sinito bene coquantur, et ubi pluerit et siccaverit, tum deligito et ponito in sole biduum aut triduum sub dio, si pluviae non erunt. Si pluvia erit, in tecto in cratibus conponito,
3 et siqua acina corrupta erunt, depurgato. Tum sumito aquam marinam Q. S. S. E,[1] in dolium quinquagenarium infundito aquae marinae Q. X. Tum acina de uvis miscellis decarpito de scopione in idem dolium, usque dum inpleveris. Manu conprimito acina, ut conbibant aquam marinam. Ubi inpleveris dolium, operculo operito, relinquito qua interspiret. Ubi triduum praeterierit, eximito de dolio et calcato in torculario et id vinum condito in dolia lauta et pura et sicca.

CXIII. Ut odoratum bene sit, sic facito. Sumito testam picatam, eo prunam lenem indito, suffito serta et schoeno et palma, quam habent unguentarii,

[1] *i.e. Quae Supra Scripta Est.*

in it. If it contains water, the wine will soak through and the water will remain, for a vessel of ivy wood will not hold wine.

CXII. Recipe for Coan wine: Take sea-water at a distance from the shore, where fresh water does not come, when the sea is calm and no wind is blowing, seventy days before vintage. After taking it from the sea, pour into a jar, filling it not fully but to within five quadrantals of the top. Cover the jar, leaving space for air, and thirty days later pour it slowly and carefully into another jar, leaving the sediment in the bottom. Twenty days later pour in the same way into a third jar, and leave until vintage. Allow the grapes from which you intend to make the Coan wine to remain on the vine, let them ripen thoroughly, and pick them when they have dried after a rain. Place them in the sun for two days, or in the open for three days, unless it is raining, in which case put them under cover in baskets; clear out any berries which have rotted. Then take the above-mentioned sea-water and pour 10 quadrantals into a jar holding 50; then pick the berries of ordinary grapes from the stem into the jar until you have filled it. Press the berries with the hand so that they may soak in the sea-water. When the jar is full, cover it, leaving space for air, and three days later remove the grapes from the jar, tread out in the pressing-room, and store the wine in jars which have been washed clean and dried.

CXIII. To impart a sweet aroma: Take a tile covered with pitch, spread over it warm ashes, and cover with aromatic herbs, rush and the palm which the perfumers keep, place in a jar and cover, so that

ponito in dolio et operito, ne odor exeat, antequam vinum indas. Hoc facito pridie quam vinum infundere voles. De lacu quam primum vinum in dolia indito, sinito dies XV operta, antequam oblinas,
2 relinquito qua interspiret, postea oblinito. Post dies XL diffundito in amphoras et addito in singulas amphoras sapae sextarium unum. Amphoras nolito inplere nimium, ansarum infimarum fini, et amphoras in sole ponito, ubi herba non siet, et amphoras operito, ne aqua accedat, et ne plus quadriennium in sole siveris. Post quadriennium in cuneum conponito et instipato.

CXIV. Vinum si voles concinnare, ut alvum bonam faciat, secundum vindemiam, ubi vites ablaqueantur, quantum putabis ei rei satis esse vini, tot vites ablaqueato et signato. Earum radices circumsecato et purgato. Veratri atri[1] radices contundito in pila, eas radices dato circum vitem et stercus vetus et cinerem veterem et duas partes terrae circumdato
2 radices vitis. Terram insuper inicito. Hoc vinum seorsum legito. Si voles servare in vetustatem ad alvum movendam, servato, ne conmisceas cum cetero vino. De eo vino cyatum sumito et misceto aqua et bibito ante cenam. Sine periculo alvum movebit.

CXV. In vinum mustum veratri atri manipulum coicito in amphoram. Ubi satis efferverit, de vino manipulum eicito. Id vinum servato ad alvum movendam.

2 Vinum ad alvum movendam concinnare. Vites cum ablaqueabuntur, signato rubrica, ne admisceas cum cetero vino. Tris fasciculos veratri atri circumponito circum radices et terram insuper inicito. Per

[1] *atri* added by Pontedera, from 115, 1.

the odour will not escape before you pour in the wine. Do this the day before you wish to pour in the wine. Pour the wine into the jars from the vat immediately, let them stand covered for fifteen days before sealing, leaving space for air, and then seal. Forty days later pour off into amphorae, and add one sextarius of boiled must to the amphora. Do not fill the amphorae higher than the bottom of the handles, and place them in the sun where there is no grass. Cover the amphorae so that water cannot enter, and let them stand in the sun not more than four years; four years later, arrange them in a wedge, and pack them closely.

CXIV. If you wish to make a laxative wine: After vintage, when the vines are trenched, expose the roots of as many vines as you think you will need for the purpose and mark them; isolate and clear the roots. Pound roots of black hellebore in the mortar, and apply around the vines. Cover the roots with old manure, old ashes, and two parts of earth, and cover the whole with earth. Gather these grapes separately; if you wish to keep the wine for some time as a laxative, do not mix it with the other wine. Take a cyathus of this wine, dilute it with water, and drink it before dinner; it will move the bowels with no bad results.

CXV. Throw in a handful of black hellebore to the amphora of must, and when the fermentation is complete, remove the hellebore from the wine; save this wine for a laxative.

To prepare a laxative wine: When the vines are trenched, mark with red chalk so that you will not mix with the rest of the wine; place three bundles of black hellebore around the roots and cover with

vindemiam de iis vitibus quod delegeris, seorsum servato, cyatum in ceteram potionem indito. Alvum movebit et postridie perpurgabit sine periculo.

CXVI. Lentim quo modo servari oporteat. Laserpicium aceto diluito, permisceto lentim aceto laserpiciato et ponito in sole. Postea lentim oleo perfricato, sinito arescat. Ita integra servabitur recte.

CXVII. Oleae albae quo modo condiantur. Antequam nigrae fiant, contundantur et in aquam deiciantur. Crebro aquam mutet. Deinde, ubi satis maceratae erunt, exprimat et in acetum coiciat et oleum addat, salis selibram in modium olearum. Feniculum et lentiscum seorsum condat in acetum. Si una admiscere voles, cito utito.[1] In orculam calcato. Manibus siccis, cum uti voles, sumito.

CXVIII. Oleam albam, quam secundum vindemiam uti voles, sic condito. Musti tantundem addito, quantum aceti. Cetera item condito ita, uti supra scriptum est.

CXIX. Epityrum album nigrum variumque sic facito. Ex oleis albis nigris variisque nuculeos eicito. Sic condito. Concidito ipsas, addito oleum, acetum, coriandrum, cuminum, feniculum, rutam, mentam. In orculam condito, oleum supra siet. Ita utito.[1]

CXX. Mustum si voles totum annum habere, in amphoram mustum indito et corticem oppicato, demittito in piscinam. Post dies XXX eximito. Totum annum mustum erit.

[1] *utito* Keil: *utitor*.

---

[1] The same as *silpium*, mentioned in Chapter 157, 7, where see note.

earth. Keep the yield from these vines separate during the vintage. Put a cyathus into another drink; it will move the bowels and the next day give a thorough purging without danger.

CXVI. To preserve lentils: Infuse asafetida[1] in vinegar, soak the lentils in the infusion of vinegar and asafetida, and expose to the sun; then rub the lentils with oil, allow them to dry, and they will keep quite sound.

CXVII. To season green olives: Bruise the olives before they become black and throw them into water. Change the water often, and when they are well soaked press out and throw into vinegar; add oil, and a half pound of salt to the modius of olives. Make a dressing of fennel and mastic steeped in vinegar, using a separate vessel. If you wish to mix them together they must be served at once. Press them out into an earthenware vessel and take them out with dry hands when you wish to serve them.

CXVIII. To season green olives which you wish to use after vintage, add as much must as vinegar; for the rest, season them as stated above.

CXIX. Recipe for a confection of green, ripe, and mottled olives. Remove the stones from green, ripe, and mottled olives, and season as follows: chop the flesh, and add oil, vinegar, coriander, cummin, fennel, rue, and mint. Cover with oil in an earthen dish, and serve.

CXX. If you wish to keep grape juice through the whole year, put the grape juice in an amphora, seal the stopper with pitch, and sink in the pond. Take it out after thirty days; it will remain sweet the whole year.

CXXI. Mustaceos sic facito. Farinae siligineae modium unum musto conspargito. Anesum, cuminum, adipis P. II, casei libram, et de virga lauri deradito, eodem addito, et ubi definxeris, lauri folia subtus addito, cum coques.

CXXII. Vinum concinnare, si lotium difficilius transibit. Capreidam vel iunipirum contundito in pila, libram indito, in duobus congiis vini veteris in vase aheneo vel in plumbeo defervefacito. Ubi refrixerit, in lagonam indito. Id mane ieiunus sumito cyatum; proderit.

CXXIII. Vinum ad isciacos sic facito. De iunipiro materiem semipedem crassam concidito minutim. Eam infervefacito cum congio vini veteris. Ubi refrixerit, in lagonam confundito et postea id utito cyatum mane ieiunus; proderit.

CXXIV. Canes interdiu clausos esse oportet, ut noctu acriores et vigilantiores sint.

CXXV. Vinum murteum sic facito. Murtam nigram arfacito in umbra. Ubi iam passa erit, servato ad vindemiam, in urnam musti contundito murtae semodium, id oblinito. Ubi desiverit fervere mustum, murtam eximito. Id est ad alvum crudam et ad lateris dolorem et ad coeliacum.

CXXVI. Ad tormina, et si alvus non consistet, et si taeniae et lumbrici molesti erunt. XXX mala Punica acerba sumito, contundito, indito in urceum et vini nigri austeri congios III. Vas oblinito. Post dies XXX aperito et utito; ieiunus heminam bibito.

CXXVII. Ad dyspepsiam et stranguriam mederi. Malum Punicum ubi florebit, conligito, tris minas in amphoram infundito, vini Q. I veteris addito et

---

[1] The plant is mentioned only in this passage, and cannot be identified.

CXXI. Recipe for must cake: Moisten 1 modius of wheat flour with must; add anise, cummin, 2 pounds of lard, 1 pound of cheese, and the bark of a laurel twig. When you have made them into cakes, put bay leaves under them, and bake.

CXXII. To blend a wine as a remedy for retention of urine: Macerate capreida[1] or juniper, add a pound of it, and boil in 2 congii of old wine in a copper or lead vessel. After it cools, pour into a bottle. Take a cyathus in the morning before eating; it will prove beneficial.

CXXIII. To blend a wine as a remedy for gout: Cut into small chips a piece of juniper wood a half-foot thick, boil with a congius of old wine, and after it cools pour into a bottle. Take a cyathus in the morning before eating; it will prove beneficial.

CXXIV. Dogs should be chained up during the day, so that they may be keener and more watchful at night.

CXXV. Recipe for myrtle wine: Dry out black myrtle in the shade, and when dried keep it until vintage. Macerate a half-modius of myrtle into an urna of must and seal it. When the must has ceased to ferment remove the myrtle. This is a remedy for indigestion, for pain in the side, and for colic.

CXXVI. For gripes, for loose bowels, for tapeworms and stomach-worms, if troublesome: Take 30 acid pomegranates, crush, place in a jar with 3 congii of strong black wine, and seal the vessel. Thirty days later open and use. Drink a hemina before eating.

CXXVII. Remedy for dyspepsia and strangury: Gather pomegranate blossoms when they open, and place 3 minae of them in an amphora. Add one quadrantal of old wine and a mina of clean crushed

feniculi radicem puram contusam minam. Oblinito amphoram et post dies XXX aperito et utito.[1] Ubi voles cibum concoquere et lotium facere, hinc bibito quantum voles sine periculo. Idem vinum taenias perpurgat et lumbricos, si sic concinnes. Incenatum 2 iubeto[2] esse. Postridie turis drachmam unam conterito et mel coctum drachmam unam et vini sextarium origaniti. Dato ieiuno, et puero pro aetate triobolum et vini heminam. Supra pilam inscendat et saliat decies et deambulet.

CXXVIII. Habitationem delutare. Terram quam maxime cretosam vel rubricosam, eo amurcam infundito, paleas indito. Sinito quadriduum fracescat. Ubi bene fracuerit, rutro concidito. Ubi concideris, delutato. Ita neque aspergo nocebit, neque mures cava facient, neque herba nascetur, neque lutamenta scindent se.

CXXIX. Aream, ubi frumentum teratur, sic facito. Confodiatur minute terra, amurca bene conspargatur et conbibat quam plurimum. Conminuito terram et cylindro aut pavicula coaequato. Ubi coaequata erit, neque formicae molestae erunt, et cum pluerit, lutum non erit.

CXXX. Codicillos oleagineos et cetera ligna amurca cruda perspargito et in sole ponito, perbibant bene. Ita neque fumosa erunt et ardebunt bene.

CXXXI. Piro florente dapem pro bubus facito. Postea verno arare incipito. Ea loca primum arato, quae rudecta harenosaque erunt. Postea uti quaeque gravissima et aquosissima erunt, ita postremo arato.

[1] *utito* Keil: *utitor*. [2] *iubeto* Keil: *iubet*.

---

[1] See Glossary.

root of fennel; seal the vessel and thirty days later open and use. You may drink this as freely as you wish without risk, when you wish to digest your food and to urinate. The same wine will clear out tapeworms and stomach-worms if it is blended in this way. Bid the patient refrain from eating in the evening, and the next morning macerate 1 drachm of pulverized incense, 1 drachm of boiled honey, and a sextarius of wine of wild marjoram. Administer to him before he eats, and, for a child, according to age, a triobulus[1] and a hemina. Have him climb a pillar and jump down ten times, and walk about.

CXXVIII. To plaster a dwelling: Take very chalky or red earth, pour amurca over it, and add chopped straw; let it soften for four days, and when it has softened thoroughly, work up with a spade; and when you have worked it up, plaster. With this treatment, the moisture will not injure the walls, nor the mice burrow in them, nor weeds grow, nor the plaster crack.

CXXIX. To make a floor for threshing grain: Break the ground fine, soak thoroughly with amurca and let it absorb as much as possible; then pulverize the dirt and level with a roller or rammer. When it is levelled the ants will not be troublesome, and there will be no mud when it rains.

CXXX. Wet olive logs and other firewood with crude amurca and expose them to the sun so that they will absorb it thoroughly; with this treatment, they will not be smoky, but will burn well.

CXXXI. Make the offering for the oxen when the pear trees bloom; then begin the spring ploughing. Plough first the spots which are dry and sandy. Then, the heavier and wetter the spots are, the later they should be ploughed.

CXXXII. Dapem hoc modo fieri oportet. Iovi dapali culignam vini quantam vis polluceto. Eo die feriae bubus et bubulcis et qui dapem facient. Cum pollucere oportebit, sic facies: " Iuppiter dapalis, quod tibi fieri oportet in domo familia mea culignam vini dapi, eius rei ergo macte hac illace dape pollucenda esto." Manus interluito, postea vinum sumito: " Iuppiter dapalis, macte istace dape pollucenda 2 esto, macte vino inferio esto." Vestae, si voles, dato. Daps Iovi assaria pecuina urna vini. Iovi caste profanato sua contagione. Postea dape facta serito milium, panicum, alium, lentim.

CXXXIII. Propagatio pomorum ceterarumque arborum. Arboribus abs terra pulli qui nati erunt, eos in terram deprimito, extollito, uti radicem capere pos- 2 sint. Inde, ubi tempus erit, effodito seritoque recte. Ficum, oleam, malum Punicum, mala strutea, cotonea aliaque mala omnia, laurum Cypriam, Delphicam, prunum, murtum coniugulum et murtum album et nigrum, nuces Abellanas, Praenestinas, platanum, haec omnia genera a capitibus propagari eximique 3 ad hunc modum oportebit. Quae diligentius seri voles, in calicibus seri oportet. In arboribus radices uti capiant, calicem pertusum sumito tibi aut quasillum; per eum ramulum trasserito; eum quasillum terra inpleto calcatoque, in arbore relinquito. Ubi bimum erit, ramum tenerum infra praecidito, cum quasillo serito. Eo modo quod vis genus arborum

---

[1] An epithet of Jupiter, derived from the old Roman custom of spreading an offering of food (*daps*) before the gods.
[2] See note 1, page 22.

CXXXII. The offering is to be made in this way: Offer to Jupiter Dapalis[1] a cup of wine of any size you wish, observing the day as a holiday for the oxen, the teamsters, and those who make the offering. In making the offering use this formula: "Jupiter Dapalis, forasmuch as it is fitting that a cup of wine be offered thee, in my house and in the midst of my people, for thy sacred feast; and to that end, be thou honoured by the offering of this food." Wash the hands, then take the wine, and say: "Jupiter Dapalis, be thou honoured by the offering of thy feast, and be thou honoured by the wine placed before thee." You may make an offering to Vesta if you wish. The feast to Jupiter consists of roasted meat and an urn of wine. Present it to Jupiter religiously, in the fitting form. After the offering is made plant millet, panic grass, garlic, and lentils.

CXXXIII. To layer fruit and other trees: Press back into the ground the scions which spring up from the ground, but raise their tips out, so that they will take root; dig up at the proper time and transplant vertically. In this way you should propagate from the crown and transplant fig, olive, pomegranate, quince, wild quince, and all other fruits, Cyprian and Delphic laurel, plum, conjugulan myrtle,[2] as well as white and black myrtle, Abellan and Praenestine nuts, and plane trees. Those which you wish to have planted more carefully should be planted in pots. To make them take root while on the tree, take a pot perforated at the bottom or a basket, run the shoot through it, fill the basket with earth, pack it, and leave it on the tree. When it is two years old cut off the tender branch below and plant

4 facere poteris uti radices bene habeant. Item vitem in quasillum propagato terraque bene operito, anno post praecidito, cum qualo serito.

CXXXIV. Priusquam messim facies, porcam praecidaneam hoc modo fieri oportet. Cereri porca praecidanea porco femina, priusquam hasce fruges condas,[1] far, triticum, hordeum, fabam, semen rapicium. Ture vino Iano Iovi Iunoni praefato, 2 priusquam porcum feminam inmolabis. Iano struem ommoveto sic: "Iane pater, te hac strue ommovenda bonas preces precor, uti sies volens propitius mihi liberisque meis domo familiaeque meae." Fertum Iovi ommoveto et mactato sic: "Iuppiter, te hoc ferto obmovendo bonas preces precor, uti sies volens propitius mihi liberisque meis domo familiaeque meae 3 mactus hoc ferto." Postea Iano vinum dato sic: "Iane pater, uti te strue ommovenda bonas preces bene precatus sum, eiusdem rei ergo macte vino inferio esto." Postea Iovi sic: "Iuppiter macte isto ferto esto, macte vino inferio esto." Postea porcam 4 praecidaneam inmolato. Ubi exta prosecta erunt, Iano struem ommoveto mactatoque item, uti prius obmoveris. Iovi fertum obmoveto mactatoque item, uti prius feceris. Item Iano vinum dato et Iovi vinum dato, item uti prius datum ob struem obmovendam et fertum libandum. Postea Cereri exta et vinum dato.

[1] *condas* Keil: *condantur*.

---

[1] The hog offered in sacrifice before the harvest. Aulus Gellius, IV, 6, 7 contrasts it with the *porca succidanea*, offered after the harvest.

[2] The *strues* was a heap of little offering-cakes, "not unlike the fingers joined together," as Festus describes them; and the *fertum* was an oblation-cake.

along with the basket. By this method you can make any variety of tree take root firmly. Vines may also be layered by thrusting them through a basket, packing firmly with earth, cutting a year later, and planting along with the basket.

CXXXIV. Before harvest the sacrifice of the *porca praecidanea*[1] should be offered in this manner: Offer a sow as *porca praecidanea* to Ceres before harvesting spelt, wheat, barley, beans, and rape seed; and address a prayer, with incense and wine, to Janus, Jupiter, and Juno, before offering the sow. Make an offering of cakes[2] to Janus, with these words: "Father Janus, in offering these cakes, I humbly beg that thou wilt be gracious and merciful to me and my children, my house and my household." Then make an offering of cake to Jupiter with these words: "In offering this cake, O Jupiter, I humbly beg that thou, pleased by this offering, wilt be gracious and merciful to me and my children, my house and my household." Then present the wine to Janus, saying: "Father Janus, as I prayed humbly in offering the cakes, so wilt thou to the same end be honoured by this wine placed before thee." And then pray to Jupiter thus: "Jupiter, wilt thou deign to accept the cake; wilt thou deign to accept the wine placed before thee." Then offer up the *porca praecidanea*. When the entrails have been removed, make an offering of cakes to Janus, with a prayer as before; and an offering of a cake to Jupiter, with a prayer as before. After the same manner, also, offer wine to Janus and offer wine to Jupiter, as was directed before for the offering of the cakes, and the consecration of the cake. Afterwards offer entrails and wine to Ceres.

CXXXV. Romae tunicas, togas, saga, centones, sculponeas; Calibus et Minturnis cuculliones, ferramenta, falces, palas, ligones, secures, ornamenta, murices, catellas; Venafro palas. Suessae et in Lucanis plostra, treblae; Albae, Romae dolia, labra; 2 tegulae ex Venafro. Aratra in terram validam Romanica bona erunt, in terram pullam Campanica; iuga Romanica optima erunt; vomeris indutilis optimus erit. Trapeti Pompeis, Nolae ad Rufri maceriam; claves, clostra Romae; hamae, urnae oleariae, urcei aquarii, urnae vinariae, alia vasa ahenea Capuae, Nolae; fiscinae Campanicae Capuae[1] utiles sunt. 3 Funes subductarios, spartum omne Capuae; fiscinas Romanicas Suessae, Casino . . .[2] optimae erunt Romae.

Funem torculum siquis faciet, Casini L. Tunnius, Venafri C. Mennius L. F. Eo indere oportet coria bona VIII nostratia, recentia quae depsta sient, quam minimum salis habeant. Ea depsere et 4 unguere unguine prius oportet, tum siccare. Funem exordiri oportet longum P. LXXII. Toros III habeat, lora in toros singulos VIIII lata digitos II. Cum tortus erit, longus P. XLVIIII. In conmissura abibit P. III, rel. erit P. XLVI. Ubi extentus erit, accedent P. V: longus erit P. LI. Funem torculum extentum longum esse oportet P. LV maximis vasis, 5 minoribus P. LI. Funem loreum in plostrum iustum P. LX, semifunium P. XLV, lora retinacula in plostrum P. XXXVI, ad aratrum P. XXVI, lora praeductoria P. XXVII S, subiugia in plostrum lora

[1] *Capuae* Hauler: *eame* MSS.: *hae hamae* Iucundus.
[2] Lacuna, Keil: *eae* Iucundus.

---

[1] Used for the windlass which raised and lowered the press.

CXXXV. Tunics, togas, blankets, smocks, and shoes should be bought at Rome; caps, iron tools, scythes, spades, mattocks, axes, harness, ornaments, and small chains at Cales and Minturnae; spades at Venafrum; carts and sledges at Suessa and in Lucania; jars and pots at Alba and at Rome; and tiles at Venafrum. Roman ploughs will be good for heavy soil, Campanian for black loam. Roman yokes are the best made. You will find detachable ploughshares the best. The following cities are the best markets for the articles named: oil mills at Pompeii, and at Rufrius's yard at Nola; nails and bars at Rome; pails, oil-urns, water-pitchers, wine-urns, other copper vessels at Capua and at Nola; Campanian baskets from Capua will be found useful; pulley ropes and all sorts of cordage at Capua; Roman baskets at Suessa and Casinum; . . . at Rome will be found best.

Lucius Tunnius, of Casinum, and Gaius Mennius, son of Lucius Mennius, of Venafrum, make the best press-ropes. Eight good native hides, freshly tanned, should be used for these, and should have very little salt; they should be tanned, rubbed down with fat, and then dried. The rope should be laid down 72 feet long, and should have 3 splices, with 9 leather thongs, 2 fingers wide, at each splice. When twisted it will be 49 feet long; 3 feet will be lost in the fastening, leaving 46 feet; when stretched, 5 feet will be added, and the length will be 51 feet. The press-rope[1] should be 55 feet long for the largest presses and 51 for the smaller when stretched. Proper length of thongs for the cart 60 feet, cords 45 feet, leather reins for the cart 36 feet and for the plough 26 feet; traces 27½ feet; yoke straps for the

P. XIX, funiculum P. XV, in aratrum subiugia lora P. XII, funiculum P. IIX.

6 Trapetos latos maximos P. IIII S, orbis altos P. III S, orbis medios, ex lapicaedinis cum eximet, crassos pedem et palmum, inter miliarium et labrum P. I[1] digitos II, labra crassa digitos V.[1] Secundarium trapetum latum P. IIII et palmum, inter miliarium et labrum pes unus digitus unus, labra crassa digitos V, orbis altos P. III et digitos V, crassos P. I et digitos III. Foramen in orbis semipedem quoquo versum facito. Tertium trapetum latum P. IIII, inter miliarium et labrum P. I, labrum digitos V, orbis altos P. III digitos III, crassos P. I et digitos II. Trapetum ubi arvectum erit, ubi statues, ibi accommodato concinnatoque.

CXXXVI. Politionem quo pacto partiario[2] dari oporteat. In agro Casinate et Venafro in loco bono parti octava corbi dividat, satis bono septima, tertio loco sexta; si granum modio dividet, parti quinta. In Venafro ager optimus nona parti corbi dividat. Si communiter pisunt, qua ex parte politori pars est, eam partem in pistrinum politor. Hordeum quinta modio, fabam quinta modio dividat.

CXXXVII. Vineam curandam partiario. Bene curet fundum, arbustum, agrum frumentarium. Partiario faenum et pabulum, quod bubus satis siet, qui illic sient. Cetera omnia pro indiviso.

[1] Schneider: *P. II digitos II, labra crassa digitum.*
[2] *partiario* added by Keil.

---

[1] Unthreshed grain was measured by the basket (*corbis*), threshed grain by the peck (*modius*).

cart 19 feet, lines 15; for the plough, yoke straps 12 feet and line 8 feet.

The largest mills are 4½ feet in diameter; the stones 3½ feet, the centre (when quarried) a foot and a palm thick. Interval between the column and the basin 1 foot, 2 fingers; basin 5 fingers thick. Those of the second size are 4 feet and a palm in diameter, interval between column and basin 1 foot, 1 finger, basin 5 fingers thick; stones 3 feet, 5 fingers in diameter, 1 foot, 3 fingers thick. Cut a hole ½ foot square in the stones. Those of the third size are 4 feet in diameter, interval between column and basin 1 foot, thickness of basin 5 fingers; stones 3 feet, 3 fingers in diameter, 1 foot, 2 fingers thick. Assemble and adjust the press after it has been brought to the place where you wish to set it up.

CXXXVI. Terms for letting the tending of the land to a share tenant: In the district of Casinum and Venafrum, on good land he should receive one-eighth of the unthreshed grain,[1] on fairly good land one-seventh, on land of third quality one-sixth; if the threshed grain is shared, one-fifth. In the district of Venafrum the division is one-ninth of the unthreshed grain on the best land. If they mill in common, the caretaker shall pay for the milling in proportion to the share he receives. He should receive one-fifth of threshed barley and one-fifth of shelled beans.

CXXXVII. [Terms for letting] the care of the vineyard to a share tenant: he must take good care of the estate, the orchard, and the grain land. The share worker is to have enough hay and fodder for the cattle on the place; everything else is in common.

CXXXVIII. Boves feriis coniungere licet. Haec licet facere: arvehant ligna, fabalia, frumentum, quod conditurus[1] erit. Mulis, equis, asinis feriae nullae, nisi si in familia sunt.

CXXXIX. Lucum conlucare Romano more sic oportet. Porco piaculo facito, sic verba concipito: "Si deus, si dea es, quoium illud sacrum est, uti tibi ius est porco piaculo facere illiusce sacri coercendi ergo harumque rerum ergo, sive ego sive quis iussu meo fecerit, uti id recte factum siet, eius rei ergo te hoc porco piaculo inmolando bonas preces precor, uti sies volens propitius mihi domo familiaeque meae liberisque meis; harumce rerum ergo macte hoc porco piaculo inmolando esto."

CXL. Si fodere voles, altero piaculo eodem modo facito, hoc amplius dicito: "operis faciundi causa." Dum opus, cotidie per partes facito. Si intermiseris aut feriae publicae aut familiares intercesserint, altero piaculo facito.

CXLI. Agrum lustrare sic oportet. Impera suovitaurilia circumagi: "Cum divis volentibus quodque bene eveniat, mando tibi, Mani, uti illace suovitaurilia fundum agrum terramque meam quota ex parte sive circumagi sive circumferenda censeas, uti cures 2 lustrare." Ianum Iovemque vino praefamino, sic dicito: "Mars pater, te precor quaesoque uti sies volens propitius mihi domo familiaeque nostrae,

[1] *conditurus* Keil: *non daturus*.

---

[1] Three victims of three kinds were offered, a swine, a ram, and a bull (*sus*, *ovis*, *taurus*).

[2] Variously taken as a slave, an overseer, a sooth-sayer, or our "John Doe."

CXXXVIII. Oxen may be yoked on feast days for these purposes: to haul firewood, bean stalks, and grain for storing. There is no holiday for mules, horses, or donkeys, except the family festivals.

CXXXIX. The following is the Roman formula to be observed in thinning a grove: A pig is to be sacrificed, and the following prayer uttered: "Whether thou be god or goddess to whom this grove is dedicated, as it is thy right to receive a sacrifice of a pig for the thinning of this sacred grove, and to this intent, whether I or one at my bidding do it, may it be rightly done. To this end, in offering this pig to thee I humbly beg that thou wilt be gracious and merciful to me, to my house and household, and to my children. Wilt thou deign to receive this pig which I offer thee to this end."

CXL. If you wish to till the ground, offer a second sacrifice in the same way, with the addition of the words: "for the sake of doing this work." So long as the work continues, the ritual must be performed in some part of the land every day; and if you miss a day, or if public or domestic feast days intervene, a new offering must be made.

CXLI. The following is the formula for purifying land: Bidding the *suovetaurilia*[1] to be led around, use the words: "That with the good help of the gods success may crown our work, I bid thee, Manius,[2] to take care to purify my farm, my land, my ground with this *suovetaurilia*, in whatever part thou thinkest best for them to be driven or carried around." Make a prayer with wine to Janus and Jupiter, and say: "Father Mars, I pray and beseech thee that thou be gracious and merciful to me, my house, and my household; to which intent I have

quoius re ergo agrum terram fundumque meum suovitaurilia circumagi iussi, uti tu morbos visos invisosque, viduertatem vastitudinemque, calamitates intemperiasque prohibessis defendas averruncesque; 3 utique tu fruges, frumenta, vineta virgultaque grandire beneque evenire siris, pastores pecuaque salva servassis duisque bonam salutem valetudinemque mihi domo familiaeque nostrae; harumce rerum ergo, fundi terrae agrique mei lustrandi lustrique faciendi ergo, sicuti dixi, macte hisce suovitaurilibus lactentibus inmolandis esto; Mars pater, eiusdem rei ergo 4 macte hisce suovitaurilibus lactentibus esto." Item cultro facito struem et fertum uti adsiet, inde obmoveto. Ubi porcum inmolabis, agnum vitulumque, sic oportet: "Eiusque rei ergo macte suovitaurilibus inmolandis esto." Nominare vetat Martem neque agnum vitulumque.[1] Si minus in omnis litabit, sic verba concipito: "Mars pater, siquid tibi in illisce suovitaurilibus lactentibus neque satisfactum est, te hisce suovitaurilibus piaculo." Si in uno duobusve dubitabit, sic verba concipito: "Mars pater, quod tibi illoc porco neque satisfactum est, te hoc porco piaculo."

CXLII. Vilici officia quae sunt, quae dominus praecepit, ea omnia quae in fundo fieri oportet quaeque emi pararique oportet, quo modoque cibaria, vestimenta familiae dari oportet, eadem uti curet faciatque moneo dominoque dicto audiens sit. Hoc

[1] The text is corrupt. Schneider suggests *porcum* for *Martem*, thus making the meaning: "It is not permitted to call the pig, the lamb, or the calf by name."

---

[1] Norden, *Antike Kunstprosa*, p. 157, calls attention to the metrical character of the passage. [2] See note 2, page 114.
[3] For the lacuna, see critical note above.

bidden this *suovetaurilia* to be led around my land, my ground, my farm; that thou keep away, ward off, and remove sickness, seen and unseen, barrenness and destruction, ruin and unseasonable influence; and that thou permit my harvests, my grain, my vineyards, and my plantations to flourish and to come to good issue, preserve in health my shepherds and my flocks, and give good health and strength to me, my house, and my household.[1] To this intent, to the intent of purifying my farm, my land, my ground, and of making an expiation, as I have said, deign to accept the offering of these suckling victims; Father Mars, to the same intent deign to accept the offering of these suckling offering." Also heap the cakes[2] with the knife and see that the oblation cake be hard by, then present the victims. When you offer up the pig, the lamb, and the calf, use this formula: "To this intent deign to accept the offering of these victims." . . .[3] If favourable omens are not obtained in response to all, speak thus: "Father Mars, if aught hath not pleased thee in the offering of those sucklings, I make atonement with these victims." If there is doubt about one or two, use these words: "Father Mars, inasmuch as thou wast not pleased by the offering of that pig, I make atonement with this pig."

CXLII. Those things which are the duty of the overseer, the instructions which the master has given, all those things which should be done on the farm and what should be bought or brought in, and how food and raiment should be issued to the servants—the same I warn that he do and perform, and that he hearken to the master's instructions. Furthermore, he must know how to manage the

amplius, quo modo vilicam uti oportet et quo modo eae imperari oportet, uti adventu domini quae opus sunt parentur curenturque diligenter.

CXLIII. Vilicae quae sunt officia, curato faciat. Si eam tibi dederit dominus uxorem, ea esto contentus. Ea te metuat facito. Ne nimium luxuriosa siet. Vicinas aliasque mulieres quam minimum utatur neve domum neve ad sese recipiat. Ad cenam nequo eat neve ambulatrix siet. Rem divinam ni faciat neve mandet, qui pro ea faciat, iniussu domini aut dominae. Scito dominum pro tota familia rem 2 divinam facere. Munda siet; villam conversam mundeque habeat; focum purum circumversum cotidie, priusquam cubitum eat, habeat. Kalendis, Idibus, Nonis, festus dies cum erit, coronam in focum indat, per eosdemque dies lari familiari pro copia supplicet. Cibum tibi et familiae curet uti coctum 3 habeat. Gallinas multas et ova uti habeat. Pira arida, sorba, ficos, uvas passas, sorba in sapa et pira et uvas in doliis et mala strutea, uvas in vinaciis et in urceis in terra obrutas et nuces Praenestinas recentes in urceo in terra obrutas habeat. Mala Scantiana in doliis et alia quae condi solent et silvatica, haec omnia quotannis diligenter uti condita habeat. Farinam bonam et far suptile sciat facere.

[1] Cf. Chap. 7, Sec. 2, and Columella, XII, 45.

housekeeper and how to give her directions, so that the master, at his coming, will find that all necessary preparations and arrangements have been made with care.

CXLIII. See that the housekeeper performs all her duties. If the master has given her to you as wife, keep yourself only to her. Make her stand in awe of you. Restrain her from extravagance. She must visit the neighbouring and other women very seldom, and not have them either in the house or in her part of it. She must not go out to meals, or be a gad-about. She must not engage in religious worship herself or get others to engage in it for her without the orders of the master or the mistress; let her remember that the master attends to the devotions for the whole household. She must be neat herself, and keep the farmstead neat and clean. She must clean and tidy the hearth every night before she goes to bed. On the Kalends, Ides, and Nones, and whenever a holy day comes, she must hang a garland over the hearth, and on those days pray to the household gods as opportunity offers. She must keep a supply of cooked food on hand for you and the servants. She must keep many hens and have plenty of eggs. She must have a large store of dried pears, sorbs, figs, raisins, sorbs in must, preserved pears and grapes and quinces. She must also keep preserved grapes in grape-pulp[1] and in pots buried in the ground, as well as fresh Praenestine nuts kept in the same way, and Scantian quinces in jars, and other fruits that are usually preserved, as well as wild fruits. All these she must store away diligently every year. She must also know how to make good flour and to grind spelt fine.

CXLIV. Oleam legendam hoc modo locare oportet. Oleam cogito recte omnem arbitratu domini, aut quem custodem fecerit, aut cui olea venierit. Oleam ne stringito neve verberato iniussu domini aut custodis. Si adversus ea quis fecerit, quod ipse eo die delegerit, pro eo nemo solvet neque debebitur. Qui oleam legerint, omnes iuranto ad dominum aut ad custodem sese oleam non subripuisse neque quemquam suo dolo malo ea oletate ex fundo L. Manli. Qui eorum non ita iuraverit, quod is legerit omne, pro eo argentum nemo dabit neque debebitur. Oleam cogi recte satis dato arbitratu L. Manli. Scalae ita uti datae erunt, ita reddito, nisi quae vetustate fractae erunt. Si non erunt redditae, 3 aequom viri boni[1] arbitratu deducetur. Siquid redemptoris opera domino damni datum erit, resolvito; id viri boni arbitratu deducetur. Legulos, quot opus erunt, praebeto et strictores. Si non praebuerit, quanti conductum erit aut locatum erit, deducetur; tanto minus debebitur. De fundo ligna et oleam ne deportato. Qui oleam legerit, qui deportarit, in singulas deportationes SS.N.[2] II deducentur neque

[1] Mommsen: *reddet eaeque.*
[2] *i.e. Sestertii nummi.*

---

[1] Keil remarks in his commentary that it is clear that many points in Chapters 144–50 are confused, and that some passages are corrupt because of the loss of words; that all the passages cannot have been written by Cato as we have them now, but that passages were changed or added by other users of the book. He has therefore attempted only to restore the reading of the archetype. Mommsen, in the appendix to Bruns, *Fontes Iuris Romani Antiqui*, 6th edition, has made certain conjectures and interpretations, some of which have been incorporated in the text and translation.

CXLIV.[1] Terms for letting the gathering of olives: The contractor will gather the whole harvest carefully, according to the directions of the owner or his representative or the purchaser of the crop. He will not pick or beat down olives without the orders of the owner or his representative. If anyone violates this rule, no one will pay or be liable for what he has picked that day. All gatherers will take an oath before the owner or his representative that they have not stolen olives, nor has anyone with their connivance stolen olives from the estate of Lucius Manlius[2] during that harvest; if any refuse to take the oath, no one will pay or be liable for what he has gathered. He must give security for the proper harvesting of the olives, satisfactory to Lucius Manlius. Ladders are to be returned in as good condition as when they were issued, except those which have been broken because of age; if they are not returned, a fair deduction will be made by arbitration of an honest man. Whatever damage is done the owner through the fault of the contractor the latter will make good, the amount to be deducted after arbitration by an honest person. The contractor will furnish as many gatherers[3] and pickers as are needed; and if he fails to do so, a deduction will be made of the cost of hiring or contracting, and the total will be less by that amount. He is not to remove firewood or olives from the farm; and if any of his gatherers carry them off, a deduction will be made of 2 sesterces for each load, and that amount will not be

[2] Lucius Manlius is our "John Doe."

[3] The *leguli* harvested the wind-fall olives, while the *strictores* picked olives from the trees.

4 id debebitur. Omnem oleam puram metietur modio oleario. Adsiduos homines L praebeto, duas partes strictorum praebeto. Nequis concedat, quo olea legunda et faciunda carius locetur, extra quam siquem socium inpraesentiarum dixerit. Siquis adversum ea fecerit, si dominus aut custos volent, iurent omnes 5 socii. Si non ita iuraverint, pro ea olea legunda et faciunda nemo dabit neque debebitur ei qui non iuraverit. Accessiones: in M̊ ∞ CC[1] accedit oleae salsae M̊ V, olei puri P. VIIII, in tota oletate aceti Q. V. quod oleae salsae non acceperint, dum oleam legent, in modios singulos SS. V dabuntur.

CXLV. Oleam faciundam hac lege oportet locare. Facito recte arbitratu domini aut custodis, qui id negotium curabit. Si sex iugis vasis opus erit, facito. Homines eos dato, qui placebunt aut custodi aut quis eam oleam emerit. Si opus erit trapetis[2] facito. Si operarii conducti erunt aut facienda locata erit, 2 pro eo resolvito, aut deducetur. Oleum ne tangito utendi causa neque furandi causa, nisi quod custos dederit aut dominus. Si sumpserit, in singulas sumptiones SS.N. XL deducentur neque debebitur. Factores, qui oleum fecerint, omnes iuranto aut ad dominum aut ad custodem sese de fundo L. Manli neque alium quemquam suo dolo malo oleum neque

[1] The abbreviation M̊ is for *modios*; ∞ for M = 1000.
[2] *em⟨erit. si opus⟩erit trapeti* (vel *trapetis*) suggested, with a query, by Goetz: *emerit trapeti.*

---

[1] The whole topic is discussed by Mommsen, *History of Rome*, Book III, Chapter 12.
[2] In Chapter 12 the usual number is given as five.

due. All olives will be measured clean in an olive measure. He is to furnish fifty active workmen, two-thirds being pickers. No one shall form a combination for the purpose of raising the contract price for harvesting and milling olives, unless he names his associate at the time; in case of a violation of this rule, if the owner or his representative wish, all the associates shall take an oath, and if anyone refuses so to swear, no one will pay or be liable for pay for the gathering or milling of the olives to one who has not so sworn. Bonuses: The extra allowance for a harvest of 1200 modii will be 5 modii of salted olives, 9 pounds of pure oil, 5 quadrantals of vinegar for the whole harvest; for that part of the salted olives which they do not take during the harvesting, an allowance of 5 sesterces per modius of the aforesaid will be made.[1]

CXLV. Terms on which contracts are to be made for the milling of olives: Mill them honestly, to the satisfaction of the owner or his representative in charge of the work. If necessary, supply six complete equipments.[2] Furnish workmen to the satisfaction of the representative of the owner or the one who has bought the olives. If a mill is necessary, set it up. If labourers are hired, or the work has to be sublet, settle for this, or let it be deducted. Do not touch any oil by way of use or pilfering beyond what the owner or his representative issues; if he takes it, 40 sesterces will be deducted for each offence, and that amount will not be due. All hands engaged in the manufacturing will take an oath before the owner or his representative that neither they nor anyone with their connivance has stolen oil or olives from the farm

3 oleam subripuisse. Qui eorum non ita iuraverit, quae eius pars erit, omne deducetur neque debebitur. Socium nequem habeto, nisi quem dominus iusserit aut custos. Siquid redemptoris opera domino damni datum erit, viri boni arbitratu deducetur. Si viride oleum opus siet, facito. Accedet oleum et sale suae usioni quod satis siet, vasarium vict. II.

CXLVI. Oleam pendentem hac lege venire oportet. Olea pendens in fundo Venafro venibit. Qui oleam emerit, amplius quam quanti emerit omnis pecuniae centesima accedet, praeconium praesens SS.L, et oleum: Romanici P. ∞ D, viridis P. CC, oleae caducae M̊ L, strictivae M̊ X modio oleario mensum dato, unguinis P. X; ponderibus modiisque domini

2 dato frugis[1] primae cotulas duas. Dies argento ex K. Nov. mensum X oleae legendae faciendae quae locata est, et si emptor locarit, Idibus solvito. Recte haec dari fierique satisque dari domino, aut cui iusserit, promittito satisque dato arbitratu domini. Donicum solutum erit aut ita satis datum erit, quae in fundo inlata erunt, pigneri sunto; nequid eorum de fundo deportato; siquid deportaverit, domini esto.

3 Vasa torcula, funes, scalas, trapetos, siquid et aliut datum erit, salva recte reddito, nisi quae vetustate

[1] Hauler: *iri pri.*

---

[1] *Vasarium* is the amount paid for the use of the mill. The *victoriatus* was one-half a denarius. [2] See Glossary.

[3] This old formula is supposed to be based on an original year of ten months; but it is denied by some scholars that the year was ever one of ten months.

of Lucius Manlius. If any one of them will not take such oath, his share of the pay will be deducted, and that amount will not be due. You will have no partner without the approval of the owner or his representative. Any damage done to the owner through the fault of the contractor will be deducted on the decision of an honest person. If green oil is required, make it. There will be an allowance of a sufficient quantity of oil and salt for his own use, and two victoriati as toll.[1]

CXLVI. Terms for the sale of olives on the tree: Olives for sale on the tree on an estate near Venafrum. The purchaser of the olives will add one per cent. of all money more than the purchase price; the auctioneer's fee of 50 sesterces; and pay 1500 pounds of Roman oil, 200 pounds of green oil, 50 modii of windfall olives, 10 modii of picked olives, all measured by olive measure, and 10 pounds of lubricating oil; and pay 2 cotylae[2] of the first pressing for the use of the weights and measures of the owner. Date of payment: within ten months[3] from the first of November he will pay the contract price for gathering and working up the olives, even if the purchaser has made a contract, on the Ides. Sign a contract and give bond to the satisfaction of the owner that such payments will be made in good faith, and that all will be done to the satisfaction of the owner or his representative. Until payment is made, or such security has been given, all property of the purchaser on the place will be held in pledge, and none of it shall be removed from the place; whatever is so removed becomes the property of the owner. All presses, ropes, ladders, mills, and whatever else has been furnished by the owner, will be returned in the

fracta erunt. Si non reddet, aequom solvito. Si emptor legulis et factoribus, qui illic opus fecerint, non solverit, cui dari oportebit, si dominus volet, solvat. Emptor domino debeto et id satis dato, proque ea re ita uti S. S. E[1] item pignori sunto.

CXLVII. Hac lege vinum pendens venire oportet. Vinaceos inlutos et faecem relinquito. Locus vinis ad K. Octob. primas dabitur. Si non ante ea exportaverit, dominus vino quid volet faciet. Cetera lex, quae oleae pendenti.

CXLVIII. Vinum in doliis hoc modo venire oportet. Vini in culleos singulos quadragenae et singulae urnae dabuntur. Quod neque aceat neque muceat, id dabitur. In triduo proxumo viri boni arbitratu degustato. Si non ita fecerit, vinum pro degustato erit. Quot dies per dominum mora fuerit, quo minus 2 vinum degustet, totidem dies emptori procedent. Vinum accipito ante K. Ian. primas. Si non ante acceperit, dominus vinum admetietur. Quod admensus erit, pro eo resolvito. Si emptor postularit, dominus ius iurandum dabit verum fecisse. Locus vinis ad K. Octobres primas dabitur. Si ante non deportaverit, dominus vino quid volet faciet. Cetera lex, quae oleae pendenti.

[1] *i.e. Supra Scriptum Est.*

same good condition, except articles broken because of age; and a fair price will be paid for all not returned. If the purchaser does not pay the gatherers and the workmen who have milled the oil, the owner may, if he wishes, pay the wages due; and the purchaser will be liable to the owner for the amount, and give bond, and his property will be held in pledge as described above.

CXLVII. Terms for the sale of grapes on the vine: The purchaser will leave unwashed lees and dregs. Storage will be allowed for the wine until the first of October next following; if it is not removed before that time, the owner will do what he will with the wine. All other terms as for the sale of olives on the tree.

CXLVIII. Terms for the sale of wine in jars: Forty-one urns to the culleus will be delivered, and only wine which is neither sour nor musty will be sold. Within three days it shall be tasted subject to the decision of an honest man, and if the purchaser fails to have this done, it will be considered tasted; but any delay in the tasting caused by the owner will add as many days to the time allowed the purchaser. The acceptance will take place before the first of January next following; and in default of acceptance by the purchaser the owner will measure the wine, and settlement will be made on the basis of such measurement; if the purchaser wishes, the owner will take an oath that he has measured it correctly. Storage will be allowed for the wine until the first of October next following; if it is not removed before that date, the owner will do what he wishes with the wine. Other terms as for olives on the tree.

CXLIX. Qua lege pabulum hibernum venire oporteat. Qua vendas fini dicito. Pabulum frui occipito ex Kal. Septembribus. Prato sicco decedat, ubi pirus florere coeperit; prato inriguo, ubi super inferque vicinus permittet, tum decedito, vel diem certam utrique facito. Cetero pabulo Kal. Martiis de-
2 cedito. Bubus domitis binis, cantherio uni, cum emptor pascet, domino pascere recipitur. Holeris, asparagis, lignis, aqua, itinere, actu domini usioni recipitur. Siquid emptor aut pastores aut pecus emptoris domino damni dederit, viri boni arbitratu resolvat. Siquid dominus aut familia aut pecus emptori damni dederit, viri boni arbitratu resolvetur. Donicum pecuniam solverit aut satisfecerit aut delegarit, pecus et familia, quae illic erit, pigneri sunto. Siquid de iis rebus controversiae erit, Romae iudicium fiat.

CL. Fructum ovium hac lege venire oportet. In singulas casei P. I S dimidium aridum, lacte feriis quod mulserit dimidium et praeterea lactis urnam unam; hisce legibus, agnus diem et noctem qui vixerit in fructum; et Kal. Iun. emptor fructu
2 decedat; si interkalatum erit, K. Mais. Agnos XXX ne amplius promittat. Oves quae non pepererint binae pro singulis in fructu cedent. Ex quo[1] die lanam et agnos vendat menses X ab coactore releget.

[1] *Ex quo* added by Mommsen.

---

[1] In the years when 22 or 23 days were added to the ordinary number of 355, they were inserted after the 23rd day of February which ended on that day. The remaining five days along with the inserted days constituted a thirteenth month, called *intercalaris*.

CXLIX. Terms for the lease of winter pasturage: The contract should state the limits of pasturage. The use of the pasturage should begin on the first of September, and should end on dry meadows when the pear trees begin to bloom, and on water meadows when the neighbours above and below begin irrigating, or on a definite date fixed for each; on all other meadows on the first of March. The owner reserves the right to pasture two yoke of oxen and one gelding while the renter pastures; the use of vegetables, asparagus, firewood, water, roads, and right of way is reserved for the owner. All damage done to the owner by the renter or his herdsmen or cattle shall be settled for according to the decision of an honest man; and all damage done to the renter by the owner or his servants or cattle shall be settled for according to the decision of a good man. Until such damage is settled for in cash or by security, or the debt is assigned, all herds and servants on the place shall be held in pledge; and if there arises any dispute over such matters, let the decision be made at Rome.

CL. Terms for the sale of the increase of the flock: The lessee will pay per head 1½ pounds of cheese, one-half dry; one-half of the milking on holy days; and an urn of milk on other days. For the purpose of this rule a lamb which lives for a day and a night is counted as increase; the lessee will end the increase on the first of June, or, if an intercalation[1] intervene, on the first of May. The lessor will not promise more than thirty lambs; ewes which have borne no lambs count in the increase two for one. Ten months after the date of the sale of wool and lambs he shall receive his money from the col-

Porcos serarios in oves denas singulos pascat. Conductor duos menses pastorem praebeat. Donec domino satisfecerit aut solverit, pignori esto.

CLI. Semen cupressi quo modo legi seri propagarique oporteat et quo pacto cupresseta seri oporteat, Minius Percennius Nolanus ad hunc modum 2 monstravit. Semen cupressi Tarentinae per ver legi oportet; materiem, ubi hordeum flavescit. Id ubi legeris, in sole ponito, semen purgato. Id aridum condito, uti aridum expostum siet. Per ver serito in loco ubi terra tenerrima erit, quam pullam vocant, ubi aqua propter siet. Eum locum stercorato primum bene stercore caprino aut ovillo, tum vortito bipalio, terram cum stercore bene permisceto, depurgato ab herba graminibusque, bene terram 3 comminuito. Areas facito pedes latas quaternos; subcavas facito, uti aquam continere possint; inter eas sulcos facito, qua herbas de areis purgare possis. Ubi areae factae erunt, semen serito crebrum, ita uti linum seri solet. Eo cribro terram incernito, dimidiatum digitum terram altam succernito. Id bene 4 tabula aut manibus aut pedibus conplanato. Siquando non pluet, uti terra sitiat, aquam inrigato leniter in areas. Si non habebis unde inriges, gerito inditoque leniter. Quotienscumque opus erit, facito uti aquam addas. Si herbae natae erunt, facito uti ab herbis purges. Quam tenerrimis herbis, et quotiens opus erit, purges. Per aestatem ita uti dictum est fieri oportet, et ubi semen satum siet,

lector. He may feed one whey-fed hog for every ten sheep. The lessee will furnish a shepherd for two months; and he shall remain in pledge until the owner is satisfied either by security or by payment.

CLI. As to cypress seed, the best method for its gathering, planting, and propagation, and for the planting of the cypress bed has been given as follows by Minius Percennius of Nola: The seed of the Tarentine cypress should be gathered in the spring, and the wood when the barley turns yellow; when you gather the seed, expose it to the sun, clean it, and store it dry so that it may be set out dry. Plant the seed in the spring, in soil which is very mellow, the so-called *pulla*, close to water. First cover the ground thick with goat or sheep dung, then turn it with the trenching spade and mix it well with the dung, cleaning out grass and weeds; break the ground fine. Form the seed-beds four feet wide, with the surface concave, so that they will hold water, leaving a footway between the beds so that you may clean out the weeds. After the beds are formed, sow the seed as thickly as flax is usually sowed, sift dirt over it with a sieve to the depth of a half-finger, and smooth carefully with a board, or the hands or feet. In case the weather is dry so that the ground becomes thirsty, irrigate by letting a stream gently into the beds; or, failing a stream, have the water brought and poured gently; see that you add water whenever it is needed. If weeds spring up, see that you free the beds of them. Clean them when the weeds are very young, and as often as is necessary. This procedure should be continued as stated throughout the summer. The seed, after

stramentis operiri; ubi germinascere coeperit, tum demi.

CLII. De scopis virgeis, Q. A. M.[1] Manlii monstraverunt. In diebus XXX, quibus vinum legeris, aliquotiens facito scopas virgeas ulmeas aridas, in asserculo alligato, eabus latera doliis intrinsecus usque bene perfricato, ne faex in lateribus adhaerescat.

CLIII. Vinum faecatum sic facito. Fiscinas olearias Campanicas duas illae rei habeto. Eas faecis inpleto sub prelumque subdito exprimitoque.

CLIV. Vinum emptoribus sine molestia quo modo admetiaris. Labrum culleare illae rei facito. Id habeat ad summum ansas IIII, uti transferri possitur. Id imum pertundito; ea fistulam subdito, uti opturarier recte possit; et ad summum, qua fini culleum capiet, pertundito. Id in suggestu inter dolia positum habeto, uti in culleum de dolio vinum[2] salire possit. Id inpleto, postea obturato.

CLV. Per hiemem aquam de agro depelli oportet. In monte fossas inciles puras habere oportet. Prima autumnitate cum pulvis est, tum maxime ab aqua periculum est. Cum pluere incipiet, familiam cum ferreis sarculisque exire oportet, incilia aperire, aquam diducere in vias et curare oportet uti fluat. 2 In villa, cum pluet, circumire oportet, sicubi perpluat, et signare carbone, cum desierit pluere, uti tegula mutetur. Per segetem in frumentis aut in segete aut

[1] Various interpretations have been proposed, of which Gesner's *quemadmodum* seems the most probable.
[2] *de dolio vinum* Pontedera: *decovinum*, PA.

---

[1] The *suggestus* was an elevated walk-way around the *lacus* into which the wine flowed from the press.

being planted, should be covered with straw, which should be removed when they begin to sprout.

CLII. Of brush-brooms, according to the directions of the Manlii: At several times during the thirty days of the vintage, make brooms of dry elm twigs bound around a stick. With these scrape continually the inner surfaces of the wine jars, to keep the wine dregs from sticking to the sides.

CLIII. To make lees-wine: Keep two Campanian olive baskets for the purpose; fill them with lees, place them under the press-beam, and force out the juice.

CLIV. A convenient method of measuring wine for buyers: Take for this purpose a cask of culleus size, with four handles at the top for easy handling; make a hole at the bottom, fitting into it a pipe so that it can be stopped tight, and also pierce near the top at the point where it will hold exactly a culleus. Keep it on the elevation[1] among the jars, so that the wine can run from the jar into the cask; and when the cask is filled close it up.

CLV. Land ought to be drained during the winter, and the drain-ditches on the hillsides kept clean. The greatest danger from water is in the early autumn, when there is dust. When the rains begin, the whole household must turn out with shovels and hoes, open the ditches, turn the water into the roads, and see that it flows off. You should look around the farmstead while it is raining, and mark all leaks with charcoal, so that the tile can be replaced after the rain stops. During the growing season, if water is standing anywhere, in the grain

in fossis sicubi aqua constat aut aliquid aquae obstat, id emittere, patefieri removerique oportet.

CLVI. De brassica quod concoquit. Brassica est quae omnibus holeribus antistat. Eam esto vel coctam vel crudam. Crudam si edes, in acetum intinguito. Mirifice concoquit, alvum bonam facit, lotiumque ad omnes res salubre est. Si voles in convivio multum bibere cenareque libenter, ante cenam esto crudam quantum voles ex aceto, et item, ubi cenaveris, comesto aliqua V folia; reddet te quasi nihil ederis, bibesque quantum voles.

2 Alvum si voles deicere superiorem, sumito brassicae quae levissima erit P. IIII inde facito manipulos aequales tres conligatoque. Postea ollam statuito cum aqua. Ubi occipiet fervere, paulisper demittito unum manipulum, fervere desistet. Postea ubi occipiet fervere, paulisper demittito ad modum 3 dum quinque numeres, eximito. Item facito alterum manipulum, item tertium. Postea conicito, contundito, item eximito in linteum, exurgeto sucum quasi heminam in pocillum fictile. Eo indito salis micam quasi ervum et cumini fricti tantum quod oleat. Postea ponito pocillum in sereno noctu. Qui poturus erit, lavet calida, bibat aquam mulsam, 4 cubet incenatus. Postea mane bibat sucum deambuletque horas IIII, agat, negoti siquid habebit. Ubi libido veniet, nausia adprehendet, decumbat

---

[1] Compare Chapter 157, 10.

or the seed-bed or in ditches, or if there is any obstruction to the water, it should be cleared, opened and removed.

CLVI. Of the medicinal value of the cabbage: It is the cabbage which surpasses all other vegetables. It may be eaten either cooked or raw; if you eat it raw, dip it into vinegar. It promotes digestion marvellously and is an excellent laxative, and the urine is wholesome for everything.[1] If you wish to drink deep at a banquet and to enjoy your dinner, eat as much raw cabbage as you wish, seasoned with vinegar, before dinner, and likewise after dinner eat some half a dozen leaves; it will make you feel as if you had not dined, and you can drink as much as you please.

If you wish to clean out the upper digestive tract, take four pounds of very smooth cabbage leaves, make them into three equal bunches and tie them together. Set a pot of water on the fire, and when it begins to boil sink one bunch for a short time, which will stop the boiling; when it begins again sink the bunch briefly while you count five, and remove. Do the same with the second and third bunches, then throw the three together and macerate. After macerating, squeeze through a cloth about a hemina of the juice into an earthen cup; add a lump of salt the size of a pea, and enough crushed cummin to give it an odour, and let the cup stand in the air through a calm night. Before taking a dose of this, one should take a hot bath, drink honey-water, and go to bed fasting. Early the next morning he should drink the juice and walk about for four hours, attending to any business he has. When the desire comes on him and he is seized with nausea, he

purgetque sese. Tantum bilis pituitaeque eiciet, uti ipse miretur, unde tantum siet. Postea ubi deorsum versus ibit, heminam aut paulo plus bibat. Si amplius ibit, sumito farinae minutae concas duas, 5 infriet in aquam, paulum bibat, constituet. Verum quibus tormina molesta erunt, brassicam in aqua macerare oportet. Ubi macerata erit, coicito in aquam calidam, coquito usque donec conmadebit bene, aquam defundito. Postea salem addito et cumini paululum et pollinem polentae eodem addito 6 et oleum. Postea fervefacito, infundito in catinum, uti frigescat. Eo interito quod volet cibi, postea edit. Sed si poterit solam brassicam esse, edit. Et si sine febre erit, dato vini atri duri aquatum bibat quam minimum; si febris erit, aquam. Id facito cotidie mane. Nolito multum dare, ne pertaedescat, uti possit porro libenter esse. Ad 7 eundem modum viro et mulieri et puero dato. Nunc de illis, quibus aegre lotium it quibusque substillum est. Sumito brassicam, coicito in aquam ferventem, coquito paulisper, uti subcruda siet. Postea aquam defundito non omnem. Eo addito oleum bene et salem et cumini paululum, infervefacito paulisper. Postea inde iusculum frigidum sorbere et ipsam brassicam esse, uti quam primum excoquatur. Cotidie id facito.

CLVII. De brassica Pythagorea, quid in ea boni sit salubritatisque. Principium te cognoscere oportet, quae genera brassicae sint et cuius modi naturam habeant. Omnia ad salutem temperat conmutatque sese semper cum calore, arida simul et umida et dulcis et amara et acris. Sed quae vocantur septem bona in conmixtura, natura omnia haec habet

should lie down and purge himself; he will evacuate such a quantity of bile and mucus that he will wonder himself where it all came from. Afterwards, when he goes to stool, he should drink a hemina or a little more. If it acts too freely, if he will take two conchas of fine flour, sprinkle it into water, and drink a little, it will cease to act. Those who are suffering from colic should macerate cabbage in water, then pour into hot water, and boil until it is quite soft. Pour off the water, add salt, a bit of cummin, barley flour dust, and oil, and boil again; turn into a dish and allow to cool. You may break any food you wish into it and eat it; but if you can eat the cabbage alone, do so. If the patient has no fever, administer a very little strong, dark wine, diluted; but if he has fever give only water. The dose should be repeated every morning, but in small quantities, so that it may not pall but continue to be eaten with relish. The treatment is the same for man, woman, and child. Now for those who pass urine with difficulty and suffer from strangury: take cabbage, place it in hot water and boil until it is half-done; pour off most of the water, add a quantity of oil, salt, and a bit of cummin, and boil for a short time. After that drink the broth of this and eat the cabbage itself, that it may be absorbed quickly. Repeat the treatment daily.

CLVII. Of Pythagoras's cabbage, what virtue and health-giving qualities it has. The several varieties of cabbage and the quality of each should first be known; it has all the virtues necessary for health, and constantly changes its nature along with the heat, being moist and dry, sweet, bitter, and acid. The cabbage has naturally all the virtues of the so-called "Seven

brassica. Nunc uti cognoscas naturam earum, prima est levis quae nominatur; ea est grandis, latis foliis, caule magno, validam habet naturam et vim 2 magnam habet. Altera est crispa, apiacon vocatur; haec est natura et aspectu bona, ad curationem validior est quam quae supra scripta est. Et item est tertia, quae lenis vocatur, minutis caulibus, tenera, et acerrima omnium est istarum, tenui suco vehementissima. Et primum scito, de omnibus brassicis nulla 3 est illius modi medicamento. Ad omnia vulnera tumores eam contritam inponito. Haec omnia ulcera purgabit sanaque faciet sine dolore. Eadem tumida concoquit, eadem erumpit, eadem vulnera putida canceresque purgabit sanosque faciet, quod aliud medicamentum facere non potest. Verum prius quam id inponas, aqua calida multa lavato; postea bis in die contritam inponito; ea omnem putorem adimet. Cancer ater, is olet et saniem spurcam mittit; albus purulentus est, sed fistulosus 4 et subtus suppurat sub carne. In ea vulnera huiusce modi teras brassicam, sanum faciet; optima est ad huiusce modi vulnus. Et luxatum siquid est, bis die aqua calida foveto, brassicam tritam opponito, cito sanum faciet; bis die id opponito, dolores auferet. Et siquid contusum est, erumpet; brassicam tritam opponito, sanum faciet. Et siquid in mammis ulceris natum et carcinoma, brassicam tritam opponito, 5 sanum faciet. Et si ulcus acrimoniam eius ferre

[1] The introduction to this chapter is not considered genuine. We are told by Plutarch and Diogenes Laertius that the philosophers, such as Solon and Pythagoras, made a point of eating uncooked food, in contrast with the luxury of others. Gesner suggests that the "Seven Blessings" were heat, cold, moisture, dryness, sweetness, bitterness, and sourness. Hörle,

Blessings" mixture.[1] To give, then, the several varieties: the first is the so-called smooth; it is large, with broad leaves and thick stem; it is hardy and has great potency. The second is the curly variety, called "parsley cabbage"; it has a good nature and appearance, and has stronger medicinal properties than the above-mentioned variety. So also has the third, the mild, with small stalk, tender, and the most pungent of all; and its juice, though scanty, has the most powerful effect. No other variety of cabbage approaches it in medicinal value. It can be used as a poultice on all kinds of wounds and swellings; it will cleanse all sores and heal without pain; it will soften and open boils; it will cleanse suppurating wounds and tumours, and heal them, a thing which no other medicine can do. But before it is applied, the surface should be washed with plenty of warm water, and then the crushed cabbage should be applied as a poultice, and renewed twice a day; it will remove all putridity. The black ulcer has a foul odour and exudes putrid pus, the white is purulent but fistulous, and suppurates under the surface; but if you macerate cabbage it will cure all such sores—it is the best remedy for sores of this kind. Dislocations will be healed quickly if they are bathed twice a day in warm water and a cabbage poultice is applied; if applied twice a day, the treatment will relieve the pain. A contusion will burst, and when bruised cabbage is applied, it will heal. An ulcer on the breast and a cancer can be healed by the application of macerated cabbage; and if the spot is too tender to endure the astringency,

suggesting *vis* for *septem* (vii), would emend to *quae vocatur vis bona, in commixta natura,* at the beginning of the sentence.

non poterit, farinam hordeaceam misceto, ita opponito. Huiusce modi ulcera omnia haec sana faciet, quod aliud medicamentum facere non potest neque purgare. Et puero et puellae si ulcus erit huiusce modi, farinam hordeaceam addito. Et si voles eam consectam lautam siccam sale aceto sparsam esse, 6 salubrius nihil est. Quo libentius edis, aceto mulso spargito; lautam siccam et rutam coriandrum sectam sale sparsam paulo libentius edes. Id bene faciet et mali nihil sinet in corpore consistere et alvum bonam faciet. Siquid antea mali intus erit, omnia sana faciet, et de capite et de oculis omnia deducet et sanum faciet. Hanc mane esse oportet 7 ieiunum. Et si bilis atra est et si lienes turgent et si cor dolet et si iecur aut pulmones aut praecordia, uno verbo omnia sana faciet intro quae dolitabunt. Eodem silpium inradito, bonum est. Nam venae omnes ubi sufflatae sunt ex cibo, non possunt perspirare in toto corpore; inde aliqui morbus nascitur. Ubi ex multo cibo alvus non it, pro portione brassica si uteris, id ut te moneo, nihil istorum usu veniet morbis. Verum morbum articularium nulla res tam purgat, quam brassica cruda, si edes concisam et rutam et coriandrum concisam siccam et sirpicium inrasum et brassicam ex aceto oxymeli et sale sparsam. 8 Haec si uteris, omnis articulos poteris experiri. Nullus sumptus est, et si sumptus esset, tamen valetudinis causa experires. Hanc oportet mane ieiu-

[1] Greek σίλφιον, an umbelliferous plant, the juice of which was used in food and medicine. Bentley thinks it is the *asafetida*, still much eaten as a relish in the East. It is now thought that the Persian sort was the *asafetida*, and that the African sort was the *Ferula tingitana* or the *Thapsia gummifera*.

[2] This seems to refer to the theory of Erasistratus, that the arteries carry air, and the veins carry blood through the body.

the cabbage should be mixed with barley-flour and so applied. All sores of this kind it will heal, a thing which no other medicine can do or cleanse. When applied to a sore of this kind on a boy or girl the barley-meal should be added. If you eat it chopped, washed, dried, and seasoned with salt and vinegar, nothing will be more wholesome. That you may eat it with better appetite, sprinkle it with grape vinegar, and you will like it a little better when washed, dried, and seasoned with rue, chopped coriander and salt. This will benefit you, allow no ill to remain in the body, and promote digestion; and will heal any ill that may be inside. Headache and eyeache it heals alike. It should be eaten in the morning, on an empty stomach. Also if you are bilious, if the spleen is swollen, if the heart is painful, or the liver, or the lungs, or the diaphragm—in a word, it will cure all the internal organs which are suffering. (If you grate silphium [1] into it, it will be good.) For when all the veins are gorged with food they cannot breathe [2] in the whole body, and hence a disease is caused; and when from excess of food the bowels do not act, if you eat cabbage proportionately, prepared as I direct above, you will have no ill effects from these. But as to disease of the joints, nothing so purges it as raw cabbage, if you eat it chopped, and rue, chopped dry coriander, grated asafetida, and cabbage out of vinegar and honey, and sprinkled with salt.[3] After using this remedy you will have the use of all your joints. There is no expense involved; and even if there were, you should try it for your health's sake. It should be eaten in the morning, on an empty stomach. One who

[3] The passage is very corrupt.

num esse. Insomnis vel siquis est seniosus, hac eadem curatione sanum facies. Verum assam brassicam et unctam caldam, salis paulum dato homini ieiuno. Quam plurimum ederit, tam citissime 9 sanus fiet ex eo morbo. Tormina quibus molesta erunt, sic facito. Brassicam macerato bene, postea in aulam coicito, defervefacito bene. Ubi cocta erit bene, aquam defundito. Eo addito oleum bene et salis paululum et cuminum et pollinem polentae. Postea ferve bene facito. Ubi ferverit, in catinum indito. Dato edit, si poterit, sine pane; si non, dato panem purum ibidem madefaciat. Et si febrim non habebit, dato vinum atrum bibat; cito sanus 10 fiet. Et hoc siquando usus venerit, qui debilis erit, haec res sanum facere potest: brassicam edit ita uti S. S. E. Et hoc amplius lotium conservato eius qui brassicam essitarit, id calfacito, eo hominem demittito, cito sanum facies hac cura; expertum hoc est. Item pueros pusillos si laves eo lotio, numquam debiles fient. Et quibus oculi parum clari sunt, eo lotio inunguito, plus videbunt. Si caput aut cervices 11 dolent, eo lotio caldo lavito, desinent dolere. Et si mulier eo lotio locos fovebit, numquam miseri [1] fient, et fovere sic oportet: ubi in scutra fervefeceris, sub sellam supponito pertusam. Eo mulier adsidat, operito, circum vestimenta eam dato.

12 Brassica erratica maximam vim habet. Eam arfacere et conterere oportet bene minutam. Siquem purgare voles, pridie ne cenet, mane ieiuno dato

[1] *miseri* Hauler: *umseri*.

is sleepless or debilitated you can make well by this same treatment. But give the person, without food, simply warm cabbage, oiled, and a little salt. The more the patient eats the more quickly will he recover from the disease. Those suffering from colic should be treated as follows: Macerate cabbage thoroughly, then put in a pot and boil well; when it is well done pour off the water, add plenty of oil, very little salt, cummin, and fine barley-flour, and let it boil very thoroughly again. After boiling turn it into a dish. The patient should eat it without bread, if possible; if not, plain bread may be soaked in it; and if he has no fever he may have some dark wine. The cure will be prompt. And further, whenever such occasion arises, if a person who is debilitated will eat cabbage prepared as I have described above, he will be cured. And still further, if you save the urine of a person who eats cabbage habitually, heat it, and bathe the patient in it, he will be healed quickly; this remedy has been tested. Also, if babies are bathed in this urine they will never be weakly; those whose eyes are not very clear will see better if they are bathed in this urine; and pain in the head or neck will be relieved if the heated urine is applied. If a woman will warm the privates with this urine, they will never become diseased. The method is as follows: when you have heated it in a pan, place it under a chair whose seat has been pierced. Let the woman sit on it, cover her, and throw garments around her.

Wild cabbage has the greatest strength; it should be dried and macerated very fine. When it is used as a purge, let the patient refrain from food the previous night, and in the morning, still fasting,

brassicam tritam, aquae cyatos IIII. Nulla res tam bene purgabit, neque elleborum neque scamonium, et sine periculo, et scito salubrem esse corpori. 13 Quos diffidas sanos facere, facies. Qui hac purgatione purgatus erit, sic eum curato. Sorbitione liquida hoc per dies septem dato. Ubi esse volet, carnem assam dato. Si esse non volet, dato brassicam coctam et panem, et bibat vinum lene dilutum, lavet raro, utatur unctione. Qui sic purgatus erit, diutina valetudine utetur, neque ullus morbus veniet nisi sua culpa. Et siquis ulcus taetrum vel recens habebit, hanc brassicam erraticam aqua spargito, 14 opponito; sanum facies. Et si fistula erit, turundam intro trudito. Si turundam non recipiet, diluito, indito in vesicam, eo calamum alligato, ita premito, in fistulam introeat; ea res sanum faciet cito. Et ad omnia ulcera vetera et nova contritam cum melle 15 opponito, sanum faciet. Et si polypus in naso intro erit, brassicam erraticam aridam tritam in manum conicito et ad nasum admoveto, ita subducito susum animam quam plurimum poteris; in triduo polypus excidet. Et ubi exciderit, tamen aliquot dies idem 16 facito, ut radices polypi persanas facias. Auribus si parum audies, terito cum vino brassicam, sucum exprimito, in aurem intro tepidum instillato; cito te intelleges plus audire. Depetigini spurcae[1] brassicam opponito, sanam faciet et ulcus non faciet.

CLVIII. Alvum deicere hoc modo oportet, si vis bene tibi deicere. Sume tibi ollam, addito eo aquae

[1] *Depetigini spurcae* Schneider: *de petiginis porcae.*

take macerated cabbage with four cyathi of water. Nothing will purge so well, neither hellebore, nor scammony; it is harmless, and highly beneficial; it will heal persons whom you despair of healing. The following is the method of purging by this treatment: Administer it in a liquid form for seven days; if the patient has an appetite, feed him on roast meat, or, if he has not, on boiled cabbage and bread. He should drink diluted mild wine, bathe rarely, and rub with oil. One so purged will enjoy good health for a long time, and no sickness will attack him except by his own fault. If one has an ulcer, whether suppurated or new, sprinkle this wild cabbage with water and apply it; you will cure him. If there is a fistula, insert a pellet; or if it will not admit a pellet, make a solution, pour into a bladder attached to a reed, and inject into the fistula by squeezing the bladder. It will heal quickly. An application of wild cabbage macerated with honey to any ulcer, old or new, will heal it. If a nasal polypus appears, pour macerated dry wild cabbage into the palm of the hand; apply to the nostril and sniff with the breath as vigorously as possible. Within three days the polypus will fall out, but continue the same treatment for several days after it has fallen out, so that the roots of the polypus may be thoroughly cleaned. In case of deafness, macerate cabbage with wine, press out the juice, and instil warm into the ear, and you will soon know that your hearing is improved. An application of cabbage to a malignant scab will cause it to heal without ulcerating.

CLVIII. Recipe for a purgative, if you wish to purge thoroughly: Take a pot and pour into it

sextarios sex et eo addito ungulam de perna. Si ungulam non habebis, addito de perna frustum P. S quam minime pingue. Ubi iam coctum incipit esse, eo addito brassicae coliculos duos, betae coliculos duos cum radice sua, feliculae pullum, herbae Mercurialis non multum, mitulorum L. II, piscem capitonem et scorpionem I, cochleas sex et lentis pugillum.
2 Haec omnia decoquito usque ad sextarios III iuris. Oleum ne addideris. Indidem sume tibi sextarium unum tepidum, adde vini Coi cyatum unum, bibe, interquiesce, deinde iterum eodem modo, deinde tertium: purgabis te bene. Et si voles insuper vinum Coum mixtum bibere, licebit bibas. Ex iis tot rebus quod scriptum est unum, quod eorum vis, alvum deicere potest. Verum ea re tot res sunt, uti bene deicias, et suave est.

CLIX. Intertrigini remedium. In viam cum ibis, apsinthi Pontici surculum sub anulo habeto.

CLX. Luxum siquod est, hac cantione sanum fiet. Harundinem prende tibi viridem P. IIII aut quinque longam, mediam diffinde, et duo homines teneant ad coxendices. Incipe cantare: "motas uaeta daries dardares astataries dissunapiter," usque dum coeant. Ferrum insuper iactato. Ubi coierint et altera alteram tetigerint, id manu prehende et dextera sinistra praecide, ad luxum aut ad fracturam alliga, sanum fiet. Et tamen cotidie cantato et luxato vel hoc modo: "huat haut haut istasis tarsis ardannabou dannaustra."

CLXI. Asparagus quo modo seratur. Locum su-

---

[1] A plant which still retains this name in botany.
[2] Unknown.
[3] *i.e.* the two halves, apparently at both ends.

six sextarii of water and add the hock of a ham, or, if you have no hock, a half-pound of ham-scraps with as little fat as possible. Just as it comes to a boil, add two cabbage leaves, two beet plants with the roots, a shoot of fern, a bit of the mercury-plant,[1] two pounds of mussels, a capito[2] fish and one scorpion, six snails, and a handful of lentils. Boil all together down to three sextarii of liquid, without adding oil. Take one sextarius of this while warm, add one cyathus of Coan wine, drink, and rest. Take a second and a third dose in the same way, and you will be well purged. You may drink diluted Coan wine in addition, if you wish. Any one of the many ingredients mentioned above is sufficient to move the bowels; but there are so many ingredients in this concoction that it is an excellent purgative, and, besides, it is agreeable.

CLIX. To prevent chafing: When you set out on a journey, keep a small branch of Pontic wormwood under the anus.

CLX. Any kind of dislocation may be cured by the following charm: Take a green reed four or five feet long and split it down the middle, and let two men hold it to your hips. Begin to chant: "motas uaeta daries dardares astataries dissunapiter" and continue until they meet. Brandish a knife over them, and when the reeds[3] meet so that one touches the other, grasp with the hand and cut right and left. If the pieces are applied to the dislocation or the fracture, it will heal. And none the less chant every day, and, in the case of a dislocation, in this manner, if you wish: "huat haut haut istasis tarsis ardannabou dannaustra."

CLXI. Method of planting asparagus: Break up

bigere oportet bene, qui habeat umorem, aut locum crassum. Ubi erit subactus, areas facito, ut possis dextra sinistraque sarire runcare, ne calcetur. Cum areas deformabis, intervallum facito inter areas semipedem latum in omnes partes. Deinde serito, ad lineam palo grana bina aut terna demittito et eodem palo cavum terra operito. Deinde supra areas stercus spargito bene. Serito secundum 2 aequinoctium vernum. Ubi erit natum, herbas crebro purgato cavetoque ne asparagus una cum herba vellatur. Quo anno severis, satum stramentis per hiemem operito, ne praeuratur. Deinde primo vero aperito, sarito runcatoque. Post annum tertium, quam severis, incendito vere primo. Deinde ne ante sarueris, quam asparagus natus erit, ne in sariendo radices laedas. Tertio aut quarto anno 3 asparagum vellito ab radice. Nam si defringes, stirpes fient et intermorientur. Usque licebit vellas, donicum in semen videris ire. Semen maturum fit ad autumnum. Ita, cum sumpseris semen, incendito, et cum coeperit asparagus nasci, sarito et stercorato. Post annos VIII aut novem, cum iam est vetus, digerito et in quo loco posturus eris terram bene subigito et stercerato. Deinde fossulas facito, quo 4 radices asparagi demittas. Intervallum sit ne minus pedes singulos inter radices asparagi. Evellito, sic circumfodito, ut facile vellere possis; caveto ne frangatur. Stercus ovillum quam plurimum fac ingeras; id est optimum ad eam rem; aliut stercus herbas creat.

CLXII. Salsura pernarum et ofellae Puteolanae.

thoroughly ground that is moist, or is heavy soil. When it has been broken, lay off beds, so that you may hoe and weed them in both directions without trampling the beds. In laying off the beds, leave a path a half-foot wide between the beds on each side. Plant along a line, dropping two or three seeds together in a hole made with a stick, and cover with the same stick. After planting, cover the beds thickly with manure; plant after the vernal equinox. When the shoots push up, weed often, being careful not to uproot the asparagus with the weeds. The year it is planted, cover the bed with straw through the winter, so that it will not be frostbitten. Then in the early spring uncover, hoe, and weed. The third year after planting burn it over in the early spring; after this do not work it before the shoots appear, so as not to injure the roots by hoeing. In the third or fourth year you may pull asparagus from the roots; for if you break it off, sprouts will start and die off. You may continue pulling until you see it going to seed. The seed ripens in autumn; when you have gathered it, burn over the bed, and when the asparagus begins to grow, hoe and manure. After eight or nine years, when it is now old, dig it up, after having thoroughly worked and manured the ground to which you are to transplant it, and made small ditches to receive the roots. The interval between the roots of the asparagus should be not less than a foot. In digging, loosen the earth around the roots so that you can dig them easily, and be careful not to break them. Cover them very deep with sheep dung; this is the best for this purpose, as other manure produces weeds.

CLXII. Method of curing hams and Puteolan

Pernas sallire sic oportet in dolio aut in seria. Cum pernas emeris, ungulas earum praecidito. Salis Romaniensis moliti in singulas semodios. In fundo dolii aut seriae sale sternito, deinde pernam ponito, 2 cutis deosum spectet, sale obruito totam. Deinde alteram insuper ponito, eodem modo obruito. Caveto ne caro carnem tangat. Ita omnes obruito. Ubi iam omnes conposueris, sale insuper obrue, ne caro appareat; aequale facito. Ubi iam dies quinque in sale fuerint, eximito omnis cum suo sale. Quae tum summae fuerint, imas facito eodemque modo 3 obruito et conponito. Post dies omnino XII pernas eximito et salem omnem detergeto et suspendito in vento biduum. Die tertio extergeto spongea bene, perunguito oleo, suspendito in fumo biduum. Tertio die demito, perunguito oleo et aceto conmixto, suspendito in carnario. Nec tinia nec vermes tangent.

ofella.[1] You should salt hams in the following manner, in a jar or large pot: When you have bought the hams cut off the hocks. Allow a half-modius of ground Roman salt to each ham. Spread salt on the bottom of the jar or pot; then lay a ham, with the skin facing downwards, and cover the whole with salt. Place another ham over it and cover in the same way, taking care that meat does not touch meat. Continue in the same way until all are covered. When you have arranged them all, spread salt above so that the meat shall not show, and level the whole. When they have remained five days in the salt remove them all with their own salt. Place at the bottom those which had been on top before, covering and arranging them as before. Twelve days later take them out finally, brush off all the salt, and hang them for two days in a draught. On the third day clean them thoroughly with a sponge and rub with oil. Hang them in smoke for two days, and the third day take them down, rub with a mixture of oil and vinegar, and hang in the meat-house. No moths or worms will touch them.

[1] Supposed to be the same as the *offulae carnis*, lumps of salted pork, mentioned by Columella (XII, 55, 4).

MARCUS TERENTIUS VARRO
ON AGRICULTURE

M. TERENTI VARRONIS

RERUM RUSTICARUM

LIBER PRIMUS

I. Otium si essem consecutus, Fundania, commodius tibi haec scriberem, quae nunc, ut potero, exponam cogitans esse properandum, quod, ut dicitur, si est homo bulla, eo magis senex. Annus enim octogesimus admonet me ut sarcinas conligam, antequam pro- 2 ficiscar e vita. Quare, quoniam emisti fundum, quem bene colendo fructuosum cum facere velis, meque ut id mihi habeam curare roges, experiar; et non solum, ut ipse quoad vivam, quid fieri oporteat ut 3 te moneam, sed etiam post mortem. Neque patiar Sibyllam non solum cecinisse quae, dum viveret, prodessent hominibus, sed etiam quae cum perisset ipsa, et id etiam ignotissimis quoque hominibus; ad cuius libros tot annis post publice solemus redire, cum desideramus, quid faciendum sit nobis ex aliquo portento; me, ne dum vivo quidem, necessariis meis 4 quod prosit facere. Quocirca scribam tibi tres libros indices, ad quos revertare, siqua in re quaeres, quem ad modum quidque te in colendo oporteat facere. Et quoniam, ut aiunt, dei facientes adiuvant, prius

MARCUS TERENTIUS VARRO

ON AGRICULTURE

BOOK I

I. HAD I possessed the leisure, Fundania,[1] I should write in a more serviceable form what now I must set forth as I can, reflecting that I must hasten; for if man is a bubble, as the proverb has it, all the more so is an old man. For my eightieth year admonishes me to gather up my pack before I set forth from life. Wherefore, since you have bought an estate and wish to make it profitable by good cultivation, and ask that I concern myself with the matter, I will make the attempt; and in such wise as to advise you with regard to the proper practice not only while I live but even after my death. And I cannot allow the Sibyl to have uttered prophecies which benefited mankind not only while she lived, but even after she had passed away, and that too people whom she never knew—for so many years later we are wont officially to consult her books when we desire to know what we should do after some portent—and not do something, even while I am alive, to help my friends and kinsfolk. Therefore I shall write for you three handbooks to which you may turn whenever you wish to know, in a given case, how you ought to proceed in farming. And since, as we are told,

[1] Varro's wife.

invocabo eos, nec, ut Homerus et Ennius, Musas, sed duodecim deos Consentis; neque tamen eos urbanos, quorum imagines ad forum auratae stant, sex mares et feminae totidem, sed illos XII deos, qui maxime 5 agricolarum duces sunt. Primum, qui omnis fructos agri culturae caelo et terra continent, Iovem et Tellurem; itaque, quod ii parentes magni dicuntur, Iuppiter pater appellatur, Tellus terra mater. Secundo Solem et Lunam, quorum tempora observantur, cum quaedam seruntur et conduntur. Tertio Cererem et Liberum, quod horum fructus maxime necessari ad victum; ab his enim cibus et potio venit 6 e fundo. Quarto Robigum ac Floram, quibus propitiis neque robigo frumenta atque arbores corrumpit, neque non tempestive florent. Itaque publice Robigo feriae Robigalia, Florae ludi Floralia instituti. Item adveneror Minervam et Venerem, quarum unius procuratio oliveti, alterius hortorum; quo nomine rustica Vinalia instituta. Nec non etiam precor Lympham ac Bonum Eventum, quoniam sine aqua omnis arida ac misera agri cultura, sine successu ac bono eventu frustratio est, non 7 cultura. Iis igitur deis ad venerationem advocatis ego referam sermones eos quos de agri cultura habuimus nuper, ex quibus quid te facere oporteat

---

[1] In the Etrusco-Romish language of religion, the twelve superior deities who formed the common council of the gods, assembled by Jupiter. The word occurs only in this phrase. Their names are given by Ennius in the hexameters:

Juno, Vesta, Minerva, Ceres, Diana, Venus, Mars,
Mercurius, Ioui', Neptunus, Vulcanus, Apollo.

[2] The festival occurred on the 19th of August; and Varro tells us (*De Ling. Lat.*, VI, 16) that it was "because at that time a temple was dedicated to Venus, and the protection of the

the gods help those who call upon them, I will first invoke them—not the Muses, as Homer and Ennius do, but the twelve councillor-gods[1]; and I do not mean those urban gods, whose images stand around the forum, bedecked with gold, six male and a like number female, but those twelve gods who are the special patrons of husbandmen. First, then, I invoke Jupiter and Tellus, who, by means of the sky and the earth, embrace all the fruits of agriculture; and hence, as we are told that they are the universal parents, Jupiter is called "the Father," and Tellus is called "Mother Earth." And second, Sol and Luna, whose courses are watched in all matters of planting and harvesting. Third, Ceres and Liber, because their fruits are most necessary for life; for it is by their favour that food and drink come from the farm. Fourth, Robigus and Flora; for when they are propitious the rust will not harm the grain and the trees, and they will not fail to bloom in their season; wherefore, in honour of Robigus has been established the solemn feast of the Robigalia, and in honour of Flora the games called Floralia. Likewise I beseech Minerva and Venus, of whom the one protects the oliveyard and the other the garden; and in her honour the rustic Vinalia has been established.[2] And I shall not fail to pray also to Lympha and Bonus Eventus, since without moisture all tilling of the ground is parched and barren, and without success and "good issue" it is not tillage but vexation. Having now duly invoked these divinities, I shall relate the conversations which we had recently about agriculture, from which you may

garden is assigned to her." But Ovid, *Fasti*, IV, 877 f. gives another explanation.

animadvertere poteris. In quis quae non inerunt et quaeres, indicabo a quibus scriptoribus repetas et Graecis et nostris.

Qui Graece scripserunt dispersim alius de alia re, 8 sunt plus quinquaginta. Hi sunt, quos tu habere in consilio poteris, cum quid consulere voles, Hieron Siculus et Attalus Philometor; de philosophis Democritus physicus, Xenophon Socraticus, Aristoteles et Theophrastus peripatetici, Archytas Pythagoreus; item Amphilochus Atheniensis, Anaxipolis Thasius, Apollodorus Lemnius, Aristophanes Mallotes, Antigonus Cymaeus, Agathocles Chius, Apollonius Pergamenus, Aristandros Atheniensis, Bacchius Milesius, Bion Soleus, Chaeresteus et Chaereas Athenienses, Diodorus Prieneus, Dion Colophonius, Diophanes Nicaeensis, Epigenes Rhodius, Euagon Thasius, Euphronii duo, unus Atheniensis, alter Amphipolites, Hegesias Maronites, Menandri duo, unus Prieneus, alter Heracleotes, Nicesius 9 Maronites, Pythion Rhodius. De reliquis, quorum quae fuerit patria non accepi, sunt Androtion, Aeschrion, Aristomenes, Athenagoras, Crates, Dadis, Dionysios, Euphiton, Euphorion, Eubulus, Lysimachus, Mnaseas, Menestratus, Plentiphanes, Persis, Theophilus. Hi quos dixi omnes soluta oratione scripserunt; easdem res etiam quidam versibus, ut 10 Hesiodus Ascraeus, Menecrates Ephesius. Hos nobilitate Mago Carthaginiensis praeteriit, poenica lingua qui res dispersas comprendit libris XXIIX,

learn what you ought to do; and if matters in which you are interested are not treated, I shall indicate the writers, both Greek and Roman, from whom you may learn them.

Those who have written various separate treatises in Greek, one on one subject, another on another, are more than fifty in number. The following are those whom you can call to your aid when you wish to consider any point: Hiero of Sicily and Attalus Philometor; of the philosophers, Democritus the naturalist, Xenophon the Socratic, Aristotle and Theophrastus the Peripatetics, Archytas the Pythagorean, and likewise Amphilochus of Athens, Anaxipolis of Thasos, Apollodorus of Lemnos, Aristophanes of Mallos, Antigonus of Cyme, Agathocles of Chios, Apollonius of Pergamum, Aristandrus of Athens, Bacchius of Miletus, Bion of Soli, Chaeresteus and Chaereas of Athens, Diodorus of Priene, Dion of Colophon, Diophanes of Nicaea, Epigenes of Rhodes, Euagon of Thasos, the two Euphronii, one of Athens and the other of Amphipolis, Hegesias of Maronea, the two Menanders, one of Priene and the other of Heraclea, Nicesius of Maronea, and Pythion of Rhodes. Among other writers, whose birthplace I have not learned, are: Androtion, Aeschrion, Aristomenes, Athenagoras, Crates, Dadis, Dionysius, Euphiton, Euphorion, Eubulus, Lysimachus, Mnaseas, Menestratus, Plentiphanes, Persis, Theophilus. All these whom I have named are prose writers; others have treated the same subjects in verse, as Hesiod of Ascra and Menecrates of Ephesus. All these are surpassed in reputation by Mago of Carthage, who gathered into twenty-eight books, written in the Punic tongue, the subjects they had dealt with

quos Cassius Dionysius Uticensis vertit libris XX ac Graeca lingua Sextilio praetori misit; in quae volumina de Graecis libris eorum quos dixi adiecit non pauca et de Magonis dempsit instar librorum VIII. Hosce ipsos utiliter ad VI libros redegit

11 Diophanes in Bithynia et misit Deiotaro regi. Quo brevius de ea re conor tribus libris exponere, uno de agri cultura, altero de re pecuaria, tertio de villaticis pastionibus, hoc libro circumcisis rebus, quae non arbitror pertinere ad agri culturam. Itaque prius ostendam, quae secerni oporteat ab ea, tum de his rebus dicam sequens naturales divisiones. Ea erunt ex radicibus trinis, et quae ipse in meis fundis colendo animadverti, et quae legi, et quae a peritis audii.

II. Sementivis feriis in aedem Telluris veneram rogatus ab aeditumo, ut dicere didicimus a patribus nostris, ut corrigimur a recentibus urbanis, ab aedituo. Offendi ibi C. Fundanium, socerum meum, et C. Agrium equitem R. Socraticum et P. Agrasium publicanum spectantes in pariete pictam Italiam. Quid vos hic? inquam, num feriae sementivae otiosos huc adduxerunt, ut patres et avos solebant

2 nostros? Nos vero, inquit Agrius, ut arbitror, eadem causa quae te, rogatio aeditumi. Itaque si ita est, ut annuis, morere oportet nobiscum, dum ille

---

[1] A village festival, following the sowing of the seed, on a date set by the Pontifices. See Ovid, *Fasti*, I, 657 f.

[2] The names are evidently chosen in the spirit of the punster. They are genuine Roman names, but all derived from *ager* or *fundus*.

separately. These Cassius Dionysius of Utica translated into Greek and published in twenty books, dedicated to the praetor Sextilius. In these volumes he added not a little from the Greek writers whom I have named, taking from Mago's writings an amount equivalent to eight books. Diophanes, in Bithynia, further abridged these in convenient form into six books, dedicated to king Deiotarus. I shall attempt to be even briefer and treat the subject in three books, one on agriculture proper, the second on animal husbandry, the third on the husbandry of the steading, omitting in this book all subjects which I do not think have a bearing on agriculture. And so, after first showing what matter should be omitted, I shall treat of the subject, following the natural divisions. My remarks will be derived from three sources: what I have myself observed by practice on my own land, what I have read, and what I have heard from experts.

II. On the festival of the Sementivae[1] I had gone to the temple of Tellus at the invitation of the *aeditumus* (sacristan), as we have been taught by our fathers to call him, or of the *aedituus*, as we are being set right on the word by our modern purists. I found there Gaius Fundanius, my father-in-law, Gaius Agrius, a Roman knight of the Socratic school, and Publius Agrasius,[2] the tax-farmer, examining a map of Italy painted on the wall. "What are you doing here?" said I. "Has the festival of the Sementivae brought you here to spend your holiday, as it used to bring our fathers and grandfathers?" "I take it," replied Agrius, "that the same reason brought us which brought you—the invitation of the sacristan. If I am correct, as your nod implies, you will have to await with us his

revertatur. Nam accersitus ab aedile, cuius procuratio huius templi est, nondum rediit et nos uti expectaremus se reliquit qui rogaret. Voltis igitur interea vetus proverbium, quod est "Romanus sedendo vincit," usurpemus, dum ille venit? Sane, inquit Agrius, et simul cogitans portam itineri dici longissimam esse ad subsellia sequentibus nobis procedit.

3 Cum consedissemus, Agrasius, Vos, qui multas perambulastis terras, ecquam cultiorem Italia vidistis? inquit. Ego vero, Agrius, nullam arbitror esse quae tam tota sit culta. Primum cum orbis terrae divisus sit in duas partes ab Eratosthene 4 maxume secundum naturam, ad meridiem versus et ad septemtriones, et sine dubio quoniam salubrior pars septemtrionalis est quam meridiana, et quae salubriora illa fructuosiora, dicendum utique Italiam magis etiam fuisse opportunam ad colendum quam Asiam, primum quod est in Europa, secundo quod haec temperatior pars quam interior. Nam intus paene sempiternae hiemes, neque mirum, quod sunt regiones inter circulum septemtrionalem et inter cardinem caeli, ubi sol etiam sex mensibus continuis non videtur. Itaque in oceano in ea parte ne navigari quidem posse dicunt propter mare congela- 5 tum. Fundanius, Em ubi tu quicquam nasci putes posse aut coli natum. Verum enim est illud Pacuvi sol si perpetuo sit aut nox, flammeo vapore aut frigore

[1] Cf. Fabius Cunctator's remark to Paulus (Livy, XXII, 39, 15): "Dubitas ergo quin sedendo superaturi simus?"

[2] The application of the proverb is that we will waste no time on preliminaries, but will begin at once.

[3] The grammarian Festus (p. 482, Lindsay) quotes a verse from Pacuvius's *Antiope*: "Flammeo vapore torrens terrae fetum exusserit."

return; he was summoned by the aedile who has supervision of this temple, and has not yet returned; and he left a man to ask us to wait for him. Do you wish us then meanwhile to follow the old proverb, 'the Roman wins by sitting still,'[1] until he returns?" "By all means," replied Agrius; and reflecting that the longest part of the journey is said to be the passing of the gate,[2] he walked to a bench, with us in his train.

When we had taken our seats Agrasius opened the conversation: "You have all travelled through many lands; have you seen any land more fully cultivated than Italy?" "For my part," replied Agrius, "I think there is none which is so wholly under cultivation. Consider first: Eratosthenes, following a most natural division, has divided the earth into two parts, one to the south and the other to the north; and since the northern part is undoubtedly more healthful than the southern, while the part which is more healthful is more fruitful, we must agree that Italy at least was more suited to cultivation than Asia. In the first place, it is in Europe; and in the next place, this part of Europe has a more temperate climate than we find farther inland. For the winter is almost continuous in the interior, and no wonder, since its lands lie between the arctic circle and the pole, where the sun is not visible for six months at a time; wherefore we are told that even navigation in the ocean is not possible in that region because of the frozen sea." "Well," remarked Fundanius, "do you think that anything can germinate in such a land, or mature if it does germinate? That was a true saying of Pacuvius,[3] that if either day or night be uninterrupted, all the fruits of the earth perish, from

terrae fructos omnis interire. Ego hic, ubi nox et dies modice redit et abit, tamen aestivo die, si non diffinderem meo insiticio somno meridie, vivere non 6 possum. Illic in semenstri die aut nocte quem ad modum quicquam seri aut alescere aut meti possit? Contra quid in Italia utensile non modo non nascitur, sed etiam non egregium fit? Quod far conferam Campano? Quod triticum Apulo? Quod vinum Falerno? Quod oleum Venafro? Non arboribus 7 consita Italia, ut tota pomarium videatur? An Phrygia magis vitibus cooperta, quam Homerus appellat ἀμπελόεσσαν, quam haec? Aut tritico Argos, quod idem poeta πολύπυρον? In qua terra iugerum unum denos et quinos denos culleos fert vini, quot quaedam in Italia regiones? An non M. Cato scribit in libro Originum sic: "ager Gallicus Romanus vocatur, qui viritim cis Ariminum datus est ultra agrum Picentium. In eo agro aliquotfariam in singula iugera dena cullea vini fiunt"? Nonne item in agro Faventino, a quo ibi trecenariae appellantur vites, quod iugerum trecenas amphoras reddat? Simul aspicit me, Certe, inquit, Libo Marcius, praefectus fabrum tuos, in fundo suo Faventiae hanc 8 multitudinem dicebat suas reddere vites. Duo in primis spectasse videntur Italici homines colendo, possentne fructus pro impensa ac labore redire et utrum saluber locus esset an non. Quorum si

---

[1] Columella, II, 6, 3, gives the several varieties of this grain, a coarse kind of wheat, of which Cato has spoken repeatedly.

[2] *Iliad*, III, 184. [3] *Iliad*, XIV, 372.

[4] See Glossary.

[5] The work survives only in fragments in quotations such as this. [6] The Ager Gallicus.

[7] A city in Cispadane Gaul, modern Faenza.

the fiery vapour or from the cold. For my part, I could not live even here, where the night and the day alternate at moderate intervals, if I did not break the summer day with my regular midday nap; but there, where the day and the night are each six months long, how can anything be planted, or grow, or be harvested? On the other hand, what useful product is there which not only does not grow in Italy, but even grow to perfection? What spelt[1] shall I compare to the Campanian, what wheat to the Apulian, what wine to the Falernian, what oil to the Venafran? Is not Italy so covered with trees that the whole land seems to be an orchard? Is that Phrygia, which Homer[2] calls 'the vine-clad,' more covered with vines than this land, or Argos, which the same poet[3] calls 'the rich in corn,' more covered with wheat? In what land does one iugerum[4] bear ten and fifteen cullei[4] of wine, as do some sections of Italy? Or does not Marcus Cato use this language in his *Origines*?[5] 'The land lying this side of Ariminum and beyond the district of Picenum, which was allotted to colonists, is called Gallo-Roman.[6] In that district, at several places, ten cullei of wine are produced to the iugerum.' Is not the same true in the district of Faventia?[7] The vines there are called by this writer *trecenariae*, from the fact that the iugerum yields three hundred amphorae." And he added, turning to me, "At least your friend, Marcius Libo, the engineer officer, used to tell me that the vines on his estate at Faventia bore this quantity. The Italian seems to have had two things particularly in view in his farming: whether the land would yield a fair return for the investment in money and labour, and whether the situation was healthful or not. If

alterutrum decolat et nihilo minus quis vult colere, mente est captus adque adgnatos et gentiles est deducendus. Nemo enim sanus debet velle impensam ac sumptum facere in cultura, si videt non posse refici, nec si potest reficere fructus, si videt eos fore ut

9 pestilentia dispereant. Sed, opinor, qui haec commodius ostendere possint adsunt. Nam C. Licinium Stolonem et Cn. Tremelium Scrofam video venire; unum, cuius maiores de modo agri legem tulerunt (nam Stolonis illa lex, quae vetat plus D iugera habere civem R.), et qui propter diligentiam culturae Stolonum confirmavit cognomen, quod nullus in eius fundo reperiri poterat stolo, quod effodiebat circum arbores e radicibus quae nascerentur e solo, quos stolones appellabant. Eiusdem gentis C. Licinius, tr. pl. cum esset, post reges exactos annis CCCLXV primus populum ad leges accipiendas in septem

10 iugera forensia e comitio eduxit. Alterum collegam tuum, viginti virum qui fuit ad agros dividendos Campanos, video huc venire, Cn. Tremelium Scrofam, virum omnibus virtutibus politum, qui de agri cultura Romanus peritissimus existimatur. An non iure? inquam. Fundi enim eius propter culturam iucundiore spectaculo sunt multis, quam regie polita aedificia aliorum, cum huius spectatum veniant villas,

---

[1] Both *Stolo* and *Scrofa* are genuine Roman names, but are clearly selected here for the sake of the pun. That on *Stolo* is explained immediately, and for *Scrofa*="a pig," cf. Book II, 4.

[2] The Licinian Law, passed in 367 B.C. Authorities differ as to its exact terms. Columella, I, 3, 11, refers to the same law.

[3] See Cicero, *de Amicitia*, XXV. Licinius was the first to address the people in the forum instead of the patricians in the comitium. The "seven iugera" was traditional for a farm, as it was the amount of land assigned each citizen after the expulsion of the kings (Pliny, *N.H.*, XVIII, 18).

either of these elements is lacking, any man who, in spite of that fact, desires to farm has lost his wits, and should be taken in charge by his kinsmen and family. For no sane man should be willing to undergo the expense and outlay of cultivation if he sees that it cannot be recouped; or, supposing that he can raise a crop, if he sees that it will be destroyed by the unwholesomeness of the situation. But, I think, there are some gentlemen present who can speak with more authority on these subjects; for I see Gaius Licinius Stolo and Gnaeus Tremelius Scrofa[1] approaching, one of them a man whose ancestors originated the bill to regulate the holding of land (for that law which forbids a Roman citizen to hold more than 500 iugera was proposed by a Stolo),[2] and who has proved the appropriateness of the family name by his diligence in farming; he used to dig around his trees so thoroughly that there could not be found on his farm a single one of those suckers which spring up from the roots and are called *stolones*. Of the same family was that Gaius Licinius who, when he was tribune of the plebs, 365 years after the expulsion of the kings, was the first to lead the people, for the hearing of laws, from the comitium into the "farm" of the forum.[3] The other whom I see coming is your colleague, who was of the Commission of Twenty for parcelling the Campanian lands, Gnaeus Tremelius Scrofa, a man distinguished by all the virtues, who is esteemed the Roman most skilled in agriculture." "And justly so," I exclaimed. "For his estates, because of their high cultivation, are a more pleasing sight to many than the country seats of others, furnished in a princely style. When people come to inspect his farmsteads,

non, ut apud Lucullum, ut videant pinacothecas, sed oporothecas. Huiusce, inquam, pomarii summa sacra via, ubi poma veneunt contra aurum, imago.

11 Illi interea ad nos, et Stolo, Num cena comessa, inquit, venimus? Nam non L. videmus Fundilium, qui nos advocavit. Bono animo este, inquit Agrius. Nam non modo ovom illut sublatum est, quod ludis circensibus novissimi curriculi finem facit quadrigis, sed ne illud quidem ovom vidimus, quod in cenali 12 pompa solet esse primum. Itaque dum id nobiscum una videatis ac venit aeditumus, docete nos, agri cultura quam summam habeat, utilitatemne an voluptatem an utrumque. Ad te enim rudem esse agri culturae nunc, olim ad Stolonem fuisse dicunt. Scrofa, Prius, inquit, discernendum, utrum quae serantur in agro, ea sola sint in cultura, an etiam 13 quae inducantur in rura, ut oves et armenta. Video enim, qui de agri cultura scripserunt et Poenice et Graece et Latine, latius vagatos, quam oportuerit. Ego vero, inquit Stolo, eos non in omni re imitandos arbitror et eo melius fecisse quosdam, qui minore pomerio finierunt exclusis partibus quae non pertinent ad hanc rem. Quare tota pastio, quae coniungitur a plerisque cum agri cultura, magis ad pastorem quam 14 ad agricolam pertinere videtur. Quocirca principes qui utrique rei praeponuntur vocabulis quoque sunt diversi, quod unus vocatur vilicus, alter magister pecoris. Vilicus agri colendi causa constitutus

[1] The "eggs," or wooden balls, were taken, one by one, from the *meta* in the race-course, at the end of each lap; and the egg traditionally began the dinner.

it is not to see collections of pictures, as at Lucullus's, but collections of fruit. The top of the Via Sacra," I added, "where fruit brings its weight in gold, is a very picture of his orchard."

While we were speaking they came up, and Stolo inquired: "We haven't arrived too late for dinner? For I do not see Lucius Fundilius, our host." "Do not be alarmed," replied Agrius, "for not only has that egg which shows the last lap of the chariot race at the games in the circus not been taken down, but we have not even seen that other egg which usually heads the procession at dinner.[1] And so, while you and we are waiting to see the latter, and our sacristan is returning, tell us what end agriculture has in view, profit, or pleasure, or both; for we are told that you are now the past-master of agriculture, and that Stolo formerly was." "First," remarked Scrofa, "we should determine whether we are to include under agriculture only things planted, or also other things, such as sheep and cattle, which are brought on to the land. For I observe that those who have written on agriculture, whether in Punic, or Greek, or Latin, have wandered too far from the subject." "For my part," replied Stolo, "I do not think that they are to be imitated in every respect, but that certain of them have acted wisely in confining the subject to narrower limits, and excluding matters which do not bear directly on this topic. Thus the whole subject of grazing, which many writers include under agriculture, seems to me to concern the herdsman rather than the farmer. For that reason the persons who are placed in charge of the two occupations have different names, the one being called *vilicus*, and the other *magister pecoris*. The *vilicus* is appointed for

atque appellatus a villa, quod ab eo in eam convehuntur fructus et evehuntur, cum veneunt. A quo rustici etiam nunc quoque viam veham appellant propter vecturas et vellam, non villam, quo vehunt et unde vehunt. Item dicuntur qui vecturis vivunt
15 velaturam facere. Certe, inquit Fundanius, aliut pastio et aliut agri cultura, sed adfinis et ut dextra tibia alia quam sinistra, ita ut tamen sit quodam modo coniuncta, quod est altera eiusdem carminis
16 modorum incentiva, altera succentiva. Et quidem licet adicias, inquam, pastorum vitam esse incentivam, agricolarum succentivam auctore doctissimo homine Dicaearcho, qui Graeciae vita qualis fuerit ab initio nobis ita ostendit, ut superioribus temporibus fuisse doceat, cum homines pastoriciam vitam agerent neque scirent etiam arare terram aut serere arbores aut putare; ab iis inferiore gradu aetatis susceptam agri culturam. Quocirca ea succinit pastorali, quod est inferior, ut tibia sinistra a dextrae
17 foraminibus. Agrius, Tu, inquit, tibicen non solum adimis domino pecus, sed etiam servis peculium, quibus domini dant ut pascant, atque etiam leges colonicas tollis, in quibus scribimus, colonus in agro surculario ne capra natum pascat; quas etiam

---

[1] Smith, *Dict. of Ant.*, s. v. *Tibia*, says:

"Varro (R.R. I, 2, 15, 16) tells us that the melody was played on the right instrument, which he calls *incentiva*, and the accompaniment on the left or the *succentiva*."

"The two pipes were tuned so that the melody played on one could be accompanied an octave lower on the other."

The dictionaries define "treble" as "the instrument that takes the upper part in concerted music."

[2] Cf. Book II, Chap. 3, 7, and Book II, Chap. 1, 7. The argument is, that if grazing is not to be included, the master is deprived of his flocks, and the slaves of their right to graze an animal of their own; and the homestead laws clearly do

the purpose of tilling the ground, and the name is derived from *villa*, the place into which the crops are hauled (*vehuntur*), and out of which they are hauled by him when they are sold. For this reason the peasants even now call a road *veha*, because of the hauling; and they call the place to which and from which they haul *vella* and not *villa*. In the same way, those who make a living by hauling are said *facere velaturam*." "Certainly," said Fundanius, "grazing and agriculture are different things, though akin; just as the right pipe of the tibia is different from the left, but still in a way united, inasmuch as the one is the treble, while the other plays the accompaniment of the same air." "You may even add this," said I, "that the shepherd's life is the treble, and the farmer's plays the accompaniment, if we may trust that most learned man, Dicaearchus. In his sketch of Greek life from the earliest times, he says that in the primitive period, when people led a pastoral life, they were ignorant even of ploughing, of planting trees, and of pruning, and that agriculture was adopted by them only at a later period. Wherefore the art of agriculture 'accompanies' the pastoral because it is subordinate, as the left pipe is to the stops of the right."[1] "You and your piping," retorted Agrius, "are not only robbing the master of his flock and the slaves of their *peculium*—the grazing which their master allows them—but you are even abrogating the homestead laws, among which we find one reciting that the settler may not graze a young orchard with the offspring of the she-goat,[2] a race which

not exclude grazing, as is evident from this prohibition. The astronomical remark which follows is quite in Varro's manner, as is the preceding play on *inferior*, later, and *inferior*, subordinate, which cannot be reproduced in English.

astrologia in caelum recepit, non longe ab tauro.
18 Cui Fundanius, Vide, inquit, ne, Agri, istuc sit ab hoc, cum in legibus etiam scribatur "pecus quoddam." Quaedam enim pecudes culturae sunt inimicae ac veneno, ut istae, quas dixisti,[1] caprae. Eae enim omnia novella sata carpendo corrumpunt, non
19 minimum vites atque oleas. Itaque propterea institutum diversa de causa ut ex caprino genere ad alii dei aram hostia adduceretur, ad alii non sacrificaretur, cum ab eodem odio alter videre nollet, alter etiam videre pereuntem vellet. Sic factum ut Libero patri, repertori vitis, hirci immolarentur, proinde ut capite darent poenas; contra ut Minervae caprini generis nihil immolarent propter oleam, quod eam quam laeserit fieri dicunt sterilem; eius enim
20 salivam esse fructuis venenum; hoc nomine etiam Athenis in arcem non inigi, praeterquam semel ad necessarium sacrificium, ne arbor olea, quae primum dicitur ibi nata, a capra tangi possit. Nec ullae, inquam, pecudes agri culturae sunt propriae, nisi quae agrum opere, quo cultior sit, adiuvare, ut eae
21 quae iunctae arare possunt. Agrasius, Si istuc ita est, inquit, quo modo pecus removeri potest ab agro, cum stercus, quod plurimum prodest, greges pecorum ministrent? Sic, inquit Agrius, venalium greges dicemus agri culturam esse, si propter istam rem

[1] *dixisti* Iucundus: *dixi*.

astrology, too, has placed in the heavens, not far from the Bull." "Be careful, Agrius," interrupted Fundanius, "that your citation be not wide of the mark; for it is also written in the law, 'a certain kind of flock.' For certain kinds of animals are the foes of plants, and even poisonous, such as the goats of which you spoke; for they destroy all young plants by their browsing, and especially vines and olives. Accordingly there arose a custom, from opposite reasons, that a victim from the goat family might be led to the altar of one god, but might not be sacrificed on the altar of another; since, because of the same hatred, the one was not willing to see a goat, while the other was pleased to see him die. So it was that he-goats were offered to Father Bacchus, the discoverer of the vine, so that they might pay with their lives for the injuries they do him; while, on the other hand, no member of the goat family was sacrificed to Minerva on account of the olive, because it is said that any olive plant which they bite becomes sterile; for their spittle is poisonous to its fruit. For this reason, also, they are not driven into the acropolis at Athens except once a year, for a necessary sacrifice—to avoid the danger of having the olive tree, which is said to have originated there, touched by a she-goat." "Cattle are not properly included in a discussion on agriculture," said I, "except those which enhance the cultivation of the land by their labour, such as those which can plough under the yoke." "If that is so," replied Agrasius, "how can cattle be kept off the land, when manure, which enhances its value very greatly, is supplied by the herds?" "By that method of reasoning," retorted Agrius, "we may assert that slave-trading is a branch of agriculture, if we decide

habendum statuerimus. Sed error hinc, quod pecus in agro esse potest et fructus in eo agro ferre, quod non sequendum. Nam sic etiam res aliae diversae ab agro erunt adsumendae, ut si habet plures in fundo textores atque institutos histonas, sic alios artifices.

Scrofa, Diiungamus igitur, inquit, pastionem a 22 cultura, et siquis quid vult aliud. Anne ego, inquam, sequar Sasernarum patris et filii libros ac magis putem pertinere, figilinas quem ad modum exerceri oporteat, quam argentifodinas aut alia metalla, quae 23 sine dubio in aliquo agro fiunt? Sed ut neque lapidicinae neque harenariae ad agri culturam pertinent, sic figilinae. Neque ideo non in quo agro idoneae possunt esse non exercendae, atque ex iis capiendi fructus; ut etiam, si ager secundum viam et opportunus viatoribus locus, aedificandae tabernae devorsoriae, quae tamen, quamvis sint fructuosae, nihilo magis sunt agri culturae partes. Non enim, siquid propter agrum aut etiam in agro profectus domino, agri culturae acceptum referre debet, sed id modo quod ex satione terra sit natum ad fruendum. 24 Suscipit Stolo, Tu, inquit, invides tanto scriptori et obstrigillandi causa figlinas reprehendis, cum praeclara quaedam, ne laudes, praetermittas, quae ad agri 25 culturam vehementer pertineant. Cum subrisisset Scrofa, quod non ignorabat libros et despiciebat, et Agrasius se scire modo putaret ac Stolonem rogasset ut diceret, coepit: Scribit cimices quem ad modum

---

[1] *Figilinae* (*fodinae*) are pits from which clay could be dug to make bricks, tiles or the like, and Varro is pointing out that though they may be useful to the farmer the working of them is not really a part of agriculture.

[2] Lit. for the sake of hindering.

to keep a gang for that purpose, The error lies in the assumption that, because cattle can be kept on the land and be a source of profit there, they are part of agriculture. It does not follow; for by that reasoning we should have to embrace other things quite foreign to agriculture; as, for instance, you might keep on your farm a number of spinners, weavers, and other artisans."

"Very well," said Scrofa, "let us exclude grazing from agriculture, and whatever else anyone wishes." "Am I, then," said I, "to follow the writings of the elder and the younger Saserna, and consider that how to manage clay-pits[1] is more related to agriculture than mining for silver or other mining such as undoubtedly is carried out on some land? But as quarries for stone or sand-pits are not related to agriculture, so too clay-pits. This is not to say that they are not to be worked on land where it is suitable and profitable; as further, for instance, if the farm lies along a road and the site is convenient for travellers, a tavern might be built; however profitable it might be, still it would form no part of agriculture. For it does not follow that whatever profit the owner makes on account of the land, or even on the land, should be credited to the account of agriculture, but only that which, as the result of sowing, is born of the earth for our enjoyment." "You are jealous of that great writer," interrupted Stolo, "and you attack his potteries carpingly,[2] while passing over the excellent observations he makes bearing very closely on agriculture, so as not to praise them." This brought a smile from Scrofa, who knew the books and despised them; and Agrasius, thinking that he alone knew them, asked Stolo to give a quotation. "This is his

interfici oporteat his verbis: "cucumerem anguinum condito in aquam eamque infundito quo voles, nulli accedent; vel fel bubulum cum aceto mixtum,
26 unguito lectum." Fundanius aspicit ad Scrofam, Et tamen verum dicit, inquit, hic, ut hoc scripserit in agri cultura. Ille, Tam hercle quam hoc, siquem glabrum facere velis, quod iubet ranam luridam coicere in aquam, usque qua ad tertiam partem decoxeris, eoque unguere corpus. Ego, Quod magis, inquam, pertineat ad Fundani valetudinem in eo libro, est satius dicas; nam huiusce pedes solent
27 dolere, in fronte contrahere rugas. Dic sodes, inquit Fundanius; nam malo de meis pedibus audire, quam quem ad modum pedes betaceos seri oporteat. Stolo subridens, Dicam, inquit, eisdem quibus ille verbis scripsit (vel Tarquennam audivi, cum homini pedes dolere coepissent, qui tui meminisset, ei mederi posse): "ego tui memini, medere meis pedibus, terra pestem teneto, salus hic maneto in meis pedibus." Hoc ter noviens cantare iubet,
28 terram tangere, despuere, ieiunum cantare. Multa, inquam, item alia miracula apud Sasernas invenies, quae omnia sunt diversa ab agri cultura et ideo repudianda. Quasi vero, inquit, non apud ceteros quoque scriptores talia reperiantur. An non in magni illius Catonis libro, qui de agri cultura est editus, scripta sunt permulta similia, ut haec, quem

---

[1] The pun on *pedes* cannot be reproduced.

[2] The passage is corrupt, as is usual with such incantations It is assumed that Tarquenna is an *anagnostes*, or trained reader, but Storr-Best suggests that Tarquenna is the mythical founder of Tarquinii, the ecclesiastical metropolis of Etruria, and that the invocation is addressed to him directly: "As I heard, O Tarquenna, that when a mortal's feet began to

recipe for killing bugs," he said: "'Soak a wild cucumber in water, and wherever you sprinkle the water the bugs will not come.' And again, 'Grease your bed with ox gall, mixed with vinegar.'" "And still it is good advice," said Fundanius, glancing at Scrofa, "even if he did write it in a book on agriculture." "Just as good, by Hercules," he replied, "as this one for the making of a depilatory: 'Throw a yellow frog into water, boil it down to one-third, and rub the body with it.'" "It would be better for you to quote from that book," said I, "a passage which bears more closely on the trouble from which Fundanius suffers; for his feet are always hurting him and bringing wrinkles to his brow." "Tell me, pray," exclaimed Fundanius; "I would rather hear about my feet than how beet-roots ought to be planted."[1] "I will tell you," said Stolo, with a smile, "in the very words in which he wrote it (at least I have heard Tarquenna say that when a man's feet begin to hurt he may be cured if he will think of you): 'I am thinking of you, cure my feet. The pain go in the ground, and may my feet be sound.'[2] He bids you chant this thrice nine times, touch the ground, spit on it, and be fasting while you chant." "You will find many other marvels in the books of the Sasernas," said I, "which are all just as far away from agriculture and therefore to be disregarded." "Just as if," said he, "such things are not found in other writers also. Why, are there not many such items in the book of the renowned Cato, which he published on the subject of agriculture,

ache by thinking of you he could be cured, I think now of you, cure my feet," etc. The words *in meis pedibus* seem clearly to be a gloss, and destroy the jingle.

ad modum placentam facere oporteat, quo pacto libum, qua ratione pernas sallere? Illud non dicis, inquit Agrius, quod scribit, "si velis in convivio multum bibere cenareque libenter, ante esse oportet brassicam crudam ex aceto aliqua folia quinque."

III. Igitur, inquit Agrasius, quae diiungenda essent a cultura cuius modi sint, quoniam discretum, de iis rebus quae scientia sit in colendo nos docete, ars id an quid aliud, et a quibus carceribus decurrat ad metas. Stolo cum aspexisset Scrofam, Tu, inquit, et aetate et honore et scientia quod praestas, dicere debes. Ille non gravatus, Primum, inquit, non modo est ars, sed etiam necessaria ac magna; eaque est scientia, quae sint in quoque agro serenda ac facienda, quo terra maximos perpetuo reddat fructus.

IV. Eius principia sunt eadem, quae mundi esse Ennius scribit, aqua, terra, anima et sol. Haec enim cognoscenda, priusquam iacias semina, quod initium fructuum oritur. Hinc profecti agricolae ad duas metas dirigere debent, ad utilitatem et voluptatem. Utilitas quaerit fructum, voluptas delectationem; priores partes agit quod utile est, quam quod

2 delectat. Nec non ea, quae faciunt cultura honestiorem agrum, pleraque non solum fructuosiorem eadem faciunt, ut cum in ordinem sunt consita arbusta atque oliveta, sed etiam vendibiliorem atque adiciunt ad fundi pretium. Nemo enim eadem

---

[1] Cf. Cato, 75, 76, 162, and for the "famous one," 156.

[2] Ennius, *Epicharmus*, frag. III, *aqua, terra, anima, sol.*

[3] *Quod initium fructuum oritur* is variously interpreted.

such as his recipes for placenta, for libum, and for the salting of hams?"[1] "You do not mention that famous one of his composing," said Agrius: "'If you wish to drink deep at a feast and to have a good appetite, eat some half-dozen leaves of raw cabbage with vinegar before dinner.'"

III. "Well, then," said Agrasius, "since we have decided the nature of the subjects which are to be excluded from agriculture, tell us whether the knowledge of those things used in agriculture is an art or not, and trace its course from starting-point to goal." Glancing at Scrofa, Stolo said: "You are our superior in age, in position, and in knowledge, so you ought to speak." And he, nothing loath, began: "In the first place, it is not only an art but an important and noble art. It is, as well, a science, which teaches what crops are to be planted in each kind of soil, and what operations are to be carried on, in order that the land may regularly produce the largest crops.

IV. "Its elements are the same as those which Ennius says are the elements of the universe—water, earth, air, and fire.[2] You should have some knowledge of these before you cast your seed, which is the first step in all production.[3] Equipped with this knowledge, the farmer should aim at two goals, profit and pleasure; the object of the first is material return, and of the second enjoyment. The profitable plays a more important rôle than the pleasurable; and yet for the most part the methods of cultivation which improve the aspect of the land, such as the planting of fruit and olive trees in rows, make it not only more profitable but also more saleable, and add to the value of the estate. For any man would rather pay

The present editors believe that *quod* must refer to the nominal idea *iacere semina* derived from the preceding clause.

utilitati non formosius quod est emere mavult pluris,
3 quam si est fructuosus turpis. Utilissimus autem is ager qui salubrior est quam alii, quod ibi fructus certus; contra in pestilenti calamitas, quamvis in feraci agro, colonum ad fructus pervenire non patitur. Etenim ubi ratio cum orco habetur, ibi non modo fructus est incertus, sed etiam colentium vita. Quare ubi salubritas non est, cultura non aliud est
4 atque alea domini vitae ac rei familiaris. Nec haec non deminuitur scientia. Ita enim salubritas, quae ducitur e caelo ac terra, non est in nostra potestate, sed in naturae, ut tamen multum sit in nobis, quo graviora quae sunt ea diligentia leviora facere possimus. Etenim si propter terram aut aquam odore, quem aliquo loco eructat, pestilentior est fundus, aut propter caeli regionem ager calidior sit, aut ventus non bonus flet, haec vitia emendari solent domini scientia ac sumptu, quod permagni interest, ubi sint positae villae, quantae sint, quo spectent
5 porticibus, ostiis ac fenestris. An non ille Hippocrates medicus in magna pestilentia non unum agrum, sed multa oppida scientia servavit? Sed quid ego illum voco ad testimonium? Non hic Varro noster, cum Corcyrae esset exercitus ac classis et omnes domus repletae essent aegrotis ac funeribus, immisso fenestris novis aquilone et obstructis pestilentibus ianuaque permutata ceteraque eius generis diligentia suos comites ac familiam incolumes reduxit?

---

[1] The incident is mentioned by Pliny, *N.H.*, VII, 123.

[2] Pompey's forces lay at Corcyra before Pharsalia, and Varro had joined him after surrendering to Caesar in Spain.

more for a piece of land which is attractive than for one of the same value which, though profitable, is unsightly. Further, land which is more wholesome is more valuable, because on it the profit is certain; while, on the other hand, on land that is unwholesome, however rich it may be, misfortune does not permit the farmer to reap a profit. For where the reckoning is with death, not only is the profit uncertain, but also the life of the farmers; so that, lacking wholesomeness, agriculture becomes nothing else than a game of chance, in which the life and the property of the owner are at stake. And yet this risk can be lessened by science; for, granting that healthfulness, being a product of climate and soil, is not in our power but in that of nature, still it depends greatly on us, because we can, by care, lessen the evil effects. For if the farm is unwholesome on account of the nature of the land or the water, from the miasma which is exhaled in some spots; or if, on account of the climate, the land is too hot or the wind is not salubrious, these faults can be alleviated by the science and the outlay of the owner. The situation of the buildings, their size, the exposure of the galleries, the doors, and the windows, are matters of the highest importance. Did not that famous physician, Hippocrates, during a great pestilence save not one farm but many cities by his skill?[1] But why do I cite him? Did not our friend Varro here, when the army and fleet were at Corcyra, and all the houses were crowded with the sick and the dead, by cutting new windows to admit the north wind, and shutting out the infected winds, by changing the position of doors, and other precautions of the same kind, bring back his comrades and his servants in good health?[2]

V. Sed quoniam agri culturae quod esset initium et finis dixi, relinquitur quot partes ea disciplina habeat ut sit videndum. Equidem innumerabiles mihi videntur, inquit Agrius, cum lego libros Theophrasti complures, qui inscribuntur *φυτῶν ἱστορίας* et 2 alteri *φυτικῶν αἰτίων*. Stolo, Isti, inquit, libri non tam idonei iis qui agrum colere volunt, quam qui scholas philosophorum; neque eo dico, quo non 3 habeant et utilia et communia quaedam. Quapropter tu potius agri culturae partes nobis expone. Scrofa, Agri culturae, inquit, quattuor sunt partes summae: e quis prima cognitio fundi, solum partesque eius quales sint; secunda, quae in eo fundo opus sint ac debeant esse culturae causa; tertia, quae in eo praedio colendi causa sint facienda; quarta, quo 4 quicque tempore in eo fundo fieri conveniat. De his quattuor generalibus partibus[1] singulae minimum in binas dividuntur species, quod habet prima ea quae ad solum pertinent terrae et quae[2] ad villas et stabula. Secunda pars, quae moventur atque in fundo debent esse culturae causa, est item bipertita, de hominibus, per quos colendum, et de reliquo instrumento. Tertia pars quae de rebus dividitur, quae ad quamque rem sint praeparanda et ubi quaeque facienda. Quarta pars de temporibus, quae ad solis circumitum annuum sint referenda et quae ad lunae menstruum cursum. De primis quattuor partibus prius dicam, deinde subtilius de octo secundis.

[1] *generalibus partibus* Keil: *generibus*.
[2] *et quae* Iucundus: *et iterum quae* Goetz: *et alterum quae*.

---

[1] It will be understood that the word covers the meaning of our "scientist."

V. "But as I have stated the origin and the limits of the science, it remains to determine the number of its divisions." "Really," said Agrius, "it seems to me that they are endless, when I read the many books of Theophrastus, those which are entitled 'The History of Plants' and 'The Causes of Vegetation.'" "His books," replied Stolo, "are not so well adapted to those who wish to tend land as to those who wish to attend the schools of the philosophers;[1] which is not to say that they do not contain matter which is both profitable and of general interest. So, then, do you rather explain to us the divisions of the subject." "The chief divisions of agriculture are four in number," resumed Scrofa: "First, a knowledge of the farm, comprising the nature of the soil and its constituents; second, the equipment needed for the operation of the farm in question; third, the operations to be carried out on the place in the way of tilling; and fourth, the proper season for each of these operations. Each of these four general divisions is divided into at least two subdivisions: the first comprises questions with regard to the soil as such, and those which pertain to housing and stabling. The second division, comprising the movable equipment which is needed for the cultivation of the farm, is also subdivided into two: the persons who are to do the farming, and the other equipment. The third, which covers operations, is subdivided: the plans to be made for each operation, and where each is to be carried on. The fourth, covering the seasons, is subdivided: those which are determined by the annual revolution of the sun, and those determined by the monthly revolution of the moon. I shall discuss first the four chief divisions, and then the eight subdivisions in more detail.

VI. Igitur primum de solo fundi videndum haec quattuor, quae sit forma, quo in genere terrae, quantus, quam per se tutus. Formae cum duo genera sint, una quam natura dat, altera quam sationes imponunt, prior, quod alius ager bene natus, alius male, posterior, quod alius fundus bene consitus est, alius male, dicam prius de naturali.
2 Igitur cum tria genera sint a specie simplicia agrorum, campestre, collinum, montanum, et ex iis tribus quartum, ut in eo fundo haec duo aut tria sint, ut multis locis licet videre, e quibus tribus fastigiis simplicibus sine dubio infimis alia cultura aptior quam summis, quod haec calidiora quam summa, sic collinis, quod ea tepidiora quam infima aut summa; haec apparent magis ita esse in latioribus regionibus,
3 simplicia cum sunt. Itaque ubi lati campi, ibi magis aestus, et eo in Apulia loca calidiora ac graviora, et ubi montana, ut in Vesuvio, quod leviora et ideo salubriora; qui colunt deorsum, magis aestate laborant, qui susum, magis hieme. Verno tempore in campestribus maturius eadem illa seruntur quae in superioribus et celerius hic quam ilic coguntur. Nec non susum quam deorsum tardius seruntur ac me-
4 tuntur. Quaedam in montanis prolixiora nascuntur ac firmiora propter frigus, ut abietes ac sappini, hic,

VI. "First, then, with respect to the soil of the farm, four points must be considered: the conformation of the land, the quality of the soil, its extent, and in what way it is naturally protected. As there are two kinds of conformation, the natural and that which is added by cultivation, in the former case one piece of land being naturally good, another naturally bad, and in the latter case one being well tilled, another badly, I shall discuss first the natural conformation. There are, then, with respect to the topography, three simple types of land—plain, hill, and mountain; though there is a fourth type consisting of a combination of these, as, for instance, on a farm which may contain two or three of those named, as may be seen in many places. Of these three simple types, undoubtedly a different system is applicable to the lowlands than to the mountains, because the former are hotter than the latter; and the same is true of hillsides, because they are more temperate than either the plains or the mountains. These qualities are more apparent in broad stretches, when they are uniform; thus the heat is greater where there are broad plains, and hence in Apulia the climate is hotter and more humid, while in mountain regions, as on Vesuvius, the air is lighter and therefore more wholesome. Those who live in the lowlands suffer more in summer; those who live in the uplands suffer more in winter; the same crops are planted earlier in the spring in the lowlands than in the uplands, and are harvested earlier, while both sowing and reaping come later in the uplands. Certain trees, such as the fir and the pine, flourish best and are sturdiest in the mountains on account of the cold climate, while the poplar and the willow thrive here

quod tepidiora, populi ac salices; susum fertiliora, ut arbutus ac quercus, deorsum, ut nuces graecae ac mariscae fici. In collibus humilibus societas maior cum campestri fructu quam cum montano, in altis contra. 5 Propter haec tria fastigia formae discrimina quaedam fiunt sationum, quod segetes meliores existimantur esse campestres, vineae collinae, silvae montanae. Plerumque hiberna iis esse meliora, qui colunt campestria, quod tunc prata ibi herbosa, putatio arborum tolerabilior; contra aestiva montanis locis commodiora, quod ibi tum et pabulum multum, quod in campis aret, et cultura arborum aptior, quod tum hic 6 frigidior aer. Campester locus is melior, qui totus aequabiliter in unam partem verget, quam is qui est ad libellam aequos, quod is, cum aquae non habet delapsum, fieri solet uliginosus; eo magis, siquis est inaequabilis, eo deterior, quod fit propter lacunas aquosus. Haec atque huiusce modi tria fastigia agri ad colendum disperiliter habent momentum.

VII. Stolo, Quod ad hanc formam naturalem pertinet, de eo non incommode Cato videtur dicere, cum scribit optimum agrum esse, qui sub radice montis situs 2 sit et spectet ad meridianam caeli partem. Subicit Scrofa, De formae cultura hoc dico, quae specie fiant venustiora, sequi ut maiore quoque fructu sint, ut qui habent arbusta, si sata sunt in quincuncem, propter ordines atque intervalla modica. Itaque maiores nostri ex arvo aeque magno male consito et minus mul-

---

[1] Cato, 1, 3.

where the climate is warmer; the arbute and the oak do better in the uplands, the almond and the mariscan fig in the lowlands. On the foothills the growth is nearer akin to that of the plains than to that of the mountains; on the higher hills the opposite is true. Owing to these three types of configuration different crops are planted, grain being considered best adapted to the plains, vines to the hills, and forests to the mountains. Usually the winter is better for those who live in the plains, because at that season the pastures are fresh, and pruning can be carried on in more comfort. On the other hand, the summer is better in the mountains, because there is abundant forage at that time, whereas it is dry in the plains, and the cultivation of the trees is more convenient because of the cooler air. A lowland farm that everywhere slopes regularly in one direction is better than one that is perfectly level, because the latter, having no outlet for the water, tends to become marshy. Even more unfavourable is one that is irregular, because pools are liable to form in the depressions. These points and the like have their differing importance for the cultivation of the three types of configuration."

VII. "So far as concerns the natural situation," said Stolo, "it seems to me that Cato was quite right when he said that the best farm was one that was situated at the foot of a mountain, facing south."[1] Scrofa continued: "With regard to the conformation due to cultivation, I maintain that the more regard is had for appearances the greater will be the profits: as, for instance, if those who have orchards plant them in quincunxes, with regular rows and at moderate intervals. Thus our ancestors, on the same amount of land but not so well

tum et minus bonum faciebant vinum et frumentum, quod quae suo quicque loco sunt posita, ea minus loci occupant, et minus officit aliud alii ab sole ac luna et 3 vento. Hoc licet coniectura videre ex aliquot rebus, ut nuces integras quas uno modio comprendere possis, quod putamina suo loco quaeque habet natura composita, cum easdem, si fregeris, vix sesquimodio concipere possis. Praeterea quae arbores in ordinem satae sunt, eas aequabiliter ex omnibus partibus sol ac luna coquunt. Quo fit ut uvae et oleae plures nascantur et ut celerius coquantur. Quas res duas sequuntur altera illa duo, ut plus reddant musti et olei et preti pluris.

5 Sequitur secundum illud, quali terra solum sit fundi, a qua parte vel maxime bonus aut non bonus appellatur. Refert enim, quae res in eo seri nascique et cuius modi possint; non enim eadem omnia in eodem agro recte possunt. Nam ut alius est ad vitem appositus, alius ad frumentum, sic de ceteris alius 6 ad aliam rem. Itaque Cretae ad Cortyniam dicitur platanus esse, quae folia hieme non amittat, itemque in Cypro, ut Theophrastus ait, una, item Subari, qui nunc Thurii dicuntur, quercus simili esse natura, quae est in oppidi conspectu; item contra atque apud nos fieri ad Elephantinen, ut neque ficus neque vites amittant folia. Propter eandem causam multa sunt

[1] Theophrastus, *Hist. Plant.*, I, 9, 5 (L.C.L. Theophrastus, Vol. I, p. 65); cf. also p. 196.

laid out, made less wine and grain than we do, and of a poorer quality; for plants which are placed exactly where each should be take up less ground and screen each other less from the sun, the moon, and the air. You may prove this by one of several experiments; for instance, a quantity of nuts which you can hold in a modius measure with their shells whole, because the shells naturally keep them compacted, you can scarcely pack into a modius and a half when they are cracked. As to the second point, trees which are planted in a row are warmed by the sun and the moon equally on all sides, with the result that more grapes and olives form, and that they ripen earlier; which double result has the double consequence that they yield more must and oil, and of greater value.

"We come now to the second division of the subject, the type of soil of which the farm is composed. It is in respect of this chiefly that a farm is considered good or bad; for it determines what crops, and of what variety, can be planted and raised on it, as not all crops can be raised with equal success on the same land. As one type is suited to the vine and another to grain, so of others—one is suited to one crop, another to another. Thus near Cortynia, in Crete, there is said to be a plane tree which does not shed its leaves in winter, and another in Cyprus, according to Theophrastus.[1] Likewise at Sybaris, which is now called Thurii, there is said to be an oak tree of like character, in sight of the town; and that near Elephantine neither the fig nor the vine sheds its leaves—which is quite the opposite of what happens with us. For the same reason there are many trees which bear two crops a year, such as the

bifera, ut vitis apud mare Zmyrnae, malus in agro 7 Consentino. Idem ostendit, quod in locis feris plura ferunt, in iis quae sunt culta meliora. Eadem de causa sunt quae non possunt vivere nisi in loco aquoso aut etiam aqua, et id discriminatim alia in lacubus, ut harundines in Reatino, alia in fluminibus, ut in Epiro arbores alni, alia in mari, ut scribit Theo- 8 phrastus palmas et squillas. In Gallia transalpina intus, ad Rhenum cum exercitum ducerem, aliquot regiones accessi, ubi nec vitis nec olea nec poma nascerentur, ubi agros stercorarent candida fossicia creta, ubi salem nec fossicium nec maritimum haberent, sed ex quibusdam lignis combustis carboni- 9 bus salsis pro eo uterentur. Stolo, Cato quidem, inquit, gradatim praeponens alium alio agrum meliorem dicit esse in novem discriminibus, quod sit primus ubi vineae possint esse bono vino et multo, secundus ubi hortus inriguus, tertius ubi salicta, quartus ubi oliveta, quintus ubi pratum, sextus ubi campus frumentarius, septimus ubi caedua silva, octavus ubi arbustum, nonus ubi glandaria silva. 10 Scrofa, Scio, inquit, scribere illum; sed de hoc non consentiunt omnes, quod alii dant primatum bonis pratis, ut ego, a quo antiqui prata parata appellarunt. Caesar Vopiscus, aedilicius causam cum ageret apud censores, campos Roseae Italiae dixit esse sumen, in

---

[1] Theophrastus, *Hist. Plant.*, I, 4, 3 (L.C.L. Theophrastus, Vol. I, p. 33). [2] Cato, 1, 7.

[3] A very fertile district near Reate, now La Roscie.

vine on the coast near Smyrna, and the apple in the district of Consentia. The fact that trees produce more fruit in uncultivated spots, and better fruit under cultivation, proves the same thing. For the same reason there are plants which cannot live except in marshy ground, or actually in the water—and not in every kind of water. Some grow in ponds, as the reeds near Reate, others in streams, as the alder trees in Epirus, and still others in the sea, as the palms and squills of which Theophrastus writes.[1] When I was in command of the army in the interior of Transalpine Gaul near the Rhine, I visited a number of spots where neither vines nor olives nor fruit trees grew; where they fertilized the land with a white chalk which they dug; where they had no salt, either mineral or marine, but instead of it used salty coals obtained by burning certain kinds of wood." "Cato,[2] you know," interjected Stolo, "in arranging plots according to the degree of excellence, formed nine categories: first, land on which the vines can bear a large quantity of wine of good quality; second, land suited for a watered garden; third, for an osier bed; fourth, for olives; fifth, for meadows; sixth, for a grain field; seventh, for a wood lot: eighth, for an orchard; ninth, for a mast grove." "I know he wrote that," replied Scrofa, "but all authorities do not agree with him on this point. There are some who assign the first place to good meadows, and I am one of them. Hence our ancestors gave the name *prata* to meadow-land as being ready (*parata*). Caesar Vopiscus, once an aedile, in pleading a case before the censors, spoke of the plains of Rosea[3] as the nursing-ground[4] of Italy, such that if a rod were left there

[4] *Sumen*: lit. "udder," the richest part.

quo relicta pertica postridie non appareret propter herbam.

VIII. Contra vineam sunt qui putent sumptu fructum devorare. Refert, inquam, quod genus vineae sit, quod sunt multae species eius. Aliae enim humiles ac sine ridicis, ut in Hispania, aliae sublimes, quae appellantur iugatae, ut pleraeque in Italia. Cuius generis[1] nomina duo, pedamenta et iuga. Quibus stat rectis vinea, dicuntur pedamenta; quae transversa iunguntur, iuga; ab eo quoque 2 vineae iugatae. Iugorum genera fere quattuor, pertica, harundo, restes, vites: pertica, ut in Falerno, harundo, ut in Arpano, restes, ut in Brundisino, vites, ut in Mediolanensi. Iugationis species duae, una derecta, ut in agro Canusino, altera compluviata in longitudinem et latitudinem iugata, ut in Italia pleraeque. Haec ubi domo nascuntur, vinea non metuit sumptum; ubi multa e propinqua 3 villa, non valde. Primum genus quod dixi maxime quaerit salicta, secundum harundineta, tertium iunceta aut eius generis rem aliquam, quartum arbusta, ubi traduces possint fieri vitium, ut Mediolanenses faciunt in arboribus, quas vocant opulos, 4 Canusini in harundulatione[2] in ficis. Pedamentum item fere quattuor generum: unum robustum, quod optimum solet afferri in vineam e querco ac iunipiro et vocatur ridica; alterum palus e pertica, meliore

[1] *generis* added by Keil. [2] Schneider: *ardulatione.*

---

[1] The trellis was constructed in the manner of a Roman house-roof, with framework sloping inward from the corners to a quadrangular opening (*compluvium*) in the centre; cf. Columella, IV, 24, 14; 26, 3; Pliny, *N.H.*, XVII, 164.

[2] See on Cato, 1, 7, note 2, p. 7.

[3] Columella, V, 7, describes the method.

overnight, it would be lost the next morning on account of the growth of the grass.

VIII. "As an argument against the vineyard, there are those who claim that the cost of upkeep swallows up the profits. In my opinion, it depends on the kind of vineyard, for there are several: for some are low-growing and without props, as in Spain; others tall, which are called 'yoked,' as generally in Italy. For this latter class there are two names, *pedamenta* and *iuga*: those on which the vine runs vertically are called *pedamenta* (stakes), and those on which it runs transversely are called *iuga* (yokes); and from this comes the name 'yoked vines.' Four kinds of 'yokes' are usually employed, made respectively of poles, of reeds, of cords, and of vines: the first of these, for example, around Falernum, the second around Arpi, the third around Brundisium, the fourth around Mediolanum. There are two forms of this trellising: in straight lines, as in the district of Canusium, or yoked lengthways and sideways in the form of the *compluvium*,[1] as is the practice generally in Italy. If the material grows on the place the vineyard does not mind the expense; and it is not burdensome if much of it can be obtained in the neighbourhood. The first class I have named requires chiefly a willow thicket, the second a reed thicket, the third a rush bed or some material of the kind. For the fourth you must have an arbustum,[2] where trellises can be made of the vines,[3] as the people of Mediolanum do on the trees which they call *opuli* (maples), and the Canusians on lattice-work in fig trees. Likewise, there are, as a rule, four types of props. The best for common use in the vineyard is a stout post, called *ridica*, made of oak or juniper. The second best is a stake made

dura, quo diuturnior; quem cum infimum terra solvit, puter evertitur et fit solum summum; tertium, quod horum inopiae subsidio misit harundinetum. Inde enim aliquot colligatas libris demittunt in tubulos fictiles cum fundo pertuso, quas cuspides appellant, qua umor adventicius transire possit. Quartum est pedamentum nativum eius generis, ubi ex arboribus in arbores traductis vitibus vinea fit,

5 quos traduces quidam rumpos appellant. Vineae altitudinis modus longitudo hominis, intervalla pedamentorum, qua boves iuncti arare possint. Ea minus sumptuosa vinea, quae sine iugo ministrat acratophoro vinum. Huius genera duo: unum, in quo terra cubilia praebet uvis, ut in Asia multis locis, quae saepe vulpibus et hominibus fit communis. Nec non si parit humus mures, minor fit vindemia, nisi totas vineas oppleris muscipulis, quod in insula

6 Pandateria faciunt. Alterum genus vineti, ubi ea modo removetur a terra vitis, quae ostendit se adferre uvam. Sub eam, ubi nascitur uva, subiciuntur circiter bipedales e surculis furcillae, ne vindemia facta denique discat pendere in palma aut funiculo aut vinctu, quod antiqui vocabant cestum. Ibi dominus simul ac vidit occipitium vindemiatoris, furcillas reducit hibernatum in tecta, ut sine sumptu harum opera altero anno uti possit. Hac consuetudine in

---

[1] The word *acratophorum*, meaning a pitcher for unmixed wine, seems to occur only here and in Cicero, *de Fin.*, III, 4, 15, where he defends its use in Latin as a convenience.

[2] Literally, an embroidered girdle; Gr. κεστός.

from a branch, and preferably from a tough one, so that it will last longer; when one end has rotted in the ground the stake is reversed, what had been the top becoming the bottom. The third, which is used only as a substitute when the others are lacking, is formed of reeds; bundles of these, tied together with bark, are planted in what they call *cuspides*, earthenware pipes with open bottoms so that the casual water can run out. The fourth is the natural prop, where the vineyard is formed of vines growing across from tree to tree; such traverses are called by some *rumpi*. The limit to the height of the vineyard is the height of a man, and the intervals between the props should be sufficient to allow a yoke of oxen to plough between. The most economical type of vineyard is that which furnishes wine to beaker[1] without the aid of trellises. There are two kinds of these: one in which the ground serves as a bed for the grapes, as in many parts of Asia. The foxes often share the harvest with man in such vineyards, and if the land breeds mice the yield is cut short unless you fill the whole vineyard with traps, as they do in the island Pandateria. In the other type only those branches are raised from the ground which give promise of producing fruit. These are propped on forked sticks about two feet long, at the time when the grapes form, so that they may not wait until the harvest is over to learn to hang in a bunch by means of a string or the fastening which our fathers called a *cestus*.[2] In such a vineyard, as soon as the master sees the back of the vintager he takes his forks back to hibernate under cover so that he may be able to enjoy their assistance without cost the next year.

7 Italia utuntur Reatini. Haec ideo varietas maxime, quod terra cuius modi sit refert. Ubi enim natura umida, ibi altius vitis tollenda, quod in partu et alimonio vinum non ut in calice quaerit aquam, sed solem. Itaque ideo, ut arbitror, primum e vinea in arbores escendit vitis.

IX. Terra, inquam, cuius modi sit refert et ad quam rem bona aut non bona sit. Ea[1] tribus modis dicitur, communi et proprio et mixto. Communi, ut cum dicimus orbem terrae et terram Italiam aut quam aliam. In ea enim et lapis et harena et cetera eius generis sunt in nominando comprensa. Altero modo dicitur terra proprio nomine, quae nullo alio vocabulo neque 2 cognomine adiecto appellatur. Tertio modo dicitur terra, quae est mixta, in qua seri potest quid et nasci, ut argillosa aut lapidosa, sic aliae, cum in hac species non minus sint multae quam in illa communi propter admixtiones. In illa enim cum sint dissimili vi ac potestate partes permultae, in quis lapis, marmor, rudus, harena, sabulo, argilla, rubrica, pulvis, creta, cinis,[2] carbunculus, id est quae sole perferve ita fit, 3 ut radices satorum comburat, ab iis quae proprio nomine dicitur terra, cum est admixta ex iis generibus aliqua re, dicitur aut cretosa . . .[3] sic ab aliis generum discriminibus mixta. Horum varietatis ita genera haec, ut praeterea subtiliora sint alia, minimum in singula facie terna, quod alia terra est valde lapidosa, alia mediocriter, alia prope pura. Sic de aliis generibus reliquis admixtae terrae tres gradus

[1] *Ea* Iucundus: *et a.*
[2] *cinis* Keil: *ignis.*
[3] A word has been lost.

---

[1] As "land," "ground," "soil."

In Italy the people of Reate practise this custom. This variation in culture is caused chiefly by the fact that the nature of the soil makes a great difference; where this is naturally humid the vine must be trained higher, because while the wine is forming and ripening it does not need water, as it does in the cup, but sun. And that is the chief reason, I think, that the vines climb up trees.

IX. "The nature of the soil, I say, makes a great difference, in determining to what it is or is not adapted. The word *terra* is used in three senses, the general, the specific, and the mixed.[1] It is used in the general sense when we speak of the *orbis terrae*, or of the *terra* of Italy or any other country; for in that designation are included rock, and sand, and other such things. The word is used specifically in the second sense when it is employed without the addition of a qualifying word or epithet. It is used in the third or mixed sense, of the element in which seed can be planted and germinate—such as clay soil, rocky soil, etc. In this last sense of the word there are as many varieties of earth as when it is used in the general sense, on account of the different combinations of substances. For there are many substances in the soil, varying in consistency and strength, such as rock, marble, rubble, sand, loam, clay, red ochre, dust, chalk, ash, carbuncle (that is, when the ground becomes so hot from the sun that it chars the roots of plants); and soil, using the word in its specific sense, is called chalky or . . . according as one of these elements predominates—and so of other types of soil. The classes of these vary in such a way that there are, besides other subdivisions, at least three for each type: rocky soil, for instance,

4 ascendunt eosdem. Praeterea hae ipsae ternae species ternas in se habent alias, quod partim sunt umidiores, partim aridiores, partim mediocres, Neque non haec discrimina pertinent ad fructus vehementer. Itaque periti in loco umidiore far adoreum potius serunt quam triticum, contra in aridiore hordeum potius quam far, in mediocri
5 utrumque. Praeterea etiam discrimina omnium horum generum subtiliora alia, ut in sabulosa terra, quod ibi refert sabulo albus sit an rubicundus, quod subalbus ad serendos surculos alienus, contra rubicundior appositus. Sic magna tria discrimina terrae, quod refert utrum sit macra an pinguis an mediocris, quod ad culturam pinguis fecundior ad multa, macra contra. Itaque in tenui,[1] ut in Pupinia, neque arbores prolixae neque vites feraces, neque stramenta videre crassa possis neque ficum mariscam et arbores
6 plerasque ac prata retorrida muscosa. Contra in agro pingui, ut in Etruria, licet videre et segetes fructuosas ac restibilis et arbores prolixas et omnia sine musco. In mediocri autem terra, ut in Tiburti, quo propius accedit ut non sit macra, quam ut sit ieiuna, eo ad omnes res commodior, quam si inclinabit
7 ad illud quod deterius. Stolo, Non male, inquit, quae sit idonea terra ad colendum aut non, Diophanes Bithynos scribit signa sumi posse aut ex ipsa aut quae nascuntur ex iis: ex ipsa, si sit terra alba, si nigra, si levis, quae cum fodiatur, facile frietur,

[1] *in tenui* Keil: *in iis.*

---

[1] A sterile strip of soil near Rome.

may be very rocky, or moderately rocky, or almost free of rocks, and in the case of other varieties of mixed soil the same three grades are distinguished. And further, each of these three grades contains three grades: one may be very wet, one very dry, one intermediate. And these distinctions are not without the greatest importance for the crops; thus the intelligent farmer plants spelt rather than wheat on wet land, and on the other hand barley rather than spelt on dry land, while he plants either on the intermediate. Furthermore, even finer distinctions are made in all these classes, as, for instance, in loamy soil it makes a difference whether the loam be white or red, as the whitish loam is not suited to nurseries, while the reddish is well adapted. Thus there are three chief distinctions in soil, according as it is poor, rich, or medium; the rich being able to produce many kinds of vegetation, and the poor quite the opposite. In thin soil, as, for instance, in Pupinia,[1] you see no sturdy trees, nor vigorous vines, nor stout straw, nor mariscan figs, and most of the trees are covered with moss, as are the parched meadows. On the other hand, in rich soil, like that in Etruria, you can see rich crops, land that can be worked steadily, sturdy trees, and no moss anywhere. In the case of medium soil, however, such as that near Tibur, the nearer it comes to not being thin than to being sterile, the more it is suited to all kinds of growth than if it inclined to the poorer type." "Diophanes of Bithynia makes a good point," remarked Stolo, "when he writes that you can judge whether land is fit for cultivation or not, either from the soil itself or from the vegetation growing on it: from the soil according as it is white or black, light and crumbling easily

natura quae non sit cineracia neve vehementer densa; ex iis autem quae enata sunt fera, si sunt prolixa atque quae ex iis nasci debent earum rerum feracia. Sed quod sequitur, tertium illut de modis dic.

X. Ille, Modos, quibus metirentur rura, alius alios constituit. Nam in Hispania ulteriore metiuntur iugis, in Campania versibus, apud nos in agro Romano ac Latino iugeris. Iugum vocant, quod iuncti boves uno die exarare possint. Versum dicunt centum 2 pedes quoquo versum quadratum. Iugerum, quod quadratos duos actus habeat. Actus quadratus, qui et latus est pedes CXX et longus totidem; is modus acnua latine appellatur. Iugeri pars minima dicitur scripulum, id est decem pedes et longitudine et latitudine quadratum. Ab hoc principio mensores non numquam dicunt in subsicivum esse unciam agri aut sextantem, sic quid aliud, cum ad iugerum pervenerunt, quod habet iugerum scripula CCLXXXVIII, quantum as antiquos noster ante bellum punicum pendebat. Bina iugera quod a Romulo primum divisa dicebantur viritim, quae heredem sequerentur, heredium appellarunt. Haec postea centum centuria. Centuria est quadrata, in omnes quattuor partes ut habeat latera longa pedum ∞ ∞ CƉ. Hae porro quattuor, centuriae coniunctae ut sint in utramque partem binae, appellantur in agris divisis viritim publice saltus.

XI. In modo fundi nonanimadverso lapsi multi, quod alii villam minus magnam fecerunt, quam modus postulavit, alii maiorem, cum utrumque sit contra rem familiarem ac fructum. Maiora enim

[1] Meaning that the surveyors used, for the fractions of land, words which properly applied to fractions of the pound.

when it is dug, of a consistency not ashy and not excessively heavy; from the wild vegetation growing on it if it is luxuriant and bearing abundantly its natural products. But proceed to your third topic, that of measurement."

X. Scrofa resumed: "Each country has its own method of measuring land. Thus in farther Spain the unit of measure is the *iugum*, in Campania the *versus*, with us here in the district of Rome and in Latium the *iugerum*. The *iugum* is the amount of land which a yoke of oxen can plough in a day; the *versus* is an area 100 feet square; the *iugerum* an area containing two square *actus*. The square *actus*, which is an area 120 feet in each direction, is called in Latin *acnua*. The smallest section of the *iugerum*, an area ten feet square, is called a *scripulum*; and hence surveyors sometimes speak of the odd fractions of land above the *iugerum* as an *uncia* or a *sextans*, or the like; for the *iugerum* contains 288 *scripula*, which was the weight of the old pound before the Punic War.[1] Two *iugera* form a *haeredium*, from the fact that this amount was said to have been first allotted to each citizen by Romulus, as the amount that could be transmitted by will. Later on 100 *haeredia* were called a *centuria*; this is a square area, each side being 2400 feet long. Further, four such *centuriae*, united in such a way that there are two on each side, are called a *saltus* in the distribution of public lands.

XI. "Many errors result from the failure to observe the measurement of the farm, some building a steading smaller and some larger than the dimensions demand—each of which is prejudicial to the estate and its revenue. For buildings which are too large cost us

tecta et aedificamus pluris et tuemur sumptu maiore. Minora cum sunt, quam postulat fundus, fructus 2 solent disperire. Dubium enim non est quin cella vinaria maior sit facienda in eo agro, ubi vineta sint, ampliora ut horrea, si frumentarius ager est.

Villa aedificanda potissimum ut intra saepta villae habeat aquam, si non, quam proxime; primum quae ibi sit nata, secundum quae influat perennis. Si omnino aqua non est viva, cisternae faciendae sub tectis et lacus sub dio, ex altero loco ut homines, ex altero ut pecus uti possit.

XII. Danda opera ut potissimum sub radicibus montis silvestris villam ponat, ubi pastiones sint laxae, item[1] ut contra ventos, qui saluberrimi in agro flabunt. Quae posita est ad exortos aequinoctiales, aptissima, quod aestate habet umbram, hieme solem. Sin cogare secundum flumen aedificare, curandum ne adversum eam ponas; hieme enim fiet vehementer 2 frigida et aestate non salubris. Advertendum etiam, siqua erunt loca palustria, et propter easdem causas, et quod crescunt animalia quaedam minuta, quae non possunt oculi consequi, et per aera intus in corpus per os ac nares perveniunt atque efficiunt difficilis morbos. Fundanius, Quid potero, inquit, facere, si istius modi mi fundus hereditati obvenerit, quo minus pestilentia noceat? Istuc vel ego possum respondere, inquit Agrius; vendas, quot assibus 3 possis, aut si nequeas, relinquas. At Scrofa, Vitan-

[1] *item* Keil: *ita.*

too much for construction and require too great a sum for upkeep; and if they are smaller than the farm requires the products are usually ruined. There is no doubt, for instance, that a larger wine cellar should be built on an estate where there is a vineyard, and larger granaries if it is a grain farm.

"The steading should be so built that it will have water, if possible, within the enclosure, or at least very near by. The best arrangement is to have a spring on the place, or, failing this, a perennial stream. If no running water is available, cisterns should be built under cover and a reservoir in the open, the one for the use of people and the other for cattle.

XII. "Especial care should be taken, in locating the steading, to place it at the foot of a wooded hill, where there are broad pastures, and so as to be exposed to the most healthful winds that blow in the region. A steading facing the east has the best situation, as it has the shade in summer and the sun in winter. If you are forced to build on the bank of a river, be careful not to let the steading face the river, as it will be extremely cold in winter, and unwholesome in summer. Precautions must also be taken in the neighbourhood of swamps, both for the reasons given, and because there are bred certain minute creatures which cannot be seen by the eyes, which float in the air and enter the body through the mouth and nose and there cause serious diseases." "What can I do," asked Fundanius, "to prevent disease if I should inherit a farm of that kind?" "Even I can answer that question," replied Agrius; "sell it for the highest cash price; or if you can't sell it, abandon it." Scrofa, however, replied:

dum, inquit, ne in eas partes spectet villa, e quibus ventus gravior afflare soleat, neve in convalli cava et ut potius in sublimi loco aedifices, qui quod perflatur, siquid est quod adversarium inferatur, facilius discutitur. Praeterea quod a sole toto die illustratur, salubrior est, quod et bestiolae, siquae prope nascuntur et inferuntur, aut efflantur aut aritudine cito 4 pereunt. Nimbi repentini ac torrentes fluvii periculosi illis, qui in humilibus ac cavis locis aedificia habent, et repentinae praedonum manus quod improvisos facilius opprimere possunt, ab hac utraque re superiora loca tutiora.

XIII. In villa facienda stabula ita, ut bubilia sint ibi, hieme quae possint esse caldiora. Fructus, ut est vinum et oleum, loco plano in cellis, item vasa vinaria et olearia potius faciendum;[1] aridus, ut est faba et faenum, in tabulatis. Familia ubi versetur providendum, si fessi opere aut frigore aut calore, ubi commo- 2 dissime possint se quiete reciperare. Vilici proximum ianuam cellam esse oportet eumque scire, qui introeat aut exeat noctu quidve ferat, praesertim si ostiarius est nemo. In primis culina videnda ut sit admota, quod ibi hieme antelucanis temporibus aliquot res conficiuntur, cibus paratur ac capitur. Faciundum etiam plaustris ac cetero instrumento omni in cohorte ut satis magna sint tecta, quibus caelum pluvium

[1] The text is corrupt. Pontedera suggests *item ubi vasa vinaria et olearia esse possint, faciendum.*

"See that the steading does not face in the direction from which the infected wind usually comes, and do not build in a hollow, but rather on elevated ground, as a well-ventilated place is more easily cleared if anything obnoxious is brought in. Furthermore, being exposed to the sun during the whole day, it is more wholesome, as any animalculæ which are bred near by and brought in are either blown away or quickly die from the lack of humidity. Sudden rains and swollen streams are dangerous to those who have their buildings in low-lying depressions, as are also the sudden raids of robber bands, who can more easily take advantage of those who are off their guard. Against both these dangers the more elevated situations are safer.

XIII. "In laying out the steading, you should arrange the stables so that the cow-stalls will be at the place which will be warmest in winter. Such liquid products as wine and oil should be set away in store-rooms on level ground, and jars for oil and wine should be provided; while dry products, such as beans and hay, should be stored in a floored space. A place should be provided for the hands to stay in when they are tired from work or from cold or heat, where they can recover in comfort. The overseer's room should be next to the entrance, where he can know who comes in or goes out at night and what he takes; and especially if there is no porter. Especially should care be taken that the kitchen be conveniently placed, because there in winter there is a great deal going on before daylight, in the preparation and eating of food. Sheds of sufficient size should also be provided in the barnyard for the carts and all other implements which are injured by rain; for if

inimicum. Haec enim si intra clausum in consaepto et sub dio, furem modo non metuunt, adversus tempes-
3 tatem nocentem non resistunt. Cohortes in fundo magno duae aptiores: una ut interdius[1] compluvium habeat lacum, ubi aqua saliat, qui intra stylobatas, cum velit, sit semipiscina. Boves enim ex arvo aestate reducti hic bibunt, hic perfunduntur, nec minus e pabulo cum redierunt anseres, sues, porci. In cohorte exteriore lacum esse oportet, ubi maceretur lupinum, item alia quae demissa in aquam ad usum
4 aptiora fiunt. Cohors exterior crebro operta stramentis ac palea occulcata pedibus pecudum fit ministra fundo, ex ea quod evehatur. Secundum villam duo habere oportet stercilina aut unum bifariam divisum. Alteram enim partem fieri oportet novam, alteram veterem tolli in agrum, quod enim quam recens quod confracuit melius. Nec non stercilinum melius illud, cuius latera et summum virgis ac fronde vindicatum a sole. Non enim sucum, quem quaerit terra, solem ante exugere oportet. Itaque periti, qui possunt, ut eo aqua influat eo nomine faciunt (sic enim maxime retinetur sucus) in eoque quidam sellas familiaricas ponunt.
5 Aedificium facere oportet, sub quod tectum totam fundi subicere possis messem, quod vocant quidam nubilarium. Id secundum aream faciendum, ubi triturus sis frumentum, magnitudine pro modo

[1] *interdius*, MSS.: *interius* Iucundus.

---

[1] The passage, evidently corrupt, has puzzled all the commentators. *interdius* lacks satisfactory explanation, despite the ingenuity of Schneider (Addenda to Col. I, 6, 2, *Scriptores Rei Rusticae*, Vol. 2, 2, p. 703) and others. Storr-Best would read *interius* and insert *aut* before *lacum*. The translation is influenced by Varro's *lacus sub dio*, for cattle, in I, 11, 2.

these are kept in an enclosure inside the walls, but in the open, they will not have to fear thieves, yet they will be exposed to injurious weather. On a large farm it is better to have two farm-yards: one, containing an outdoor[1] reservoir—a pond with running water, which, surrounded by columns, if you like, will form a sort of fish-pond; for here the cattle will drink, and here they will bathe themselves when brought in from ploughing in the summer, not to mention the geese and hogs and pigs when they come from pasture; and in the outer yard there should be a pond for the soaking of lupines and other products which are rendered more fit for use by being immersed in water. As the outer yard is often covered with chaff and straw trampled by the cattle, it becomes the handmaid of the farm because of what is cleaned off it. Hard by the steading there should be two manure pits, or one pit divided into two parts; into one part should be cast the fresh manure, and from the other the rotted manure should be hauled into the field; for manure is not so good when it is put in fresh as when it is well rotted. The best type of manure pit is that in which the top and sides are protected from the sun by branches and leaves; for the sun ought not to dry out the essence which the land needs. It is for this reason that experienced farmers arrange it, when possible, so that water will collect there, for in this way the strength is best retained; and some people place the privies for the servants on it. You should build a shed large enough to store the whole yield of the farm under cover. This shed, which is sometimes called a *nubilarium*, should be built hard by the floor on which you are to thresh the grain; it should be of a

fundi, ex una parti apertum, et id ab area, quo et in trituram proruere facile possis et, si nubilare coepit, inde ut rursus celeriter reicere. Fenestras habere oportet ex ea parti, unde commodissime perflari

6 possit. Fundanius, Fructuosior, inquit, est certe fundus propter aedificia, si potius ad anticorum diligentiam quam ad horum luxuriam derigas aedificationem. Illi enim faciebant ad fructum rationem, hi faciunt ad libidines indomitas. Itaque illorum villae rusticae erant maioris preti quam urbanae, quae nunc sunt pleraque contra. Illic laudabatur villa, si habebat culinam rusticam bonam, praesepis laxas, cellam vinariam et oleariam ad modum agri aptam et pavimento proclivi in lacum, quod saepe, ubi conditum novum vinum, orcae in Hispania fervore musti ruptae neque non dolea in Italia. Item cetera ut essent in villa huiusce modi, quae cultura quaereret,[1]

7 providebant. Nunc contra villam urbanam quam maximam ac politissimam habeant dant operam ac cum Metelli ac Luculli villis pessimo publico aedificatis certant. Quo hi laborant ut spectent sua aestiva triclinaria ad frigus orientis, hiberna ad solem occidentem, potius quam, ut antiqui, in quam partem cella vinaria aut olearia fenestras haberet, cum fructus in ea vinarius quaerat ad dolia aera frigidiorem, item olearia calidiorem. Item videre oportet, si est collis, nisi quid impedit, ut ibi potissimum ponatur villa.

[1] *quaereret* Iucundus: *quaeret*.

---

[1] Columella, I, 6, describes the three sections of the villa, the dwelling (*villa urbana*), the barn (*villa rustica*), and the store-room (*villa fructuaria*). He also emphasizes the kitchen in the *villa rustica*.

[2] *i.e.* the old name, *villa*, is still retained, but applied to a palace.

size proportioned to that of the farm, and open only on one side, that next to the threshing floor, so that you can easily throw out the grain for threshing, and quickly throw it back again, if it begins to 'get cloudy.' You should have windows on the side from which it can be ventilated most easily." "A farm is undoubtedly more profitable, so far as the buildings are concerned," said Fundanius, "if you construct them more according to the thrift of the ancients than the luxury of the moderns; for the former built to suit the size of their crops, while the latter build to suit their unbridled luxury. Hence their farms cost more than their dwelling-houses,[1] while now the opposite is usually the case. In those days a steading was praised if it had a good kitchen, roomy stables, and cellars for wine and oil in proportion to the size of the farm, with a floor sloping to a reservoir, because often, after the new wine is laid by, not only the butts which they use in Spain but also the jars which are used in Italy are burst by the fermentation of the must. In like manner they took care that the steading should have everything else that was required for agriculture; while in these times, on the other hand, the effort is to have as large and handsome a dwelling-house as possible; and they vie with the 'farm houses'[2] of Metellus and Lucullus, which they have built to the great damage of the state. What men of our day aim at is to have their summer dining-rooms face the cool east and their winter dining-rooms face the west, rather than, as the ancients did, to see on what side the wine and oil cellars have their windows; for in a cellar wine requires cooler air on the jars, while oil requires warmer. Likewise you should see that, if there be a hill, the house, unless something prevents, should be placed there by preference."

XIV. Nunc de saeptis, quae tutandi causa fundi aut partis fiant, dicam. Earum tutelarum genera IIII, unum naturale, alterum agreste, tertium militare, quartum fabrile. Horum unum quodque species habet plures. Primum naturale saepimentum, quod opseri solet virgultis aut spinis, quod habet radices ac vivit, praetereuntis lascivi non metuet facem 2 ardentem. Secunda saeps est agrestis e ligno, sed non vivit: fit aut palis statutis crebris et virgultis implicatis aut latis perforatis et per ea foramina traiectis longuris fere binis aut ternis aut ex arboribus truncis demissis in terram deinceps constitutis. Tertium militare saepimentum est fossa et terreus agger. Sed fossa ita idonea, si omnem aquam, quae e caelo venit, recipere potest aut fastigium habet, ut 3 exeat e fundo. Agger is bonus, qui intrinsecus iunctus fossa aut ita arduus, ut eum transcendere non sit facile. Hoc genus saepes fieri secundum vias publicas solent et secundum amnes. Ad viam Salariam in agro Crustumino videre licet locis aliquot coniunctos aggeres cum fossis, ne flumen agris noceat. Aggeres faciunt sine fossa: eos quidam vocant 4 muros, ut in agro Reatino. Quartum fabrile saepimentum est novissimum, maceria. Huius fere species quattuor, quod fiunt a lapide, ut in agro Tusculano, quod e lateribus coctilibus, ut in agro

---

[1] Scrofa resumes the discussion.

[2] This type seems to differ from the third chiefly in the absence of the ditch; as the third type probably was also planted.

[3] Such primitive fences may still be seen. The limbs are lopped off, and their stumps thrust into the ground.

[4] So-called because it resembled the defence of a camp.

XIV. "Now I shall speak[1] of the enclosures which are constructed for the protection of the farm as a whole, or its divisions. There are four types of such defences: the natural, the rustic, the military, and the masonry type; and each of these types has several varieties. The first type, the natural, is a hedge, usually planted with brush or thorn, having roots and being alive, and so with nothing to fear from the flaming torch of a mischievous passer-by.[2] The second type, the rustic, is made of wood, but is not alive. It is built either of stakes planted close and intertwined with brush; or of thick posts with holes bored through, having rails, usually two or three to the panel, thrust into the openings; or of trimmed trees placed end to end, with the branches driven into the ground.[3] The third, or military type,[4] is a trench and bank of earth; but the trench is adequate only if it can hold all the rain water, or has a slope sufficient to enable it to drain the water off the land. The bank is serviceable which is close to the ditch on the inside, or so steep that it is not easy to climb. This type of enclosure is usually built along public roads and along streams. At several points along the Via Salaria, in the district of Crustumeria, one may see banks combined with trenches to prevent the river from injuring the fields. Banks built without trenches, such as occur in the district of Reate, are sometimes called walls. The fourth and last type of fence, that of masonry, is a wall, and there are usually four varieties: that which is built of stone, such as occurs in the district of Tusculum; that of burned brick, such as occurs in the Ager Gallicus;[5] that

[5] A strip of land lying along the coast of Umbria.

Gallico, quod e lateribus crudis, ut in agro Sabino, quod ex terra et lapillis compositis in formis, ut in Hispania et agro Tarentino.

XV. Praeterea sine saeptis fines praedi satione[1] arborum tutiores fiunt, ne familiae rixent cum vicinis ac limites ex litibus iudicem quaerant. Serunt alii circum pinos, ut habet uxor in Sabinis, alii cupressos, ut ego habui in Vesuvio, alii ulmos, ut multi habent in Crustumino; ubi id pote, ut ibi, quod est campus, nulla potior serenda, quod maxime fructuosa, quod et sustinet saepe ac cogit[2] aliquot corbulas uvarum et frondem iucundissimam ministrat ovibus ac bubus ac virgas praebet saepibus et foco ac furno.

Scrofa, Igitur primum haec, quae dixi, quattuor videnda agricolae, de fundi forma, de terrae natura, de modo agri, de finibus tuendis.

XVI. Relinquitur altera pars, quae est extra fundum, cuius appendices et vehementer pertinent ad culturam propter adfinitatem. Eius species totidem: si vicina regio est infesta; si quo neque fructus nostros exportare expediat neque inde quae opus sunt adportare; tertium, si viae aut fluvii, qua portetur, aut non sunt aut idonei non sunt; quartum, siquid ita est in confinibus fundis, ut nostris agris 2 prosit aut noceat. E quis quattuor quod est

[1] *satione* Schneider: *sationis notis.*
[2] *cogit* Nonius: *colit.*

of sun-dried brick, such as occurs in the Sabine country; and that formed of earth and gravel in moulds, such as occurs in Spain and the district of Tarentum.

XV. "Furthermore, if there are no enclosures, the boundaries of the estate are made more secure by the planting of trees, which prevent the servants from quarrelling with the neighbours, and make it unnecessary to fix the boundaries by lawsuits. Some plant pines around the edges, as my wife has done on her Sabine farms; others plant cypresses, as I did on my place on Vesuvius; and still others plant elms, as many have done near Crustumeria. Where that is possible, as it is there because it is a plain, there is no tree better for planting; it is extremely profitable, as it often supports and gathers many a basket of grapes, yields a most agreeable foliage for sheep and cattle, and furnishes rails for fencing, and wood for hearth and furnace.

"These points, then, which I have discussed," continued Scrofa, "are the four which are to be observed by the farmer: the topography of the land, the nature of the soil, the size of the plot, and the protection of the boundaries.

XVI. "It remains to discuss the second topic, the conditions surrounding the farm, for they too vitally concern agriculture because of their relation to it. These considerations are the same in number: whether the neighbourhood is unsafe; whether it is such that it is not profitable to transport our products to it, or to bring back from it what we need; third, whether roads or streams for transportation are either wanting or inadequate; and fourth, whether conditions on the neighbouring farms are such as to benefit or injure our land. Taking up the

primum, refert infesta regio sit necne. Multos enim agros egregios colere non expedit propter latrocinia vicinorum, ut in Sardinia quosdam, qui sunt prope Oeliem, et in Hispania prope Lusitaniam. Quae vicinitatis invectos habent idoneos, quae ibi nascuntur ubi vendant, et illinc invectos opportunos quae in fundo opus sunt, propter ea fructuosa. Multi enim habent in praediis, quibus frumentum aut vinum aliudve quid desit importandum; contra non pauci, 3 quibus aliquid sit exportandum. Itaque sub urbe colere hortos late expedit, sic violaria ac rosaria, item multa quae urps recipit, cum eadem in longinquo praedio, ubi non sit quo deferri possit venale, non expediat colere. Item si ea oppida aut vici in vicinia[1] aut etiam divitum copiosi agri ac villae, unde non care emere possis quae opus sunt in fundum, quibus quae supersint venire possint, ut quibusdam pedamenta aut perticae aut harundo, fructuosior fit fundus, quam si longe sint importanda, non numquam etiam, quam si colendo in tuo ea parare possis. 4 Itaque in hoc genus coloni potius anniversarios habent vicinos, quibus imperent, medicos, fullones, fabros, quam in villa suos habeant, quorum non numquam unius artificis mors tollit fundi fructum.

[1] *vici in vicinia* Keil: *viciniae*.

---

[1] The text is corrupt, and the place cannot be identified.
[2] Physicians were usually of the servile class.

first of the four: the safety or lack of safety of the neighbourhood is important; for there are many excellent farms which it is not advisable to cultivate because of the brigandage in the neighbourhood, as in Sardinia certain farms near . . . . ,[1] and in Spain on the borders of Lusitania. Farms which have near by suitable means of transporting their products to market and convenient means of transporting thence those things needed on the farm, are for that reason profitable. For many have among their holdings some into which grain or wine or the like which they lack must be brought, and on the other hand not a few have those from which a surplus must be sent away. And so it is profitable near a city to have gardens on a large scale; for instance, of violets and roses and many other products for which there is a demand in the city; while it would not be profitable to raise the same products on a distant farm where there is no market to which its products can be carried. Again, if there are towns or villages in the neighbourhood, or even well-furnished lands and farmsteads of rich owners, from which you can purchase at a reasonable price what you need for the farm, and to which you can sell your surplus, such as props, or poles, or reeds, the farm will be more profitable than if they must be fetched from a distance; sometimes, in fact, more so than if you can supply them yourself by raising them on your own place. For this reason farmers in such circumstances prefer to have in their neighbourhood men whose services they can call upon under a yearly contract—physicians,[2] fullers, and other artisans—rather than to have such men of their own on the farm; for sometimes the death of one artisan wipes out the profit of a farm. This department of

Quam partem lati fundi divites domesticae copiae mandare solent. Si enim a fundo longius absunt oppida aut vici, fabros parant, quos habeant in villa, sic ceteros necessarios artifices, ne de fundo familia ab opere discedat ac profestis diebus ambulet feriata potius, quam opere faciendo agrum fructuosiorem 5 reddat. Itaque ideo Sasernae liber praecipit, nequis de fundo exeat praeter vilicum et promum et unum, quem vilicus legat; siquis contra exierit, ne impune abeat; si abierit, ut in vilicum animadvertatur. Quod potius ita praecipiendum fuit, nequis iniussu vilici exierit, neque vilicus iniussu domini longius, quam ut eodem die rediret, neque id crebrius, quam 6 opus esset fundo. Eundem fundum fructuosiorem faciunt vecturae, si viae sunt, qua plaustra agi facile possint, aut flumina propinqua, qua navigari possit, quibus utrisque rebus evehi atque invehi ad multa praedia scimus. Refert etiam ad fundi fructus, quem ad modum vicinus in confinio consitum agrum habeat. Si enim ad limitem querquetum habet, non possis recte secundum eam silvam serere oleam, quod usque eo est contrarium natura, ut arbores non solum minus ferant, sed etiam fugiant, ut introrsum in fundum se reclinent, ut vitis adsita ad holus facere solet. Ut quercus, sic iugulandes magnae et crebrae finitimae fundi oram faciunt sterilem.

---

[1] Columella, V, 8, 7, says that the olive must not be planted where an oak has grown: "for even when it has been removed, the oak leaves roots which are harmful to the olive, as their poison kills it." And Pliny (*N.H.*, XXIV, 1) repeats the remark about the vine.

a great estate rich owners are wont to entrust to their own people; for if towns or villages are too far away from the estate, they supply themselves with smiths and other necessary artisans to keep on the place, so that their farm hands may not leave their work and lounge around holiday-making on working days, rather than make the farm more profitable by attending to their duties. It is for this reason, therefore, that Saserna's book lays down the rule that no person shall leave the farm except the overseer, the butler, and one person whom the overseer may designate; if one leaves against this rule he shall not go unpunished, and if he does, the overseer shall be punished. The rule should rather be stated thus: that no one shall leave the farm without the direction of the overseer, nor the overseer without the direction of the master, on an errand which will prevent his return the same day, and that no oftener than is necessary for the farm business. A farm is rendered more profitable by convenience of transportation: if there are roads on which carts can easily be driven, or navigable rivers near by. We know that transportation to and from many farms is carried on by both these methods. The manner in which your neighbour keeps the land on the boundary planted is also of importance to your profits. For instance, if he has an oak grove near the boundary, you cannot well plant olives alongside such a forest; for it is so hostile in its nature[1] that your trees will not only be less productive, but will actually bend so far away as to lean inward toward the ground, as the vine is wont to do when planted near the cabbage. As the oak, so large numbers of large walnut trees close by render the border of the farm sterile.

XVII. De fundi quattuor partibus, quae cum solo haerent, et alteris quattuor, quae extra fundum sunt et ad culturam pertinent, dixi. Nunc dicam, agri quibus rebus colantur. Quas res alii dividunt in duas partes, in homines et adminicula hominum, sine quibus rebus colere non possunt; alii in tres partes, instrumenti genus vocale et semivocale et mutum, vocale, in quo sunt servi, semivocale, in 2 quo sunt boves, mutum, in quo sunt plaustra. Omnes agri coluntur hominibus servis aut liberis aut utrisque: liberis, aut cum ipsi colunt, ut plerique pauperculi cum sua progenie, aut mercennariis, cum conducticiis liberorum operis res maiores, ut vindemias ac faenisicia, administrant, iique quos obaerarios nostri vocitarunt et etiam nunc sunt in Asia atque Aegypto 3 et in Illyrico complures. De quibus universis hoc dico, gravia loca utilius esse mercennariis colere quam servis, et in salubribus quoque locis opera rustica maiora, ut sunt in condendis fructibus vindemiae aut messis. De iis, cuius modi esse oporteat, Cassius scribit haec: operarios parandos esse, qui laborem ferre possint, ne minores annorum XXII et ad agri culturam dociles. Eam coniecturam fieri posse ex aliarum rerum imperatis, et in eo eorum e noviciis requisitione,[1] ad priorem dominum quid factitarint.

Mancipia esse oportere neque formidulosa neque 4 animosa. Qui praesint esse oportere, qui litteris

[1] *requisitione* Ursinus: *requisito*.

---

[1] Those who work off a debt by labour.
[2] That Cassius Dionysius of Utica who translated Mago; see Chapter 1, 10.

XVII. "I have now discussed the four divisions of the estate which are concerned with the soil, and the second four, which are exterior to the soil but concern its cultivation; now I turn to the means by which land is tilled. Some divide these into two parts: men, and those aids to men without which they cannot cultivate; others into three: the class of instruments which is articulate, the inarticulate, and the mute; the articulate comprising the slaves, the inarticulate comprising the cattle, and the mute comprising the vehicles. All agriculture is carried on by men—slaves, or freemen, or both; by freemen, when they till the ground themselves, as many poor people do with the help of their families; or hired hands, when the heavier farm operations, such as the vintage and the haying, are carried on by the hiring of freemen; and those whom our people called *obaerarii*,[1] and of whom there are still many in Asia, in Egypt, and in Illyricum. With regard to these in general this is my opinion: it is more profitable to work unwholesome lands with hired hands than with slaves; and even in wholesome places it is more profitable thus to carry out the heavier farm operations, such as storing the products of the vintage or harvest. As to the character of such hands Cassius [2] gives this advice: that such hands should be selected as can bear heavy work, are not less than twenty-two years old, and show some aptitude for farm labour. You may judge of this by the way they carry out their other orders, and, in the case of new hands, by asking one of them what they were in the habit of doing for their former master.

"Slaves should be neither cowed nor high-spirited. They ought to have men over them who know how

atque aliqua sint humanitate imbuti, frugi, aetate maiore quam operarios, quos dixi. Facilius enim iis quam qui[1] minore natu sunt dicto audientes. Praeterea potissimum eos praeesse oportere,[2] qui periti sint rerum rusticarum. Non solum enim debere imperare, sed etiam facere, ut facientem imitetur et ut animadvertat eum cum causa sibi
5 praeesse, quod scientia praestet. Neque illis concedendum ita imperare, ut verberibus coerceant potius quam verbis, si modo idem efficere possis. Neque eiusdem nationis plures parandos esse; ex eo enim potissimum solere offensiones domesticas fieri. Praefectos alacriores faciendum praemiis dandaque opera ut habeant peculium et coniunctas conservas, e quibus habeant filios. Eo enim fiunt firmiores ac coniunctiores fundo. Itaque propter has cognationes Epiroticae familiae sunt illustriores
6 ac cariores. Inliciendam voluntatem praefectorum honore aliquo habendo, et de operariis qui praestabunt alios, communicandum quoque cum his, quae facienda sint opera, quod, ita cum fit, minus se putant despici atque aliquo numero haberi a domino.
7 Studiosiores ad opus fieri liberalius tractando aut cibariis aut vestitu largiore aut remissione operis concessioneve, ut peculiare aliquid in fundo pascere

[1] *iis quam qui* Keil: *ii quam.*
[2] *oportere* Keil: *oportet.*

---

[1] Compare Chapter 2, 17, Section 7 below, and Chapter 19, 3.

to read and write and have some little education, who are dependable and older than the hands whom I have mentioned; for they will be more respectful to these than to men who are younger. Furthermore, it is especially important that the foremen be men who are experienced in farm operations; for the foreman must not only give orders but also take part in the work, so that his subordinates may follow his example, and also understand that there is good reason for his being over them—the fact that he is superior to them in knowledge. They are not to be allowed to control their men with whips rather than with words, if only you can achieve the same result. Avoid having too many slaves of the same nation, for this is a fertile source of domestic quarrels. The foremen are to be made more zealous by rewards, and care must be taken that they have a bit of property of their own,[1] and mates from among their fellow-slaves to bear them children; for by this means they are made more steady and more attached to the place. Thus, it is on account of such relationships that slave families of Epirus have the best reputation and bring the highest prices. The good will of the foremen should be won by treating them with some degree of consideration; and those of the hands who excel the others should also be consulted as to the work to be done. When this is done they are less inclined to think that they are looked down upon, and rather think that they are held in some esteem by the master. They are made to take more interest in their work by being treated more liberally in respect either of food, or of more clothing, or of exemption from work, or of permission to graze some cattle of their own on the farm, or other things of this kind; so that, if

liceat, huiusce modi rerum aliis, ut quibus quid gravius sit imperatum aut animadversum qui, consolando eorum restituat voluntatem ac benevolentiam in dominum.

XVIII. De familia Cato derigit ad duas metas, ad certum modum agri et genus sationis, scribens de olivetis et vineis ut duas formulas: unam, in qua praecipit, quo modo olivetum agri iugera CCXL instruere oporteat. Dicit enim in eo modo haec mancipia XIII habenda, vilicum, vilicam, operarios V, bubulcos III, asinarium I, subulcum I, opilionem I. Alteram formulam scribit de vinearum iugeribus C, ut dicat haberi oportere haec XV mancipia, vilicum, vilicam, operarios X, bubulcum, asinarium, 2 subulcum. Saserna scribit satis esse ad iugera VIII hominem unum; ea debere eum confodere diebus XLV, tametsi quaternis operis singula iugera possit; sed relinquere se operas XIII valetudini, tempestati, 3 inertiae, indiligentiae. Horum neuter satis dilucide modulos reliquit nobis, quod Cato si voluit, debuit sic, ut pro portione ad maiorem fundum et minorem adderemus et demeremus. Praeterea extra familiam debuit dicere vilicum et vilicam. Neque enim, si minus CCXL iugera oliveti colas, non possis minus uno vilico habere, nec, si bis tanto ampliorem fundum aut eo plus colas, ideo duo vilici aut tres habendi. 4 Fere operarii modo et bubulci pro portione addendi

[1] Cato, 10–11.

some unusually heavy task is imposed, or punishment inflicted on them in some way, their loyalty and kindly feeling to the master may be restored by the consolation derived from such measures.

XVIII. "With regard to the number of slaves required, Cato has in view two bases of calculation: the size of the place, and the nature of the crop grown. Writing of oliveyards and vineyards,[1] he gives two formulas. The first is one in which he shows how an oliveyard of 240 iugera should be equipped; on a place of this size he says that the following thirteen slaves should be kept: an overseer, a housekeeper, five labourers, three teamsters, one muleteer, one swineherd, one shepherd. The second he gives for a vineyard of 100 iugera, on which he says should be kept the following fifteen slaves: an overseer, a housekeeper, ten labourers, a teamster, a muleteer, a swineherd. Saserna states that one man is enough for eight iugera, and that he ought to dig over that amount in forty-five days, although he can dig over a single iugerum with four days' work; but he says that he allows thirteen days extra for such things as illness, bad weather, idleness, and laxness. Neither of these writers has left us a very clearly expressed rule. For if Cato wished to do this, he should have stated it in such a way that we add or subtract from the number proportionately as the farm is larger or smaller. Further, he should have named the overseer and the housekeeper outside of the number of slaves; for if you cultivate less than 240 iugera of olives you cannot get along with less than one overseer, nor if you cultivate twice as large a place or more will you have to keep two or three overseers. It is only the labourers and

ad maioris modos fundorum, ii quoque, si similis est ager. Sin est ita dissimilis, ut arari non possit, quod sit confragosus atque arduis clivis, minus multi opus sunt boves et bubulci. Mitto illut, quod modum neque unum nec modicum proposuit CCXL iugerum
5 (modicus enim centuria, et ea CC iugerum), e quo quom sexta pars sit ea XL, quae de CCXL demuntur, non video quem ad modum ex eius praecepto demam sextam partem et de XIII mancipiis, nihilo magis, si vilicum et vilicam removero, quem ad modum ex XI sextam partem demam. Quod autem ait in C iugeribus vinearum opus esse XV mancipia, siquis habebit centuriam, quae dimidium vineti, dimidium oliveti, sequetur ut duo vilicos et duas vilicas habeat,
6 quod est deridiculum. Quare alia ratione modus mancipiorum generatim est animadvertendus et magis in hoc Saserna probandus, qui ait singula iugera quaternis operis uno operario ad conficiendum satis esse. Sed si hoc in Sasernae fundo in Gallia satis fuit, non continuo idem in agro Ligusco montano. Itaque de familiae magnitudine et reliquo instru-
7 mento commodissime scies quantam pares, si tria animadverteris diligenter: in vicinitate praedia cuius modi sint et quanta, et quot quaeque hominibus colantur, et quot additis operis aut demptis melius aut deterius habeas cultum. Bivium nobis enim ad culturam dedit natura, experientiam et imitationem. Antiquissimi agricolae temptando pleraque constituerunt, liberi eorum magnam partem imitando.
8 Nos utrumque facere debemus, et imitari alios et

---

[1] See Chapter 10, Section 2.

teamsters that are to be added proportionately to larger bodies of land; and even then only if the land is uniform. But if it is so varied that it cannot all be ploughed, as, for instance, if it is very broken or very steep, fewer oxen and teamsters will be needed. I pass over the fact that the 240 iugera instanced is a plot which is neither a unit nor standard (the standard unit is the century, containing 200 iugera);[1] when one-sixth, or 40 iugera, is deducted from this 240, I do not see how, according to his rule, I shall take one-sixth also from thirteen slaves, or, if I leave out the overseer and the housekeeper, how I shall take one-sixth from the eleven. As to his saying that on 100 iugera of vineyard you should have fifteen slaves; if one has a century, half vineyard and half oliveyard, it will follow that he should have two overseers and two housekeepers, which is absurd. Wherefore the proper number and variety of slaves must be determined by another method, and Saserna is more to be approved in this matter; he says that each iugerum is enough to furnish four days 'work for one hand. But if this applied to Saserna's farm in Gaul, it does not necessarily follow that the same would hold good for a farm in the mountains of Liguria. Therefore you will most accurately determine the number of slaves and other equipment which you should provide if you observe three things carefully: the character of the farms in the neighbourhood and their size; the number of hands employed on each; and how many hands should be added or subtracted in order to keep your cultivation better or worse. For nature has given us two routes to agriculture, experiment and imitation. The most ancient farmers determined many of the

aliter ut faciamus experientia temptare quaedam, sequentes non aleam, sed rationem aliquam: ut si altius repastinaverimus aut minus quam alii, quod momentum ea res habeat, ut fecerunt ii in sariendo iterum et tertio, et qui insitiones ficulnas ex verno tempore in aestivum contulerunt.

XIX. De reliqua parte instrumenti, quod semivocale appellavi, Saserna ad iugera CC arvi boum iuga duo satis esse scribit, Cato in olivetis CCXL iugeris boves trinos. Ita fit ut, si Saserna dicit verum, ad C iugera iugum opus sit, si Cato, ad octogena. Sed ego neutrum modum horum omnem ad agrum convenire puto et utrumque ad aliquem. Alia enim 2 terra facilior aut difficilior est; aliam terram boves proscindere nisi magnis viribus non possunt et saepe fracta bura relinquunt vomerem in arvo. Quo sequendum nobis in singulis fundis, dum sumus novicii, triplici regula, superioris domini instituto et 3 vicinorum et experientia quadam. Quod addit asinos qui stercus vectent tres, asinum molarium, in vinea iugerum C iugum boum, asinorum iugum, asinum molendarium; in hoc genere semivocalium adiciendum de pecore ea sola quae agri colendi causa erunt et quae solent esse peculiaria pauca habenda,

---

[1] "Here, in a few words," remarks Mr. Fairfax Harrison, "is the whole doctrine of intelligent agriculture." The neglect of this injunction is what has retarded the advance of the science.

[2] Cato, 10, 1. This interpretation of *boves trinos* as three yoke of oxen and not as three oxen is accepted by most editors and seems necessary from the context. The next sentence makes it clear that Varro so interprets it.

practices by experiment, their descendants for the most part by imitation. We ought to do both—imitate others and attempt by experiment to do some things in a different way, following not chance but some system:[1] as, for instance, if we plough a second time, more or less deeply than others, to see what effect this will have. This was the method they followed in weeding a second and third time, and those who put off the grafting of figs from spring-time to summer.

XIX. "With regard to the second division of equipment, to which I have given the name of inarticulate, Saserna says that two yoke of oxen are enough for 200 iugera of cultivated land, while Cato states that three yoke[2] are needed for 240 iugera of olive-yard. Hence, if Saserna is right, one yoke is needed for every 100 iugera; if Cato is right, one to every 80. My own opinion is that neither of these standards will fit every piece of land, and that each will fit some particular piece. One piece, for instance, may be easier or harder to work than another, and there are places which oxen cannot break unless they are unusually powerful, and frequently they leave the plough in the field with broken beam. Wherefore on each farm, so long as we are unacquainted with it, we should follow a threefold guide: the practice of the former owner, the practice of neighbouring owners, and a degree of experimentation. As to his addition of three donkeys to haul manure and one for the mill (for a vineyard of 100 iugera, a yoke of oxen, a pair of donkeys, and one for the mill); under this head of inarticulate equipment it is to be added that of other animals only those are to be kept which are of service in agriculture, and the few which are usually

quo facilius mancipia[1] se tueri et assidua esse possint. In eo numero non modo qui prata habent, ut potius oves quam sues habeant curant, sed etiam qui non solum pratorum causa habent, propter stercus. De canibus vero utique, quod villa sine iis parum tuta.

XX. Igitur de omnibus quadripedibus prima est probatio, qui idonei sint boves, qui arandi causa emuntur. Quos rudis neque minoris trimos neque maioris quadrimos parandum; ut viribus magnis sint ac pares, ne in opere firmior imbecilliorem conficiat; amplis cornibus et nigris potius quam aliter ut sint, lata fronte, naribus simis, lato pectore, crassis 2 coxendicibus. Hos veteranos ex campestribus locis non emendum in dura ac montana, nec non contra si incidit, ut sit vitandum. Novellos cum quis emerit iuvencos, si eorum colla in furcas destitutas incluserit ac dederit cibum, diebus paucis erunt mansueti et ad domandum proni. Tum ita subigendum, ut minutatim assuefaciant et ut tironem cum veterano adiungant (imitando enim facilius domatur), et primum in aequo loco et sine aratro, tum eo levi, principio per 3 harenam aut molliorem terram. Quos ad vecturas, item instituendum ut inania primum ducant plaustra et, si possis, per vicum aut oppidum; creber crepitus ac varietas rerum consuetudine celeberrima ad

[1] *mancipia se* Keil: *mancipia quae solent se.*

---

[1] You must not buy from the mountains for the plains.

allowed as the private property of the slaves for their more comfortable support and to make them more diligent in their work. Of such animals, not only owners who have meadows prefer to keep sheep rather than swine because of their manure, but also those who keep animals for other reasons than the benefit of the meadows. As to dogs, they must be kept as a matter of course, for no farm is safe without them.

XX. "The first consideration, then, in the matter of quadrupeds, is the proper kind of ox to be purchased for ploughing. You should purchase them unbroken, not less than three years old and not more than four; they should be powerful and equally matched, so that the stronger will not exhaust the weaker when they work together; they should have large horns, black for choice, a broad face, flat nose, deep chest, and heavy quarters. Oxen that have reached maturity on level ground should not be bought for rough and mountainous country; moreover, if the opposite happens to be the case, it should be avoided.[1] When you have bought young steers, if you will fasten forked sticks loosely around their necks and give them food, within a few days they will grow gentle and fit for breaking to the plough. This breaking should consist in letting them grow accustomed to the work gradually, in yoking the raw ox to a broken one (for the training by imitation is easier), and in driving them first on level ground without a plough, then with a light one, and at first in sandy or rather light soil. Draught cattle should be trained in a similar way, first drawing an empty cart, and if possible through a village or town. The constant noise and the variety of objects, by frequent

utilitatem adducit. Neque pertinaciter, quem feceris dextrum, in eo manendum, quod, si alternis fit sinister, fit laboranti in alterutra parte requies.
4 Ubi terra levis, ut in Campania, ibi non bubus gravibus, sed vaccis aut asinis quod arant, eo facilius ad aratrum leve adduci possunt, ad molas et ad ea, siquae sunt, quae in fundo convehuntur. In qua re alii asellis, alii vaccis ac mulis utuntur, exinde ut pabuli facultas est; nam facilius asellus quam
5 vacca alitur, sed fructuosior haec. In eo agricolae hoc spectandum, quo fastigio sit fundus. In confragoso enim haec ac difficili valentiora parandum et potius ea quae per se fructum reddere possint, cum idem operis faciant.

XXI. Canes potius cum dignitate et acres paucos habendum quam multos, quos consuefacias potius noctu vigilare et interdiu clausos dormire. De indomitis quadripedibus ac pecore faciendum; si prata sunt in fundo neque pecus habet, danda opera ut pabulo vendito alienum pecus in suo fundo pascat ac stabulet.

XXII. De reliquo instrumento muto, in quo sunt corbulae, dolia, sic alia, haec praecipienda. Quae nasci in fundo ac fieri a domesticis poterunt, eorum nequid ematur, ut fere sunt quae ex viminibus et materia rustica fiunt, ut corbes, fiscinae, tribula, valli, rastelli; sic quae fiunt de cannabi, lino, iunco,

---

[1] Varro is illogical. He clearly means that these lighter animals can be used for more purposes than heavy steers.

repetition, accustom them to their work. The ox which you have put on the right should not remain continuously on that side, because if he is changed in turn to the left, he finds rest by working on alternate sides. In light soils, as in Campania, the ploughing is done, not with heavy steers, but with cows or donkeys; and hence they can more easily be adapted to a light plough or a mill, and to doing the ordinary hauling of the farm.[1] For this purpose some employ donkeys, others cows or mules, according to the fodder available; for a donkey requires less feed than a cow, but the latter is more profitable. In this matter the farmer must keep in mind the conformation of his land; in broken and heavy land stronger animals must be got, and preferably those which, while doing the same amount of work, can themselves return some profit.

XXI. "As to dogs, you should keep a few active ones of good traits rather than a pack, and train them rather to keep watch at night and sleep indoors during the day. With regard to unbroken animals and flocks; if the owner has meadow-lands on the farm and no cattle, the best practice is, after selling the forage, to feed and fold the flocks of a neighbour on the farm.

XXII. "With regard to the rest of the equipment —'the mute', a term which includes baskets, jars, and the like—the following rules may be laid down: nothing should be bought which can be raised on the place or made by the men on the farm, in general articles which are made of withes and of wood, such as hampers, baskets, threshing-sledges, fans, and rakes; so too articles which are made of hemp, flax,

2 palma, scirpo, ut funes, restes, tegetes. Quae e fundo sumi non poterunt, ea si empta erunt potius ad utilitatem quam ob speciem, sumptu fructum non extenuabunt; eo magis, si inde empta erunt potissimum, ubi ea et bona et proxime et vilissimo emi poterunt. Cuius instrumenti varia discrimina ac multitudo agri magnitudine finitur, quod plura opus 3 sunt, si fines distant late. Itaque, Stolo inquit, proposita magnitudine fundi de eo genere Cato scribit, oliveti iugera CCXL qui coleret, eum instruere ita oportere, ut faceret vasa olearia iuga quinque, quae membratim enumerat, ut ex aere ahenea, urceos, nassiternam, item alia; sic e ligno et ferro, ut plostra maiora tria, aratra cum vomeribus sex, crates stercorarias quattuor, item alia; sic de ferramentis quae sint et qua opus multitudine, ut ferreas octo, sarcula totidem, dimidio minus palas, item alia. 4 Item alteram formulam instrumenti fundi vinarii fecit, in qua scribit, si sit C iugerum, habere oportere vasa torcularia instructa trina, dolia cum operculis culleorum octingentorum, acinaria viginti, frumentaria viginti, item eius modi alia. Quae minus multa quidem alii, sed tantum numerum culleorum scripsisse puto, ne cogeretur quotannis vendere vinum. Vetera enim quam nova et eadem alio 5 tempore quam alio pluris. Item sic de ferramentorum varietate scribit permulta, et genere et multitudine qua sint, ut falces, palas, rastros, sic

[1] Cato, 10, 2. [2] Cato, 11, 1.

rush, palm fibre, and bulrush, such as ropes, cordage, and mats. Articles which cannot be got from the place, if purchased with a view to utility rather than for show, will not cut too deeply into the profits; and the more so if care is taken to buy them where they can be had of good quality, near by and at the lowest price. The several kinds of such equipment and their number are determined by the size of the place, more being needed if the farm is extensive. Accordingly," said Stolo, "under this head Cato,[1] fixing a definite size for his farm, writes that one who has under cultivation 240 iugera of olive land should equip it by assembling five complete sets of oil-pressing equipment; and he itemizes such equipment, as, copper kettles, pots, a pitcher with three spouts, and so forth; then implements made of wood and iron, as three large carts, six ploughs and ploughshares, four manure hampers, and so forth; then the kind and number of iron tools needed, as eight forks, as many hoes, half as many shovels, and so forth. He likewise gives [2] a second schedule for a vineyard, in which he writes that if it be one of 100 iugera it should have three complete pressing equipments, vats and covers to hold 800 cullei, twenty grape hampers, twenty grain hampers, and other like implements. Other authorities, it is true, give smaller numbers, but I imagine he fixed the number of cullei so high in order that the farmer might not be forced to sell his wine every year; for old wine brings a better price than new, and the same wine a better price at one time than at another. He likewise says much of the several kinds of tools, giving the kind and number needed, such as hooks, shovels, harrows, and so forth; some classes of which have

alia, quorum non nulla genera species habent plures, ut falces. Nam dicuntur ab eodem scriptore vineaticae opus esse XL, sirpiculae V, arborariae III, 6 rustariae X. Hic haec. At Scrofa, Instrumentum et supellectilem rusticam omnem oportet habere scriptam in urbe et rure dominum, vilicum contra ea ruri omnia certo suo quoque loco ad villam esse posita; quae non possunt esse sub clavi, quam maxime facere ut sint in conspectu oportet, eo magis ea quae in rariore sunt usu, ut quibus in vindemia utuntur corbulae, et sic alia. Quae enim res cotidie videntur, minus metuunt furem.

XXIII. Suscipit Agrasius, Et quoniam habemus illa duo prima ex divisione quadripertita, de fundo et de instrumento, quo coli solet, de tertia parte expecto. Scrofa, Quoniam fructum, inquit, arbitror esse fundi eum qui ex eo satus nascitur utilis ad aliquam rem, duo consideranda, quae et quo quidque loco maxime expediat serere. Alia enim loca apposita sunt ad faenum, alia ad frumentum, alia ad vinum, alia ad oleum, sic ad pabulum quae pertinent, in quo est ocinum, farrago, vicia, medica, cytisum, lupinum. 2 Neque in pingui terra omnia seruntur recte neque in macra nihil. Rectius enim in tenuiore terra ea quae non multo indigent suco, ut cytisum et legumina praeter cicer; hoc enim quoque legumen, ut cetera quae velluntur e terra, non subsecantur, quae, quod ita leguntur, legumina dicta. In pingui rectius quae

several subdivisions, such as the hooks—thus the same author says there will be needed forty pruning-hooks for vines, five for rushes, three for trees, ten for brambles." So far Stolo; and Scrofa resumed: "The master should keep, both in town and on the place, a complete inventory of tools and equipment of the farm, while the overseer on the place should keep all tools stored near the steading, each in its own place. Those that cannot be kept under lock and key he should manage to keep in sight so far as possible, and especially those that are used only at intervals; for instance, the implements which are used at vintage, such as baskets and the like; for articles which are seen every day run less risk from the thief."

XXIII. Agrasius remarked: "And since we have the first two of the fourfold division, the farm and the equipment with which it is usually worked, I am waiting for the third topic." "Since I hold," continued Scrofa, "that the profit of the farm is that which arises from it as the result of planting for a useful purpose, two items are to be considered: what it is most expedient to plant and in what place. For some spots are suited to hay, some to grain, others to vines, others to olive, and so of forage crops, including clover, mixed forage, vetch, alfalfa, snail clover, and lupines. It is not good practice to plant every kind of crop on rich soil, nor to plant nothing on poor soil; for it is better to plant in thinner soil those crops which do not need much nutriment, such as clover and the legumes, except the chick pea, which is also a legume, as are all those plants which are pulled from the ground and not mowed, and are called legumes from the fact that they are 'gathered'

cibi[1] sunt maioris, ut holus, triticum, siligo, linum.
3 Quaedam etiam serenda non tam propter praesentem fructum quam in annum prospicientem, quod ibi subsecta atque relicta terram faciunt meliorem. Itaque lupinum, cum minus[2] siliculam cepit, et non numquam fabalia, si ad siliquas non ita pervenit, ut fabam legere expediat, si ager macrior est, pro
4 stercore inarare solent. Nec minus ea discriminanda in conserendo quae sunt fructuosa, propter voluptatem, ut quae pomaria ac floralia appellantur, item illa quae ad hominum victum ac sensum delectationemque non pertinent neque ab agri utilitate sunt diiuncta. Idoneus locus eligendus, ubi facias
5 salictum et harundinetum, sic alia quae umidum locum quaerunt, contra ubi segetes frumentarias, ubi fabam potissimum seras, item alia quae arida loca secuntur; sic ut umbrosis locis alia seras, ut corrudam, quod ita petit asparagus; aprica, ut ibi seras violam et hortos facias, quod ea sole nutricantur, sic alia. Et alio loco virgulta serenda, ut habeas vimina, unde viendo quid facias, ut sirpeas, vallus, crates; alio loco ut seras ac colas silvam
6 caeduam, alio ubi aucupere, sic ubi cannabim, linum, iuncum, spartum, unde nectas bubus soleas,[1] lineas, restis, funes. Quaedam loca eadem alia ad serendum

[1] *cibi* Victorius: *ibi*. [2] *minus* Keil: *dominus*.

---

[1] Columella, VI, 12, 2, refers to such shoes made of broom. Our horse-shoes, nailed on the hoof, were not known.

(*leguntur*) in this way. In rich soil it is better to plant those requiring more food, as cabbage, wheat, winter wheat, and flax. Some crops are also to be planted not so much for the immediate return as with a view to the year later, as when cut down and left on the ground they enrich it. Thus, it is customary to plough under lupines as they begin to pod—and sometimes field beans before the pods have formed so far that it is profitable to harvest the beans—in place of dung, if the soil is rather thin. And also in planting selection should be made of those things which are profitable for the pleasure they afford, such as those plots which are called orchards and flower gardens, and also of those which do not contribute either to the sustenance of man or to the pleasure of his senses, but are not without value to the farm. So a suitable place is to be chosen for planting a willow bed and a reed thicket, together with other plants which prefer humid ground; and on the other hand places best suited for planting grain crops, beans, and other plants which like dry ground. Similarly, you should plant some crops in shady spots, as, for instance, the wild asparagus, because the asparagus prefers that type; while sunny ground should be chosen for planting violets and laying out gardens, as these flourish in the sun, and so forth. In still another place should be planted thickets, so that you may have withes with which to weave such articles as wicker wagon bodies, winnowing baskets, and hampers; and in another plant and tend a wood-lot, in another a wood for fowling; and have a place for hemp, flax, rush, and Spanish broom, from which to make shoes for cattle,[1] thread, cord, and rope. Some

idonea. Nam et in recentibus pomariis dissitis seminibus in ordinemque arbusculis positis primis annis, antequam radices longius procedere possint, alii conserunt hortos, alii quid aliud, neque cum convaluerunt arbores idem faciunt, ne violent radices.

XXIV. Stolo, Quod ad haec pertinet, Cato non male, quod scribit de sationibus, ager crassus et laetus si sit sine arboribus, eum agrum frumentarium fieri oportere; idem ager si nebulosus sit, rapa, raphanos, milium, panicum; in agro crasso et calido oleam conditaneam, radium maiorem, Sallentinam, orcitem, poseam, Sergianam, Colminiam, albicerem, quam earum in iis locis optimam dicant esse, eam maxime serere. Agrum oliveto conserendo, nisi qui in ventum favonium spectet et soli ostentus sit, 2 alium bonum nullum esse. Qui ager frigidior et macrior sit, ibi oleam licinianam seri oportere. Si in loco crasso aut calido posueris, hostum nequam fieri et ferendo arborem perire et muscum rubrum 3 molestum esse. Hostum vocant quod ex uno facto olei reficitur. Factum dicunt quod uno tempore conficiunt, quem alii CLX aiunt esse modiorum, alii ita minus magnum, ut ad CXX descendat, exinde ut vasa olearia quot et quanta habeant, quibus conficiunt illut. Quod Cato ait, circum fundum ulmos et populos, unde fros ovibus et bubus sit et materies, seri oportere (sed hoc neque in omnibus fundis opus est neque, in quibus est opus, propter frondem

---

[1] Cato, 6, 1–2.

places are suitable at the same time for the planting of other crops; thus in young orchards, when the seedlings have been planted and the young trees have been set in rows, during the early years before the roots have spread very far, some plant garden crops, and others plant other crops; but they do not do this after the trees have gained strength, for fear of injuring the roots."

XXIV. "What Cato says[1] about planting," said Stolo, "is very much to the point on this subject: 'Soil that is heavy, rich, and treeless should be used for grain; and the same soil, if subject to fogs, should preferably be planted in rape, turnips, millet, and panic-grass. In heavy, warm soil plant olives—those for pickling, the long variety, the Sallentine, the orcites, the posea, the Sergian, the Colminian, and the waxy; choose especially the varieties which are commonly agreed to be the best for these districts. Land which is suitable for olive planting is that which faces the west and is exposed to the sun; no other will be good. In colder and thinner soil the Licinian olive should be planted. If you plant it in rich or warm soil the yield will be worthless, the tree will exhaust itself in bearing, and a reddish scale will injure it.' A *hostus* is what they call the yield of oil from one *factus*; and a *factus* ('making') is the amount they make up at one time. Some say this is 160 *modii*, others reduce it so far as 120 *modii*, according to the number and size of the equipment they have for making it. As to Cato's remark that elms and poplars should be planted around the farm to supply leaves for sheep and cattle, and timber (but this is not necessary on all farms, and where it is necessary it is not chiefly for the forage), they

maxime), sine detrimento ponuntur a septemtrionali parte, quod non officiunt soli.

4 Ille adicit ab eodem scriptore, si locus umectus sit, ibi cacumina populorum serenda et harundinetum. Id prius bipalio verti, ibi oculos harundinis pedes ternos alium ab alio seri, . . .[1] aptam esse utrique eandem fere culturam. Salicem Graecam circum harundinetum seri oportere, uti sit qui vitis alligari possit.

XXV. Vinea quo in agro serenda sit, sic observandum. Qui locus optimus vino sit et ostentus soli, Aminneum minusculum et geminum eugeneum, helvium minusculum seri oportere. Qui locus crassior sit aut nebulosus, ibi Aminneum maius aut Murgentinum, Apicium, Lucanum seri. Ceteras vites, et de iis miscellas maxime, in omne genus agri convenire.

XXVI. In omni vinea diligenter observant ut ridica vitis ad septemtrionem versus tegatur; et si cupressos vivas pro ridicis (quas) inserunt, alternos ordines imponunt, neque eos crescere altius quam ridicas patiuntur, neque propter eos ut adserant vites, quod inter se haec inimica.

Agrius Fundanio, Vereor, inquit, ne ante aeditumus veniat huc, quam hic ad quartum actum. Vindemiam enim expecto. Bono animo es, inquit Scrofa, ac fiscinas expedi et urnam.

XXVII. Et quoniam tempora duorum generum sunt, unum annale, quod sol circuitu suo finit, alterum menstruum, quod luna circumiens comprendit, prius

[1] The lacuna might be supplied from Cato, 6, 3–4, which see.

---

[1] Cato, 6, 3–4.

[2] There is no other mention of the use of cypress for this purpose, and no statement that it is hostile to the vine. Pontedera, on the basis of Chapter 16, 6, would read *propter olus*, "near the cabbage."

may safely be planted on the northern edge, because there they do not cut off the sun."

Scrofa gave the following advice from the same author:[1] "'Wherever there is wet ground, poplar cuttings and a reed thicket should be planted. The ground should first be turned with the mattock and then the eyes of the reed should be planted three feet apart; . . . the same cultivation is adapted pretty much to each. The Greek willow should be planted along the border of the thicket, so that you may have withes for tying up vines.

XXV. 'Soil for laying out a vineyard should be chosen by the following rules: In soil which is best adapted for grapes and which is exposed to the sun the small Aminnian, the double eugeneum, and the small parti-coloured should be planted; in soil that is heavy or more subject to fogs the large Aminnian, the Murgentian, the Apician, and the Lucanian. The other varieties, and especially the hybrids, grow well anywhere.'

XXVI. "In every vineyard they are careful to see that the vine is protected toward the north by the prop; and if they plant live cypresses to serve as props they plant them in alternate rows, yet do not allow the rows to grow higher than the props, and are careful not to plant vines near them, because they are hostile to each other."[2]

"I am afraid," remarked Agrius to Fundanius, "that the sacristan will come back before our friend comes to the fourth act; for I am awaiting the vintage." "Be of good cheer," replied Scrofa, "and get ready the baskets and jar.

XXVII. "And since we have two measures of time, one annual which the sun bounds by its circuit, the other monthly which the moon embraces as it circles,

dicam de sole. Eius cursus annalis primum fere circiter ternis mensibus ad fructus est divisus in IIII partis, et idem subtilius sesquimensibus in IIX; in IIII, quod dividitur in ver et aestatem et autumnum et hiemem.

2 Vere sationes quae fiunt, terram rudem proscindere oportet, quae sunt ex ea enata, priusquam ex iis quid seminis cadat, ut sint exradicata; et simul glaebis ab sole percalefactis aptiores facere ad accipiendum imbrem et ad opus faciliores relaxatas;

3 neque eam minus binis arandum, ter melius. Aestate fieri messes oportere, autumno siccis tempestatibus vindemias, ac silvas excoli commodissime tunc, praecidi arbores oportere secundum terram; radices autem primoribus imbribus ut effodiantur, nequid ex iis nasci possit. Hieme putari arbores dumtaxat his temporibus, cum gelu cortices ex imbribus careant et glacie.

XXVIII. Dies primus est veris in aquario, aestatis in tauro, autumni in leone, hiemis in scorpione. Cum unius cuiusque horum IIII signorum dies tertius et vicesimus IIII temporum sit primus et efficiat ut ver dies habeat XCI, aestas XCIV, autumnus XCI, hiems XXCIX, quae redacta ad dies civiles nostros, qui nunc sunt, primi verni temporis ex a. d. VII id. Febr., aestivi ex a. d. VII id. Mai., autumnalis ex

2 a. d. III id. Sextil., hiberni ex a. d. IV id. Nov., suptilius descriptis temporibus observanda quaedam sunt, eaque in partes VIII dividuntur: primum a

---

[1] This seems to be a Latin variation of the Greek sevenfold division given by Hippocrates and Galen.

[2] *i.e.* the beginning of each of the four seasons is the twenty-third day after the sun enters the sign.

[3] The Julian calendar, which took effect on January 1st, 45 B.C., eight years before this was written.

I shall speak first of the sun. Its annual course is divided first into four periods of about three months each up to its completion, and more narrowly into eight periods of a month and a half each;[1] the fourfold division embraces spring, summer, autumn, and winter. For the spring plantings the untilled ground should be broken up so that the weeds which have sprung from it may be rooted up before any seed falls from them; and at the same time, when the clods have been thoroughly dried by the sun, to make them more accessible to the rain and easier to work when they have been thus broken up; and there should be not less than two ploughings, and preferably three. In summer the grain should be gathered, and in autumn, when the weather is dry, the grapes; and this is the best time for the woods to be cleared, the trees being cut close to the ground, while the roots should be dug out at the time of the early rains, so that they cannot sprout again. In winter trees should be pruned, provided it is done when the bark is free from the chill of rain and ice.

XXVIII. "The first day of spring occurs [when the sun is] in Aquarius, that of summer when it is in Taurus, of autumn when it is in Leo, of winter when it is in Scorpio. As the twenty-third day of each one of these four signs is the first day of the four seasons,[2] this makes spring contain 91 days, summer 94, autumn 91, winter 89, which numbers, reduced to the official calendar now in force,[3] fix the beginning of spring on February 7, of summer on May 9, of autumn on August 11, of winter on November 10. But in the more exact divisions certain things are to be taken into account, which cause an eightfold division: the first from the rising of the

favonio ad aequinoctium vernum dies XLV, hinc ad vergiliarum exortum dies XLIV, ab hoc ad solstitium dies XLIIX, inde ad caniculae signum dies XXVII, dein ad aequinoctium autumnale dies LXVII, exin ad vergiliarum occasum dies XXXII, ab hoc ad brumam dies LVII, inde ad favonium dies XLV.

XXIX. Primo intervallo inter favonium et aequinoctium vernum haec fieri oportet. Seminaria omne genus ut serantur, putari arbusta, stercorari[1] in pratis, circum vites ablacuari, radices quae in summa terra sunt praecidi, prata purgari, salicta seri, segetes sariri. Seges dicitur quod aratum satum est, arvum quod aratum necdum satum est, novalis, ubi satum fuit, antequam secunda aratione novatur rursus.
2 Terram cum primum arant, proscindere appellant, cum iterum, offringere dicunt, quod prima aratione glaebae grandes solent excitari; cum iteratur, offringere vocant. Tertio cum arant iacto semine, boves lirare dicuntur, id est cum tabellis additis ad vomerem simul et satum frumentum operiunt in porcis et sulcant fossas, quo pluvia aqua delabatur. Non nulli postea, qui segetes non tam latas habent, ut in Apulia et id genus praediis, per sartores occare solent, siquae in porcis relictae grandiores sunt
3 glaebae. Qua aratrum vomere lacunam striam fecit, sulcus vocatur. Quod est inter duos sulcos elata

[1] *arbusta, stercorari* added by Keil.

---

[1] *Novalis* is "fallow land," and Varro explains the term as "land waiting to be renewed" by being ploughed *again*, for this seems to be the meaning of *secunda aratione*.

west wind to the vernal equinox, 45 days, thence to the rising of the Pleiades 44 days, thence to the solstice 48 days, thence to the rising of the Dog Star 27 days, thence to the autumnal equinox 67 days, from there to the setting of the Pleiades 32 days, hence to the winter solstice 57 days, and back to the rising of the west wind 45 days.

XXIX. "These are the things which should be done in the first period, from the rising of the west wind to the vernal equinox:—All kinds of nurseries should be set out, orchards pruned, meadows manured, vines trenched and outcropping roots removed, meadows cleared, willow beds planted, grain-land weeded. The word *seges* is used of ploughed land which has been sowed, *arvum* of ploughed land not yet sowed, *novalis* of land where there has been a crop before it is "renewed"[1] (*novatur*) by a second ploughing. When they plough the first time they say they are 'breaking up,' the second time that they are 'breaking down,' because at the first ploughing large clods are usually turned up, and when the ploughing is repeated they call it 'breaking down.' When they plough the third time, after the seed has been broadcast, the oxen are said to 'ridge'; that is, with mould boards attached to the ploughshare they both cover the broadcast seed in ridges, and at the same time cut ditches to let the rain-water drain off. Some farmers, who have fields which are not very large, as in Apulia and farms of that kind, have the custom later on of breaking up with hoes any large clods which have been left on the ridges. Where the plough makes a hollow or channel with the share, it is called a 'furrow.' The space between two furrows, the raised dirt, is called *porca*, because that part of the field presents

terra dicitur porca, quod ea seges frumentum porricit. Sic quoque exta deis cum dabant, porricere dicebant.

XXX. Secundo intervallo inter vernum aequinoctium et vergiliarum exortum haec fieri. Segetes runcari, id est[1] herbam e segetibus expurgari, boves terram proscindere, salicem caedi, prata defendi. Quae superiore tempore fieri oportuerit et non sunt absoluta, antequam gemmas agant ac florescere incipiant, fieri, quod, si quae folia amittere solent ante frondere inceperunt, statim ad serendum idonea non sunt. Oleam seri interputarique oportet.

XXXI. Tertio intervallo inter vergiliarum exortum et solstitium haec fieri debent. Vineas novellas fodere aut arare et postea occare, id est comminuere, ne sit glaeba. Quod ita occidunt, occare dictum. Vites pampinari, sed a sciente (nam id quam putare maius),
2 neque in arbusto, sed in vinea fieri. Pampinare est e sarmento coles qui nati sunt, de iis, qui plurimum valent, primum ac secundum, non numquam etiam tertium, relinquere, reliquos decerpere, ne relictis colibus sarmentum nequeat ministrare sucum. Ideo in vitiario primitus cum exit vitis, tota resicari solet, ut firmiore sarmento e terra exeat atque in pariendis
3 colibus vires habeat maiores. Eiuncidum enim sarmentum propter infirmitatem sterile neque ex se

[1] *id est* added by Keil.

(*porricit*) the grain; so they also used to employ the word *porricere* when they offered the entrails to the gods.

XXX. "In the second period, between the vernal equinox and the rising of the Pleiades, these operations should be carried out:—Crops should be weeded, that is, the grass cleared from the crops, oxen should break up the ground, willows should be cut, and meadows fenced. What should have been done in the former period but was not completed should be done before the plants begin to bud and flower, because if those which are deciduous once begin to frond, they are at once unsuited for planting. Olives should be planted and pruned.

XXXI. "In the third period, between the rising of the Pleiades and the solstice, these operations should be carried out:—Digging or ploughing the young vines and then forking them, that is, breaking the ground so that there will be no clods. This is called *occare* because they crush (*occidunt*) the ground. The vines should be thinned (for that is better than pruning), but by an expert, and this should be done not in the orchard but in the vineyard. Thinning consists in leaving the first and second, sometimes even the third of the strongest shoots which spring from the stock and picking off the rest, lest the stock be not strong enough to furnish sap to the shoots if all are left. For that reason, in the vine-nursery when the vine first comes out, it is the practice for the whole to be cut back, so that it may come from the ground with a sturdier stock and have greater strength in sending out shoots. For a slender stock, on account of weakness, is sterile and cannot put out the vine, which, when it is smaller, they call

potest eicere vitem, quam vocant minorem flagellum, maiorem et iam unde uvae nascuntur palmam. Prior littera una mutata declinata a venti flatu, similiter ac flabellum flagellum. Posterior, quod ea vitis immittitur ad uvas pariendas, dicta primo 4 videtur a pariendo parilema; exin mutatis litteris, ut in multis, dici coepta palma. Ex altera parte parit capreolum. Is est coliculus viteus intortus, ut cincinnus. Hi sunt enim[1] vitis quibus teneat id quo serpit ad locum capiendum, a quo capiendo capreolus dictus. Omne pabulum, primum ocinum farraginem viciam, novissime faenum, secari. Ocinum dictum a graeco verbo ὠκέως, quod valet cito, similiter quod ocimum in horto. Hoc amplius dictum ocinum, quod citat alvom bubus et ideo iis datur, ut purgentur. Id est ex fabali segete viride sectum, antequam genat 5 siliquas. Farrago[2] contra ex segete ubi sata admixta hordeum et vicia et legumina pabuli causa viride, aut quod ferro caesa, ferrago, dicta, aut inde, quod primum in farracia segete seri coepta. Eo equi et iumenta cetera verno tempore purgantur ac saginantur. Vicia dicta a vinciendo, quod item capreolos habet, ut vitis, quibus, cum susum versus serpit, ad scapum lupini aliumve quem ut haereat, id solet vincire. Si prata inrigua habebis, simulac faenum sustuleris, inrigare. In poma, quae insita erunt, siccitatibus aquam addi cotidie vesperi. A quo, quod indigent potu, poma dicta esse possunt.

[1] *Hi sunt enim* Keil: *is enim*. [2] *Farrago* added by Goetz.

---

[1] Varro's *penchant* for derivations, based on no scientific principles, leads to many etymological absurdities. Cf. Isidore, *Orig.*, XVII, 5: "summitates vitium et fruticum flagella nuncupantur, eo quod flatu agitentur;" also Servius ad Virg. *Georg.*, II, 299.

flagellum, while the larger from which the grapes spring they call *palma*. The first word by the change of one letter is derived from *flatus*,[1] the blowing of the wind, so that they call it *flagellum* instead of *flabellum*. The second word seems at first to have been called *parilema*, from *parere*, to bear, because the shoot is sent out to bear grapes; thence, by change of letters, as in many words, it got to be called *palma*. On the other side, it bears a tendril; this is a vine twig twisted like a curl. It is by means of these tendrils that the vine holds the support on which it creeps to grasp a place, from which grasping (*capere*) it is called *capreolus*. All fodder crops should be cut, first clover, mixed fodder, and vetch, and last hay. *Ocinum* is derived, as is the garden clover (*ocimum*), from the Greek word ὠκέως, which means 'quickly.' It is called *ocinum* for the further reason that it moves (*citat*) the bowels of cattle, and is fed to them on that account, as a purgative. It is cut green from the bean crop before it forms pods. *Farrago*, on the other hand, is so called from a crop where a mixture of barley, vetch, and legumes has been sowed for green feed, either because it is cut with the steel (*ferrum*, *ferrago*) or for the reason that it was first sowed in a spelt (*far*) field. It is with this that horses and other animals are purged and fattened in the spring. Vetch is so called from *vincire*, to bind, because it also has tendrils as the vine has, with which, when it creeps up to cling to the stalk of the lupine or some other plant, it usually binds (*vincit*) it. If you have meadows to be irrigated, as soon as you have gathered the hay, irrigate them. During droughts water should be given every evening to the fruit trees that are grafted. It may be that they are called *poma* from the fact that they need drink (*potus*).

XXXII. Quarto intervallo inter solstitium et caniculam plerique messem faciunt, quod frumentum dicunt quindecim diebus esse in vaginis, quindecim florere, quindecim exarescere, cum sit maturum. Arationes absolvi, quae eo fructuosiores fiunt, quo caldiore terra aratur. Si proscideris, offringi oportet, id est iterare, ut frangantur glaebae; prima enim

2 aratione grandes glaebae ex terra scinduntur. Serendum viciam, lentem, cicerculam, ervilam ceteraque, quae alii legumina, alii, ut Gallicani quidam, legarica appellant, utraque dicta a legendo, quod ea non secantur, sed vellendo leguntur. Vineas veteres iterum occare, novellas etiam tertio, si sunt etiam tum glaebae.

XXXIII. Quinto intervallo inter caniculam et aequinoctium autumnale oportet stramenta desecari et acervos constitui, arata offringi, frondem caedi, prata inrigua iterum secari.

XXXIV. Sexto intervallo ab aequinoctio autumnali incipere scribunt oportere serere usque ad diem nonagensimum unum. Post brumam, nisi quae necessaria causa coegerit, non serere, quod tantum intersit, ut ante brumam sata quae septimo die, post brumam sata quadragesimo die vix existant. Neque ante aequinoctium incipi oportere putant, quod, si minus idoneae tempestates sint consecutae, putescere semina soleant. Fabam optime seri in vergiliarum

2 occasu; uvas autem legere et vindemiam facere inter

[1] *i.e.* its rising.

XXXII. "In the fourth period, between the solstice and the Dog Star,[1] most farmers harvest, because it is a saying that the grain is in the sheath for fifteen days, blooms for fifteen days, dries for fifteen days, and is then ripe. Ploughing should be completed, and it will be more valuable in proportion as it is done in warm ground; if you are 'breaking up' the ground it should be crushed, that is, gone over a second time so that the clods may be broken; for in the first ploughing large clods are cut from the ground. You should sow vetch, lentils, small peas, pulse, and other plants, which some call legumes, and others, like some Gallic farmers, call *legarica*, both words being derived from *legere*, because these are not reaped, but are gathered by pulling. Hoe old vines a second time, young ones even a third time, if there are clods still left.

XXXIII. "In the fifth period, between the Dog Star and the autumnal equinox, the straw should be cut and stacks built, ploughed land harrowed, leaf-fodder gathered, and irrigated meadows mowed a second time.

XXXIV. "In the sixth period, from the autumnal equinox, the authorities state that sowing should begin and continue up to the ninety-first day. After the winter solstice, unless necessity requires, there should be no sowing—a point of such importance that seeds which, when planted before the solstice, sprout in seven days, hardly sprout in forty if sowed after the solstice. And they hold that sowing should not begin before the equinox, because, if unfavourable seasons follow, the seeds usually rot. Beans are sowed to best advantage at the time of the setting of the Pleiades: while the gathering of the grapes

aequinoctium autumnale et vergiliarum occasum; dein vites putare incipere et propagare et serere poma. Haec aliquot regionibus, ubi maturius frigora fiunt asperiora, melius verno tempore.

XXXV. Septimo intervallo inter vergiliarum occasum et brumam haec fieri oportere dicunt: Serere lilium et crocum. Quae[1] iam egit radicem rosa, ea conciditur radicitus in virgulas palmares et obruitur; haec eadem postea transfertur facta viviradix. Violaria in fundo facere non est utile, ideo quod necesse est terra adruenda pulvinos fieri, quos irrigationes et pluviae tempestates abluunt et agrum faciunt 2 macriorem. A favonio usque ad arcturi exortum recte serpillum e seminario transferri, quod dictum ab eo, quod serpit. Fossas novas fodere, veteres tergere, vineas arbustumque putare, dum in XV diebus ante et post brumam, ut pleraque, ne facias. Nec non tum aliquid recte seritur, ut ulmi.

XXXVI. Octavo intervallo inter brumam et favonium haec fieri oportet. De segetibus, siqua est aqua, deduci; sin siccitates sunt et terra teneritudinem habet, sarire. Vineas arbustaque putare. Cum in agris opus fieri non potest, quae sub tecto possunt tunc conficienda antelucano tempore hiberno. Quae

[1] *quae* Keil: *quod*.

[1] Compare Cato, 39.

and the making of the vintage falls between the autumnal equinox and the setting of the Pleiades; then the pruning and layering of vines and the planting of fruit trees should begin. In some localities, where severe frosts come earlier, these operations are best carried on in spring.

XXXV. "In the seventh period, between the setting of the Pleiades and the winter solstice, they say that these operations should be carried out:—Planting of lilies and crocus. A rose which has already formed a root is cut from the root up into twigs a palm-breadth long and planted; later on the same twig is transplanted when it has made a living root. It is not profitable to plant violet beds on a farm for the reason that beds must be formed by heaping up the soil, and irrigation and heavy rains wash these away and thus make the ground poorer. From the beginning of the west wind to the rising of Arcturus, it is proper to transplant from the nursery wild thyme, which gets its name (*serpillum*) from the fact that it 'creeps' (*serpit*). Dig new ditches, clear old ones, prune vineyards and orchards, provided you do not do this, or in fact most things, during the fifteen days preceding and following the solstice. And yet some trees, such as the elm, are properly planted at that time.

XXXVI. "In the eighth period, between the solstice and the beginning of the west wind, these operations should be carried out:—Any water in the grain lands should be drained, but if there is a drought and the land is friable, harrow. Prune vineyards and orchards. When work cannot be carried out on the land, indoor tasks should be completed then in the early winter mornings.[1] You should keep the rules

dixi scripta et proposita habere in villa oportet, maxime ut vilicus norit.

XXXVII. Dies lunares quoque observandi, qui quodam modo bipertiti, quod a nova luna crescit ad plenam et inde rursus ad novam lunam decrescit, quaad veniat ad intermenstruum, quo die dicitur luna esse extrema et prima; a quo eum diem Athenis appellant ἕνην καὶ νέαν, τριακάδα alii. Quaedam facienda in agris potius crescente luna quam senescente, quaedam contra quae metas, ut frumenta et caeduas silvas.
2 Ego istaec, inquit Agrasius, non solum in ovibus tondendis, sed in meo capillo a patre acceptum servo, ni crescente luna tondens calvos fiam. Agrius, Quem ad modum, inquit, luna quadripertita? Et
3 quid ea divisio ad agros pollet? Tremelius, Numquam rure audisti, inquit, octavo Ianam lunam et crescentem et contra senescentem, et quae crescente luna fieri oporteret, tamen quaedam melius fieri post octavo Ianam lunam quam ante? Et siquae senescente fieri conveniret, melius, quanto minus haberet ignis id astrum? Dixi de quadripertita forma in cultura agri.

4 Stolo, Est altera, inquit, temporum divisio coniuncta quodam modo cum sole et luna sexpertita,

---

[1] Defined elsewhere by Varro (*de Ling. Lat.*, VI, 10) as "diem, quem diligentius Attici ἕνην καὶ νέαν appellarunt ab eo quod eo die potest videri extrema et prima luna." "The day was named more carefully 'old and new' by the Athenians, because on that day can be seen the last (of the waning) and the first (of the waxing) moon."

[2] Authorities, both ancient and modern, are undecided whether *Iana* is a variant form of *Diana*, or the feminine of *Ianus*.

[3] As enumerated by Scrofa in Chapter 5, Section **3**.

I have laid down written and posted in the farmstead, in order that the overseer particularly may know them.

XXXVII. "The lunar periods also must be taken into account; these are roughly twofold, as the moon waxes from the new to the full and then wanes again toward the new, until it reaches the *intermenstruum*, or time 'between two months,' on which day the moon is said to be 'last and first'; hence, at Athens they call this day ἔνην καὶ νέαν,[1] or 'old and new,' while others call it τριακάδα, or the 'thirtieth.' Some operations should be carried out on the land during the waxing rather than the waning of the moon, while there are certain crops which you should gather in the opposite phase, such as grain and firewood." "I learned this rule from my father," said Agrasius, "and I keep it not only in shearing my sheep but in cutting my hair, for fear that if I have it done when the moon is waxing I may become bald." "After what method is the moon divided into quarters," asked Agrius, "and what influence has that division on farming?" "Have you never heard in the country," replied Tremelius, "the expressions 'eight days before the waxing of the moon,'[2] and 'eight days before the waning of the moon,' and that of the things which should be done when the moon is waxing some are nevertheless better done after this 'eight days before the waxing' than before it; and that the things which should be done when she is waning are better done the less light that heavenly body has? I have discussed the fourfold division in agriculture."[3]

"There is," said Stolo, "a second, a sixfold-division of seasons which may be said to bear a relation

quod omnis fere fructus quinto denique gradu pervenit ad perfectum ac videt in villa dolium ac modium, unde sexto prodit ad usum. Primo praeparandum, secundo serendum, tertio nutricandum, quarto legendum, quinto condendum, sexto promendum. Ad alia in praeparando faciendi scrobes aut repastinandum aut sulcandum, ut si arbustum aut pomarium facere velis; ad alia 5 arandum aut fodiendum, ut si segetes instituas; ad quaedam bipalio vertenda terra plus aut minus. Aliae enim radices angustius diffundunt, ut cupressi, aliae latius, ut platani, usque eo ut Theophrastus scribat Athenis in Lyceo, cum etiam nunc platanus novella esset, radices trium et triginta cubitorum egisse. Quaedam si bubus et aratro prosciderís, et iterandum, antequam semen iacias.[1] Item praeparatio siquae fit in pratis, id est ut defendantur a pastione, quod fere observant a piro florente; si inrigua sunt, ut tempestive inrigentur.

XXXVIII. Quae loca in agro stercoranda, videndum, et qui et quo genere potissimum facias: nam discrimina eius aliquot. Stercus optimum scribit esse Cassius volucrium praeter palustrium ac nantium. De hisce praestare columbinum, quod sit calidissimum ac fermentare possit terram. Id ut semen

[1] *iacias* Keil: *iactas.*

---

[1] See Cato, 1, 7, note 2, page 7.
[2] *Hist. Plant.*, I, 7, 1.

to the sun and moon, because almost every product comes to perfection in five stages and reaches jar and basket in the farmstead, and from these is brought forth for use in the sixth. The first stage is the preparation, the second the planting, the third the cultivation, the fourth the harvesting, the fifth the storing, the sixth the marketing. In the matter of preparation: for some crops you must make trenches or dig thoroughly or draw furrows, as when you wish to make an *arbustum*[1] or an orchard; for others you must plough or spade, as when you are starting a grain field; for some the earth must be turned more or less deeply with the trenching spade. For some trees, such as the cypress, spread their roots less, and others, such as the plane, more; so much, indeed, that Theophrastus[2] mentions a plane tree in the Lyceum at Athens which, even when it was quite young, had thrown out its roots to a spread of thirty-three cubits. Certain land, when you have broken it with oxen and plough, must be worked a second time before you broadcast the seed. Any preparation that is made in the matter of meadows consists in closing them from grazing, a practice which is usually observed from the time of the blooming of the pear trees; and if they are irrigated, in turning in the water at the proper time.

XXXVIII. "We must observe what parts of the land must be manured, how the manure is to be applied, and the best kind to use; for there are several varieties. Cassius states that the best manure is that of birds, except marsh- and sea-fowl; and that the dung of pigeons is the best of these, because it has the most heat and causes the ground to ferment. This should be broadcast on the land like seed, and not

aspargi oportere in agro, non ut de pecore acervatim
2 poni. Ego arbitror praestare ex aviariis turdorum ac merularum, quod non solum ad agrum utile, sed etiam ad cibum ita bubus ac subus, ut fiant pingues. Itaque qui aviaria conducunt, si cavet dominus stercus ut in fundo maneat, minoris conducunt, quam ii quibus id accedit. Cassius secundum columbinum scribit esse hominis, tertio caprinum et
3 ovillum et asininum, minime bonum equinum, sed in segetes; in prata enim vel optimum, ut ceterarum veterinarum, quae hordeo pascuntur, quod multam facit herbam. Stercilinum secundum villam facere oportet, ut quam paucissimis operis egeratur. In eo, si in medio robusta aliqua materia sit depacta, negant serpentem nasci.

XXXIX. Sationis autem gradus, secundus, hanc habet curam: natura ad quod tempus cuiusque seminis apta sit ad serendum. Nam refert in agro ad quam partem caeli quisque locus spectet, sic ad quod tempus quaeque res facillime crescat. Nonne videmus alia florere verno tempore, alia aestivo, neque eadem
2 autumnali, quae hiberno? Itaque alia seruntur atque inseruntur et metuntur ante aut post quam alia; et cum pleraque vere quam autumno inserantur, circiter solstitium inseri ficos nec non brumalibus diebus
3 cerasos. Quare cum semina sint fere quattuor generum, quae natura dedit, quae transferuntur e terra in terram viva radice, quae ex arboribus dempta demittuntur in humum, quae inseruntur ex arboribus in arbores, de singulis rebus videndum, quae quoque tempore locoque facias.

---

[1] Swine, especially, eat it eagerly; but the use of bird-dung for feed has no place in modern practice.

[2] Compare Columella, II, 14, for a full discussion of this topic.

placed in piles like cattle dung. My own opinion is that the best dung is from aviaries of thrushes and blackbirds, as it is not only good for the land, but is excellent food both for cattle and swine,[1] to fatten them. Hence those who lease aviaries with the owner's stipulation that the dung shall remain on the place pay less rent than those who have the use of it. Cassius states that next to pigeon dung human excrement is the best, and in the third place goat, sheep, and ass dung; that horse dung is least valuable, but good on grain land; for on meadows it is the most valuable of all, as is that of all draught animals which feed on barley, because it produces a quantity of grass.[2] The farmer should make a dung-hill near the steading, so that the manure may be cleared out with the least labour. They say that if an oak stake is driven into the middle of it no serpent will breed there.

XXXIX. "The second step, that of planting, requires care as to the season of planting which is suited to the nature of each seed. For in a field it is important to note the exposure of every section, and also the season at which each plant grows best. Do we not observe that some blossom in spring, some in summer, and that the autumn growth is not the same as the winter? Thus some plants are sown and grafted and harvested earlier or later than others; and while most are grafted in spring rather than in autumn, figs are grafted near the solstice, and cherries actually in mid-winter. Now as seeds are, in general, of four kinds—those furnished by nature, those which are transplanted from one piece of ground to another as rooted slips, cuttings from trees planted in the ground, and grafts from tree to tree—you should observe what separate operation should be carried out at each season and in each locality.

XL. Primum semen, quod est principium genendi, id duplex, unum quod latet nostrum sensum, alterum quod apertum. Latet, si sunt semina in aere, ut ait physicos Anaxagoras, et si aqua, quae influit in agrum, inferre solet, ut scribit Theophrastus. Illud quod apparet ad agricolas, id videndum diligenter. Quaedam enim ad genendum propensa [1] usque adeo parva, ut sint obscura, ut cupressi. Non enim galbuli qui nascuntur, id est tamquam pilae parvae corticiae, 2 id semen, sed in iis intus. Primigenia semina dedit natura, reliqua invenit experientia coloni. Prima quae sine colono, priusquam sata, nata; secunda quae ex iis collecta neque, priusquam sata, nata. Prima semina videre oportet ne vetustate sint exsucta aut ne sint admixta aut ne propter similitudinem sint adulterina. Semen vetus tantum valet in quibusdam rebus, ut naturam commutet. Nam ex semine brassicae vetere sato nasci aiunt rapa et contra 3 ex raporum brassicam. Secunda semina videre oportet ne, unde tollas, nimium cito aut tarde tollas. Tempus enim idoneum, quod scribit Theophrastus, vere et autumno et caniculae exortu, neque omnibus locis ac generibus idem. In sicco et macro loco et argilloso vernum tempus idoneum, quo minus habet umoris: in terra bona ac pingui autumno, quod vere multus umor, quam sationem

[1] *propensa* Schöll; *propterea.*

---

[1] *Hist. Plant.*, III, 1, 4: "Anaxagoras says that the air contains the seeds of all things, and that these, carried along by water, produce plants." (L.C.L., Vol. I, p. 163).

[2] So Virgil, *Georg.*, II, 20 and 22:

hos Natura modos primum dedit . . .
sunt alii, quos ipse via sibi repperit usus.

[3] Varro here causes confusion by reverting to the four kinds of seeds named in 39, 3, so giving to *prima* and

XL. "In the first place, the seed, which is the origin of growth, is of two kinds, one being invisible, the other visible. There are invisible seed, if, as the naturalist Anaxagoras holds, they are in the air, and if the water which flows on the land carries them, as Theophrastus writes.[1] The seed which can be seen should be carefully watched by the farmer; for some seed, such as that of the cypress, though capable of generating, is so small that it can hardly be seen; (for the pods which it bears, that look like little balls of bark, are not the seeds but contain them). The original seeds were given by nature, while the later were discovered in the experiments of the farmer.[2] The first are those which, without the aid of the farmer, grow without being sown; the second are those which, derived from these, do not grow without being sown. In the case of the first,[3] care should be taken to see that they are not dried out from age, and that they are clean and not mixed with seed of similar appearance. The age of the seed is of such importance in the case of some plants that it alters their nature; thus from the planting of old cabbage seed it is said that rape grows, and on the other hand that cabbage grows from old rape seed. In the case of the second class of seed,[3] you should be careful not to transplant them too early or too late. The proper time is that given by Theophrastus[4]—spring, autumn, and at the rising of the Dog Star—but the time is not the same for all localities and all species. In ground that is dry, thin, or clayey, spring is the proper season, because it is less humid; in good, rich land the autumn, because in spring it is very wet.

secunda a different meaning from what they have in the preceding sentence.

[4] *Caus. Plant.*, I, 6, 3.

4 quidam metiuntur fere diebus XXX. Tertium genus seminis, quod ex arbore per surclos defertur in terram, si in humum demittitur, in quibusdam est videndum ut eo tempore sit deplantatum, quo oportet (id fit tum, antequam gemmare aut florere quid incipit); et quae de arbore transferas ut ea deplantes potius quam defringas, quod plantae solum stabilius, quo latius aut radices facilius mittit. Ea celeriter, antequam sucus exarescat, in terram demittunt. In oleagineis seminibus videndum ut sit de tenero ramo ex utraque parte aequabiliter praecisum, quas alii clavolas, alii taleas appellant ac 5 faciunt circiter pedales. Quartum genus seminis, quod transit ex arbore in aliam, videndum qua ex arbore in quam transferatur et quo tempore et quem ad modum obligetur. Non enim pirum recipit quercus, neque enim si malus pirum. Hoc secuntur multi, qui haruspices audiunt multum, a quibus proditum, in singulis arboribus quot genera insita sint, uno ictu tot fulmina fieri illut quod fulmen concepit. Si in pirum silvaticam inserueris pirum quamvis bonam, non fore tam iucundam, quam si in eam quae 6 silvestris non sit. In quamcumque arborem inseras, si eiusdem generis est, dumtaxat ut sit utraque malus, ita inserere oportet referentem ad fructum,

---

[1] Virgil likewise prescribes this method in *Georg.*, II, 23:

> hic plantas tenero abscindens de corpore matrum
> deposuit sulcis.

Cf. also Pliny, *Nat. Hist.*, XVII, 67, and (for the meaning of *deplantare*) Columella, II, 2, 26.

[2] The latter term is used by Cato, Chapter 45. Both words mean "small stick" or truncheon. Theophrastus uses θαλία for θαλλός, a young shoot.

[3] The passage is quite unintelligible, though much ingenuity

Some authorities allow about thirty days for such planting. In the third method, which consists in transferring shoots from a tree into the ground, if the shoot is buried in the earth, you must be careful, in the case of some, that the shoot be removed at the proper time—that is, before it shows any sign of budding or blossoming; and that what you transplant from the tree you tear from the stock [1] rather than break off a limb, as the heel of a shoot is steadier, or the wider it is the more easily it puts out roots. They are thrust into the ground at once, before the sap dries out. In the case of olive cuttings, care must be taken that they be from a tender branch, sharpened evenly at both ends. Such cuttings, about a foot in length, are called by some *clavolae*, and by others *taleae*.[2] In the fourth method, which consists in running a shoot from one tree to another, the points to be observed are the nature of the tree, the season, and the method of fastening. You cannot, for instance, graft a pear on an oak, even though you can on an apple. This is a matter of importance to many people who pay considerable attention to soothsayers; for these have a saying that when a tree has been grafted with several varieties, the one that attracts the lightning turns into as many bolts as there are varieties, though the stroke is a single one.[3] No matter how good the pear shoot which you graft on a wild pear, the fruit will not be as well flavoured as if you graft it on a cultivated pear. It is a general rule in grafting, if the shoot and the tree are of the same species, as, for instance, if both are of the apple family, that for the effect on the fruit the grafting

has been spent on it. Sensible people laughed at the *haruspex* long before this. Cf. Cicero, *de Divin.*, II, 24.

meliore genere ut sit surculus, quam est quo veniat arbor. Est altera species ex arbore in arborem inserendi nuper animadversa in arboribus propinquis. Ex arbore, qua vult habere surculum, in eam quam inserere vult ramulum traducit et in eius ramo praeciso ac diffisso implicat, eum locum qui contingit, ex utraque parti quod intro est falce extenuatum, ita ut ex una parti quod caelum visurum est corticem cum cortice exaequatum habeat. Eius ramuli, quem inseret, cacumen ut derectum sit ad caelum curat. Postero anno, cum comprendit, unde propagatum est, ab altera arbore praecidit.

XLI. Quo tempore quaeque transferas, haec in primis videnda, quae prius verno tempore inserebantur, nunc etiam solstitiali, ut ficus, quod densa materia non est et ideo sequitur caldorem. A quo fit ut in locis frigidis ficeta fieri non possint. Aqua recenti insito inimica; tenellum enim cito facit 2 putre. Itaque caniculae signo commodissime existimatur ea inseri. Quae autem natura minus sunt mollia, vas aliquod supra alligant, unde stillet lente aqua, ne prius exarescat surculus, quam colescat. Cuius surculi corticem integrum servandum et eum sic exacuendum, ut non denudes medullam. Ne extrinsecus imbres noceant aut nimius calor, argilla 3 oblinendum ac libro obligandum. Itaque vitem

---

[1] Compare Columella, V, 11, 13.

should be of such a nature that the shoot is of a better type than the tree on which it is grafted. There is a second method of grafting from tree to tree which has recently been developed,[1] under conditions where the trees stand close to each other, From the tree from which you wish to take the shoot a small branch is run to the tree on which you wish to graft and is inserted in a branch of the latter which has been cut off and split; the part which fits into the branch having first been sharpened on both sides with the knife so that on one side the part which will be exposed to the weather will have bark fitted accurately to bark. Care is taken to have the tip of the grafted shoot point straight up. The next year, after it has taken firm hold, it is cut off the parent stem.

XLI. "As to the proper season for grafting, this must be especially observed: that some plants which formerly were grafted in spring are now grafted in mid-summer also, such as the fig, which, as the wood is not hard, requires warm weather; it is for this reason that fig groves cannot be planted in cold localities. Moisture is harmful to a fresh graft, for it causes the tender shoot to decay quickly, and hence it is the common view that this tree is best grafted in the dog days. In the case of plants which are not so soft, however, a vessel is fastened above the graft in such a way that water may drip slowly to keep the shoot from drying out before it unites with the tree. The bark of the shoot must be kept uninjured, and the shoot itself be sharpened in such a way as not to bare the pith. To prevent moisture or excessive heat from injuring it on the surface it should be smeared with clay and tied up with bark. For this

triduo antequam inserant desecant, ut qui in ea nimius est umor defluat, antequam inseratur; aut in quam inserunt, in ea paulo infra, quam insitum est, incidunt, qua umor adventicius effluere possit. Contra in fico et malo punica, et siquae etiam horum natura aridiora, continuo. In aliis translationibus videndum ut quod transferat cacumen habeat gemmam, ut in ficis.

4 De his primis quattuor generibus seminum quaedam quod tardiora, surculis potius utendum, ut in ficetis faciunt. Fici enim semen naturale intus in ea fico; quam edimus, quae sunt minuta grana; e quibus parvis quod enasci coliculi vix queunt — omnia enim minuta et arida ad crescendum tarda, ea quae laxiora, et fecundiora, ut femina quam mas et pro portione in virgultis item; itaque ficus, malus punica et vitis propter femineam mollitiam ad crescendum prona, contra palma et cupressus et olea in crescendo 5 tarda: in hoc enim umidiora quam aridiora — quare [ex terra] potius in seminariis surculos de ficeto quam grana de fico expedit obruere, praeter si aliter nequeas, ut siquando quis trans mare semina mittere aut inde petere vult. Tum enim resticulam per ficos, quas edimus, maturas perserunt et eas, cum inaruerunt, complicant ac quo volunt mittunt, ubi obrutae 6 in seminario pariant. Sic genera ficorum, Chiae ac Chalcidicae et Lydiae et Africanae, item cetera

[1] The whole passage may be taken from Theophrastus, *Caus. Plant.*, I, 8; but the last clause is so obscure that Ursinus would reject it, together with *ex terra* below.

reason the vine is cut off three days before grafting, so that any excessive moisture in it may run out before it is grafted; or else a cut is made in the branch on which the graft is made a little lower than the graft, so as to allow casual water to run off. On the other hand, figs, pomegranates, and plants of a drier nature are grafted at once. In other graftings, such as of figs, care must be taken that the shoot contains a bud.

"Of these four forms of propagation it is better to use quicksets in the case of some slow-growing plants, as is the practice in fig groves; for the natural seed of the fig is on the inside of the fruit which we eat, in the form of very small grains. As the seedling can scarcely spring from these small grains—for all things which are small and dry grow slowly, while those which are of looser texture are also of more rapid growth, as, for instance, the female grows more rapidly than the male, a rule which holds good also in plants to some extent, the fig, the pomegranate, and the vine being, on account of their feminine softness, of rapid growth, while, on the other hand, the palm, the cypress, and the olive are of slow growth; for in this respect the humid [are quicker] than the dry[1]—it is therefore better to plant in the nursery shoots from the fig tree than grains from the fruit; unless this is impracticable, as when you wish to ship seeds overseas or import them thence. In this case we pass a string through the figs when they are ripe for eating, and after they have dried they are tied in bundles and may be sent where we will; and there they are planted in a nursery and reproduce. It was in this manner that the Chian, Chalcidian, Lydian, African, and other varieties of over-sea figs

transmarina in Italiam perlata. Simili de causa, oleae semen cum sit nuculeus, quod ex eo tardius enascebatur colis quam ex aliis, ideo potius in seminariis taleas, quas dixi, serimus.

XLII. In primis observes ne in terram nimium aridam aut variam, sed temperatam, semen demittas. In iugerum unum, si est natura temperata terra, scribunt opus esse medicae sesquimodium. Id seritur ita, ut semen iaciatur, quem ad modum cum pabulum et frumentum seritur.

XLIII. Cytisum seritur in terra bene subacta tamquam semen brassicae. Inde differtur et in sesquipedem ponitur, aut etiam de cytiso duriore virgulae deplantantur, et ita pangitur in serendo.

XLIV. Seruntur fabae modii IIII in iugero, tritici V, hordei VI, farris X, sed non nullis locis paulo amplius aut minus. Si enim locus crassus, plus; si macer, minus. Quare observabis, quantum in ea regione consuetudo erit serendi, ut tantum facias, quod tantum valet regio ac genus terrae, ut ex eodem semine aliubi cum decimo redeat, aliubi cum quinto 2 decimo, ut in Etruria locis aliquot. In Italia in Subaritano dicunt etiam cum centesimo redire solitum, in Syria ad Gadara et in Africa ad Byzacium item ex modio nasci centum. Illut quoque multum interest, in rudi terra, an in ea seras, quae quotannis obsita sit, quae vocatur restibilis, an in vervacto 3 quae interdum requierit. Cui Agrius, In Olynthia

---

[1] It is interesting to compare St. Matt. 13, 8.

[2] A province in Africa Propria, between the river Triton and the Lesser Syrtis.

[3] *Vervactum* is called by Columella, II, 10, 5 and elsewhere, *veteretum*. The word means soil which has been broken (*vervago*) and allowed to lie fallow.

were imported into Italy. For a similar reason, the seed of the olive being a nut, we prefer to plant in our nurseries the cuttings which I have described, as the stem was found to spring more slowly from the olive nut than from others.

XLII. "Be especially careful not to plant in ground that is very dry or very wet, but rather in moderately moist ground. Authorities state that the proper amount of alfalfa is a modius and a half to the iugerum, if the ground is by nature moderately moist; the method of sowing is to broadcast the seed, as is done in sowing forage crops and grain.

XLIII. "Snail-clover is sowed on land that has been thoroughly worked, like cabbage seed; then it is transplanted at intervals of a foot and a half, or, when the plant is more mature, shoots are detached, and it is set out in planting as above.

XLIV. "Beans are sowed 4 modii to the iugerum, wheat 5, barley 6, spelt 10, the amount being a little more or less in some localities; more being sowed on rich ground and less on thin. You should therefore note the amount that is usually sowed in the district and follow this practice; for the locality and the type of soil is so important that the same seed in one district yields tenfold and in another fifteen-fold —as at some places in Etruria. Around Sybaris in Italy the normal yield is said to be even a hundred to one, and a like yield is reported near Gadara [1] in Syria, and for the district of Byzacium [2] in Africa. It also makes a great difference whether the planting is on virgin soil or on what is called *restibilis*—land cultivated every year—or on *vervactum*,[3] which is allowed sometimes to lie fallow between crops." "In Olynthia," remarked Agrius, "they say that the land

restibilia esse dicunt, sed ita ut tertio quoque anno uberiores ferant fructos. Licinius, Agrum alternis annis relinqui oportet paulo levioribus sationibus, id est quae minus sugunt terram.

Dicetur, inquit Agrius, de tertio gradu, de nutri-
4 cationibus atque alimoniis eorum. Ille, Quae nata sunt, inquit, in fundo alescunt, adulta concipiunt, praegnatia, cum sunt matura, pariunt poma aut spicam, sic alia. A quo profectum, redit semen. Itaque si florem acerbumve pirum aliudve quid decerpseris, in eodem loco eodem anno nihil renascitur, quod praegnationis idem bis habere non potest. Ut enim mulieres habent ad partum dies certos, sic arbores ac fruges.

XLV. Primum plerumque e terra exit hordeum diebus VII, nec multo post triticum; legumina fere quadriduo aut quinque diebus, praeterquam faba; ea enim serius aliquanto prodit. Item milium et sesima et cetera similiter aequis fere diebus, praeterquam siquid regio aut tempestas viti attulit, quo
2 minus ita fiat. Quae in seminario nata, si loca erunt frigidiora, quae molli natura sunt, per brumalia tempora tegere oportet fronde aut stramentis. Si erunt imbres secuti, videndum necubi aqua consistat;
3 venenum enim gelum radicibus tenellis. Sub terra et supra virgulta non eodem tempore aeque crescunt; nam radices autumno aut hieme magis sub terra quam supra alescunt, quod tectae terrae tepore propagantur, supra terram aere frigidiore coguntur.[1]

[1] The text is corrupt. Victorius conj. *teguuntur*, Jackson *tinguntur*, Scaliger *restinguntur*, Pontedera *stringuntur*, Schneider *cingunter*, and Pontedera and Schneider read in the preceding line *quam quae supra adolescunt*. Varro is translating Theophrastus, *Caus. Plant.*, I, 12, 7 : "because the upper parts are checked by the surrounding air, which is chillier."

is cropped every year, but in such a way that a richer crop is produced every other year." "Land ought to be left every other year with somewhat lighter crops," rejoined Licinius; "I mean by that crops which are less exhausting to the land."

"Tell us now," said Agrius, "of the third step, the nurture and feeding of the plant." "All plants," resumed Stolo, "grow in the soil, and when mature conceive, and when the time of gestation is complete bear fruit or ear, or the like; and the seed returns whence it came.[1] Thus, if you pluck the blossom or an unripe pear, or the like, no second one will grow on the same spot in the same year, as the same plant cannot have two periods of gestation. For trees and plants, just as women, have a definite period from conception to birth.

XLV. "Barley usually appears in seven days, wheat not much later; legumes usually in four or five days, except the bean, which is somewhat slower in appearing. Millet also, and sesame, and similar plants appear in about the same number of days, except in cases where the locality or the weather prevents this from occurring. Seedlings in the nursery should be covered with leaves or straw during the winter if the locality is at all cold, as they are tender; and if rains follow, water must not be allowed to stand anywhere, for frost is baneful to the tender rootlets. Plants do not grow at the same rate below and above ground; thus roots grow faster below than does the part above ground in autumn and in winter because, being covered, they are nourished by the heat of the earth, while the part above ground is checked by the colder air. Wild plants which have

[1] *i.e.* the cycle begins again.

Itaque ita esse docent silvestria, ad quae sator non accessit. Nam prius radices, quam ex iis quod solet nasci, crescunt. Neque radices longius procedunt, nisi quo tepor venit solis. Duplex causa radicium, quod et materiem aliam quam aliam longius proicit natura, et quod alia terra alia facilius viam dat.

XLVI. Propter cuius modi res admiranda discrimina sunt naturae aliquot, ex quibusdam foliis propter eorum versuram, quod sit anni tempus, ut dici possit, ut olea et populus alba et salix. Horum enim folia cum converterunt se, solstitium dicitur fuisse. Nec minus admirandum quod fit in floribus, quos vocant heliotropia ab eo, quod ad solis ortum mane spectant et eius iter ita secuntur ad occasum, ut ad eum semper spectent.

XLVII. In seminario quae surculis consita et eorum molliora erunt natura cacumina, ut olea ac ficus, ea summa integenda binis tabellis dextra et sinistra deligatis [1] herbaeque eligendae. Eae, dum tenerae sunt, vellendae. Post [2] enim aridae factae rixantur ac celerius rumpuntur, quam secuntur. Contra herba in pratis ad spem faenisiciae nata non modo non evellenda in nutricatu, sed etiam non calcanda. Quo et pecus ab prato ablegandum et omne iumentum, etiam hominem. Solum enim hominis exitium herbae et semitae fundamentum.

XLVIII. In segetibus autem frumentum quo culmus extulit, spicam. Ea quae mutilata non est, in

[1] *deligatis* Iucundus, Schneider: *deligata.*
[2] *post* Goetz: *prius.*

not been touched by the planter show that this is true; for roots grow before the plant which comes from them, but they go no deeper than the point to which the sun's warmth reaches. The growth of roots is determined by two factors: that nature thrusts one kind of wood to a greater distance than another; and that one kind of soil yields more readily than another.

XLVI. "As a result of factors of this kind there are several remarkable differences of character; so that, for instance, the season may be told from the leaves of such trees as the olive, the silver poplar, and the willow, by the direction in which they lie; thus when the leaves of these trees turn over it is said that the summer solstice has passed. No less remarkable is the behaviour of the flowers which are called 'heliotropes' from the fact that they face the rising sun in the morning and follow his course until the setting, facing him the whole time.

XLVII. "Plants such as the olive and fig, which, reared in the nursery from shoots and naturally somewhat delicate on top, should be protected at the top by two boards tied right and left; and the weeds should be cleared. These should be pulled while they are young; for after they become dry they resist more strongly, and break off more readily than they yield. On the other hand, growth that springs up on a meadow for haying must not only not be plucked while it is maturing, but also must not be trampled. For this reason flocks, and every sort of animal, including even man, must be kept off a meadow; for the foot of man is death to grass, and marks the beginning of a path.

XLVIII. "Now, in the case of grain crops, that by which the stalk puts forth the grain is the head. If

hordeo et tritico, tria habet continentia, granum, glumam, aristam et etiam, primitus spica cum oritur, vaginam. Granum dictum quod est intimum soldum; gluma qui est folliculus eius; arista quae ut acus tenuis longa eminet e gluma, proinde ut grani 2 apex sit gluma et arista. Arista et granum omnibus fere notum, gluma paucis. Itaque id apud Ennium solum scriptum scio esse in Euhemeri libris versis. Videtur vocabulum etymum habere a glubendo, quod eo folliculo deglubitur granum. Itaque eodem vocabulo appellant fici eius, quam edimus, folliculum. Arista dicta, quod arescit prima. Granum a gerendo; id enim ut gerat spica, seritur frumentum, non ut glumam aut aristam gerat, ut vitis seritur, non ut pampinum ferat, sed ut uvam. Spica autem, quam rustici, ut acceperunt antiquitus, vocant specam, a spe videtur nominata; eam enim 3 quod sperant fore, serunt. Spica mutila dicitur, quae non habet aristam; ea enim quasi cornua sunt spicarum. Quae primitus cum oriuntur neque plane apparent, qua sub latent herba, ea vocatur vagina, ut in qua latet conditum gladium. Illut autem summa in spica iam matura, quod est minus quam granum, vocatur frit; quod in infima[1] spica ad culmum stramenti summum item minus quam granum est, appellatur urru.

XLIX. Cum conticuisset nec interrogaretur, de

[1] *quod in infima* Iucundus: *quod in firma.*

---

[1] Lit. "mutilated." The term is explained in § 3 below. Oats and spelt have no beard.

[2] Francken, in *Mnemosyne*, 28, p. 286, note, hazards the conjecture that these words are peasant corruptions of φορυτός, "chaff," and ὄρρος, "rump."

is this "hornless,"[1] as in barley and wheat, it has three components: the grain, the husk, and the beard—and the sheath, also, when the ear first appears. The hard inner part is called the grain; the husk is its envelope; and the beard is the part which rises from the husk like a long, slender needle, just as if the husk and beard formed a peaked cap for the grain. 'Beard' and 'grain' are familiar words to most people, but 'husk' (*gluma*) to few; thus the only place where it occurs, to my knowledge, is in Ennius, in his translation of Euhemerus. The word (*gluma*) seems to be derived from *glubere*, 'strip,' because the grain is stripped (*deglubitur*) from this envelope; so the same word is used for the envelope of the edible fruit of the fig tree. The beard is called *arista* from the fact that it is the first part to dry (*arescere*). The grain is so called from *gerere*; for the seed is planted that the ear may 'bear' (*gerat*) the grain, not the husk or the beard; just as the vine is planted not to bear leaves but grapes. The ear, however, which the peasants, in their old-fashioned way, call *speca*, seems to have got its name from *spes*; for it is because they hope (*sperant*) to have this grow that they plant. An ear which has no beard is said to be 'hornless,' as the beard may be said to be the 'horns' of the ear. When these are just forming and are not yet quite visible, the green envelope under which they are hidden is called the sheath, being like the sheath in which a sword is encased. The part at the top of the full-grown ear, which is smaller than the grain, is called *frit*;[2] while the part, also smaller than the grain, at the bottom of the ear where it joins the top of the stalk is called *urru*."[2]

XLIX. Stolo paused at this point, and, judging from

nutricatu credens nihil desiderari, Dicam, inquit, de fructibus maturis capiendis. Primum de pratis summissis herba, cum crescere desiit et aestu arescit, subsecari falcibus debet et, quaad perarescat, furcillis versari; cum peraruit, de his manipulos fieri ac vehi ad villam; tum de pratis stipulam rastellis
2 eradi atque addere faenisiciae cumulum. Quo facto sicilienda prata, id est falcibus consectanda quae faenisices praeterierunt ac quasi herba tuberosum reliquerunt campum. A qua sectione arbitror dictum sicilire pratum.

L. Messis proprio nomine dicitur in iis quae metimur, maxime in frumento, et ab eo esse vocabulo declinata. Frumenti tria genera sunt messionis, unum, ut in Umbria, ubi falce secundum terram succidunt stramentum et manipulum, ut quemque subsicuerunt, ponunt in terra. Ubi eos fecerunt multos, iterum eos percensent ac de singulis secant inter spicas et stramentum. Spicas coiciunt in corbem atque in aream mittunt, stramenta relincunt
2 in segete, unde tollantur in acervum. Altero modo metunt, ut in Piceno, ubi ligneum habent incurvum bacillum, in quo sit extremo serrula ferrea. Haec cum comprendit fascem spicarum, desecat et stramenta stantia in segeti relinquit, ut postea subsecentur. Tertio modo metitur, ut sub urbe Roma et locis plerisque, ut stramentum medium subsicent, quod manu sinistra summum prendunt; a quo medio

---

[1] Cf. Columella, II, 20, 3.

the fact that no questions were asked that no further discussion of nutrition was desired, he continued: "I shall discuss next the subject of harvesting the ripe crops. First the grass on the hay-meadows should be cut close with the sickle when it ceases to grow and begins to dry from the heat, and turned with the fork while it is drying out; when it is quite dry it should be made into bundles and hauled to the barn. Then the loose hay from the meadows should be raked up and added to the hay-pile. After doing this you should 'sickle' the meadows—that is, cut with the sickle what the mowers have passed over, leaving the field humped, as it were, with tufts of grass. I suppose the verb *sicilire*, used with meadow as object, is derived from this cutting (*sectio*).

L. "The word *messis* is properly employed of the crops which we 'measure' (*metimur*), especially of grain; and this, I suggest, is the derivation of the word. There are three methods of harvesting grain:[1] the first, employed in Umbria, in which the stalk is cut close to the ground with the hook, and each bundle, as it is cut, is laid on the ground. When a number of bundles are formed, they go over them again, and cut the ears from each close to the stalk. The ears are cast into the basket and carried to the threshing-floor, while the straw is left in the field and afterwards stacked. In the second method, employed in Picenum, they use a curved handle of wood with a small iron saw attached to the end; when this catches a bundle of ears it cuts them off and leaves the straw standing in the field to be cut later. In the third method, employed near Rome and in numerous other places, seizing the top with the left hand they cut the straw in the middle; and I suggest

messem dictam puto. Infra manum stramentum cum terra haeret, postea subsecatur; contra quod 3 cum spica stramentum haeret, corbibus in aream defertur. Ibi discedit in aperto loco palam; a quo potest nominata esse palea. Alii stramentum ab stando, ut stamen, dictum putant; alii ab stratu, quod id substernatur pecori. Cum est matura seges, metendum, cum in ea in iugerum fere una opera propemodum in facili agro satis esse dicatur. Messas spicas corbibus in aream deferre debent.

LI. Aream esse oportet in agro sublimiori loco, quam perflare possit ventus; hanc esse modicam pro magnitudine segetis, potissimum rutundam et mediam paulo extumidam, ut, si pluerit, non consistat aqua et quam brevissimo itinere extra aream defluere possit; omne porro brevissimum in rutundo e medio ad extremum. Solida terra pavita, maxime si est argilla, ne, aestu peminosa si sit,[1] in rimis eius grana oblitescant et recipiant aquam et ostia aperiant muribus ac formicis. Itaque amurca solent perfundere, ea enim herbarum et formicarum et talparum 2 venenum. Quidam aream ut habeant soldam, muniunt lapide aut etiam faciunt pavimentum. Non nulli etiam tegunt areas, ut in Bagiennis, quod ibi saepe id temporis anni oriuntur nimbi. Ubi ea retecta et

[1] *si sit* added by Gesner.

---

[1] See Cato, Chapter 91.

[2] Possibly Pliny's Vagienni (*N.H.*, III, 135) in the mountains of Liguria.

that the word *messis* is derived from this middle (*medium*) which they cut. The part of the stalk below the hand remains attached to the ground, and is cut later; while the part which is attached to the ear is carried to the threshing-floor in baskets. The name for straw, *palea*, may be derived from the fact that there, in an uncovered place, it is detached 'openly' (*palam*); some derive the other word, *stramentum*, from *stare*, as they do also the word *stamen*, and others from *stratus*, because it is 'spread' (*substernitur*) under cattle. The grain should be cut when it is ripe; and on easily worked land it is held that the reaping of one *iugerum* is approximately a day's work for one man—this should include the carrying of the reaped ears to the threshing-floor in baskets.

LI. "The threshing-floor should be on the place, in a somewhat elevated spot, so that the wind can sweep over it; the size should be determined by the size of the harvest. It should preferably be round, with a slight elevation at the centre, so that, if it rains, the water will not stand but be able to run off the floor in the shortest line—and of course in a circle the shortest line is from the centre to the circumference. It should be built of solid dirt, well packed, and especially if it is of clay, so that it may not crack in the heat and allow the grain to hide, or take in water and open the door to mice and ants. For this reason it is customary to coat it with amurca,[1] which is poison to weeds, ants, and moles. Some farmers build up the floor with stone to make it solid, or even pave it. Others, such as the Bagienni,[2] go so far as to build a shelter over the floors, because in that country rain-storms frequently occur at threshing time. When the floor

loca calda, prope aream faciundum umbracula, quo succedant homines in aestu tempore meridiano.

LII. Quae seges grandissima atque optima fuerit, seorsum in aream secerni oportet spicas, ut semen optimum habeat; e spicis in area excuti grana. Quod fit apud alios iumentis iunctis ac tribulo. Id fit e tabula lapidibus aut ferro asperata, quae cum imposito auriga aut pondere grandi trahitur iumentis iunctis, discutit[1] e spica grana; aut ex axibus dentatis cum orbiculis, quod vocant plostellum poenicum; in eo quis sedeat atque agitet quae trahant iumenta, ut in Hispania citeriore et aliis locis faciunt. 2 Apud alios exteritur grege iumentorum inacto et ibi agitato perticis, quod ungulis e spica exteruntur grana. Iis tritis oportet e terra subiectari vallis aut ventilabris, cum ventus spirat lenis. Ita fit ut quod levissimum est in eo atque appellatur acus ac palea evannatur foras extra aream ac frumentum, quod est ponderosum, purum veniat ad corbem.

LIII. Messi facta spicilegium venire oportet aut domi legere stipulam aut, si sunt spicae rarae et operae carae, compasci. Summa enim spectanda, ne in ea re sumptus fructum superet.

LIV. In vinetis uva cum erit matura, vindemiam ita fieri oportet, ut videas, a quo genere uvarum et a quo loco vineti incipias legere. Nam et praecox et

[1] *discutit* Keil: *aut discutit* (*discunt*, A) MSS.: *ut discutiat* Iucundus, Schneider.

---

[1] *vallus* seems to be a contraction for *vannulus*, diminutive of *vannus*. One MS. (R.) has *vallus* for *vannus* in *Georgics*, I, 166, and Servius quotes this passage. Columella (II, 20, 5) says that the *vannus* or fan is used when there is no wind.

is without a roof and the climate is hot, a shelter should be built hard by, to which the hands may go at midday in hot weather.

LII. "On the threshing-floor the largest and best ears should be placed apart, to furnish the best seed, and the grain should be threshed on the floor. This is done in some districts by means of a yoke of steers and a sledge. The latter is constructed either of a board made rough with stones or iron, which separates the grain from the ear when it is dragged by a yoke of steers with the driver or a heavy weight on it; or of a toothed axle running on low wheels, called a Punic cart, the driver sitting on it and driving the steers which drag it—a contrivance in use in Hither Spain and other places. Among other peoples the threshing is done by turning in cattle and driving them around with goads, the grain being separated from the beards by their hoofs. After the threshing the grain should be tossed from the ground when the wind is blowing gently, with winnowing fans [1] or forks. The result is that the lightest part of it, called *acus* and *palea*, is fanned outside the floor, while the grain, being heavy, comes clean to the basket.

LIII. "When the harvest is over the gleaning should be let, or the loose stalks gathered with your own force, or, if the ears left are few and the cost of labour high, it should be pastured. For the thing to be kept in view in this matter is that the expense shall not exceed the profit.

LIV. "As to vineyards, the vintage should begin when the grapes are ripe; and you must choose the variety of grapes and the part of the vineyard with which to begin. For the early grapes, and the hybrids,

miscella, quam vocant nigram, multo ante coquitur, quo prior legenda, et quae pars arbusti ac vineae 2 magis aprica, prius debet descendere de vite. In vindemia diligentis uva non solum legitur sed etiam eligitur; legitur ad bibendum, eligitur ad edendum. Itaque lecta defertur in forum vinarium, unde in dolium inane veniat; electa in secretam corbulam, unde in ollulas addatur et in dolia plena vinaciorum contrudatur, alia quae in piscinam in amphoram picatam descendat, alia quae in aream in carnarium escendat. Quae calcatae uvae erunt, earum scopi cum folliculis subiciendi sub prelum, ut, siquid reliqui 3 habeant musti, exprimatur in eundem lacum. Cum desiit sub prelo fluere, quidam circumcidunt extrema et rursus premunt et, rursus cum expressum, circumsicium appellant ac seorsum quod expressum est servant, quod resipit ferrum. Expressi acinorum folliculi in dolia coiciuntur, eoque aqua additur; ea vocatur lora, quod lota acina, ac pro vino operariis datur hieme.

LV. De oliveto oleam, quam manu tangere possis e terra ac scalis, legere oportet potius quam quatere, quod ea quae vapulavit macescit nec dat tantum olei. Quae manu stricta, melior ea quae digitis nudis, quam illa quae cum digitabulis, durities enim 2 eorum quod non solum stringit bacam, sed etiam ramos glubit ac relinquit ad gelicidium retectos. Quae manu tangi non poterunt, ita quati debent, ut

---

[1] *i.e.* "cut around." This is what Cato (23, 4) calls *circumcidaneum*; and Columella (XII, 36) *mustum tortivum*.

the so-called black, ripen much earlier and so must be gathered sooner; and the part of the plantation and the vineyard which is sunnier should have its vines stripped first. At the vintage the careful farmer not only gathers but selects his grapes; he gathers for drinking and selects for eating. So those gathered are carried to the wine-yard, thence to go into the empty jar; those selected are carried to a separate basket, to be placed thence in small pots and thrust into jars filled with wine dregs, while others are plunged into the pond in a jar sealed with pitch, and still others go up to their place in the larder. When the grapes have been trodden, the stalks and skins should be placed under the press, so that whatever must remains in them may be pressed out into the same vat. When the flow ceases under the press, some people trim around the edges of the mass and press again; this second pressing is called *circumsicium*,[1] and the juice is kept separate because it tastes of the knife. The pressed grape-skins are turned into jars and water is added; this liquid is called *lora*, from the fact that the skins are washed (*lota*), and it is issued to the labourers in winter instead of wine.

LV. "With regard to the olive harvest: the olives which can be reached from the ground or by ladders should be picked rather than shaken down, because the fruit which has been bruised dries out and does not yield so much oil. Those picked with bare fingers are better than those picked with gloves, as the hard gloves not only bruise the berry but also tear the bark from the branches and leave them exposed to the frost. Those which cannot be reached with the hand should be beaten down; but a reed

harundine potius quam pertica feriantur; gravior enim plaga medicum quaerit. Qui quatiet, ne adversam
3 caedat; saepe enim ita percussa olea secum defert de ramulo plantam, quo facto fructum amittunt posteri anni. Nec haec non minima causa, quod oliveta dicant alternis annis non ferre fructus aut non aeque
4 magnos. Olea ut uva per idem bivium redit in villam, alia ad cibum, alia ut eliquescat ac non solum corpus intus unguat sed etiam extrinsecus. Itaque dominum et in balneas et gymnasium sequitur.
5 Haec, de qua fit oleum, congeri solet acervatim in dies singulos in tabulata, ut ibi mediocriter fracescat, ac primus quisque acervos demittatur per serias ac vasa olearia ad trapetas, quae res molae oleariae ex
6 duro et aspero lapide. Olea lecta si nimium diu fuit in acervis, caldore fracescit et oleum foetidum fit. Itaque si nequeas mature conficere, in acervis
7 iactando ventilare oportet. Ex olea fructus duplex: oleum, quod omnibus notum, et amurca, cuius utilitatem quod ignorant plerique, licet videre e torculis oleariis fluere in agros ac non solum denigrare terram, sed multitudine facere sterilem; cum is umor modicus cum ad multas res tum ad agri culturam pertineat vehementer, quod circum arborum radices infundi solet, maxime ad oleam, et ubicumque in agro herba nocet.

should be used rather than a pole, as the heavier blow renders necessary the work of the tree-doctor. The one who is beating should not strike the olive directly; for an olive struck in this way often tears away the shoot with it, and the fruit of the next year is lost. This is not the least reason for the saying that the olive fails to bear a crop every other year, or does not bear so full a crop. The olive reaches the steading by the same two roads as the grape, one portion for food, and one to gush forth and anoint the body not only within but also without, thus following the master into the bath and into the gymnasium. The portion from which oil is made is usually heaped on a flooring in piles as it comes in from day to day, so that it may mellow a little; and the piles pass in the same order through the jars and the olive vessels to the *trapeta*, which is an olive-mill fitted with hard stones roughened on the surface. If the olives, after being picked, lie too long in the piles, they spoil from the heat and the oil becomes rancid; hence, if you cannot work them up promptly they should be aired by moving them about in the piles. The olive yields two products: oil, well known to all, and *amurca*. As most people are ignorant of the value of the latter, you may see it flowing out from the olive presses on to the fields, and not only blackening the ground but rendering it barren when there is a large quantity of it; whereas, in moderate quantities, this fluid is not only extremely valuable for many purposes, but is especially valuable in agriculture, as it is usually poured around the roots of trees, chiefly olive trees, and wherever noxious weeds grow in the fields."

LVI. Agrius, Iamdudum, inquit, in villa sedens expecto cum clavi te, Stolo, dum fructus in villam referas. Ille, Em quin adsum, venio, inquit, ad limen, fores aperi. Primum faenisiciae conduntur melius sub tecto quam in acervis, quod ita fit iucundius pabulum. Ex eo intellegitur, quod pecus utroque posito libentius est.

LVII. Triticum condi oportet in granaria sublimia, quae perflentur vento ab exortu ac septemtrionum regione, ad quae nulla aura umida ex propinquis locis adspiret. Parietes et solum opere tectorio 2 marmorato loricandi; si minus, ex argilla mixta acere e frumento et amurca, quod murem et vermem non patitur esse et grana facit solidiora ac firmiora. Quidam ipsum triticum conspargunt, cum addant in circiter mille modium quadrantal amurcae. Item alius aliut adfriat aut aspargit, ut Chalcidicam aut Caricam cretam aut absinthium, item huius generis alia. Quidam granaria habent sub terris speluncas, quas vocant sirus, ut in Cappadocia ac Thracia; alii, ut in agro Carthaginiensi et Oscensi in Hispania citeriore, puteos. Horum solum paleis substernunt et curant ne umor aut aer tangere possit, nisi cum promitur ad usum; quo enim spiritus non pervenit, ibi non oritur curculio. Sic conditum triticum manet 3 vel annos L, milium vero plus annos C. Supra terram granaria in agro quidam sublimia faciunt, ut in Hispania citeriore et in Apulia quidam, quae non solum a lateribus per fenestras, sed etiam subtus a

[1] *i.e.* near Osca in Hispania Tarraconensis, now Huesca in Aragon.

LVI. "I have been sitting in the steading for a long time," exclaimed Agrius, "waiting, key in hand, Stolo, for you to bring the crops into the barn." "Well," replied Stolo, "here I am. I am coming up to the threshold; open the doors. First, it is better to stow the hay crop under cover than in stacks, as by this method it makes better fodder—as is proved by the fact that when both kinds are offered them, cattle prefer the former.

LVII. "Wheat should be stored in granaries, above ground, open to the draught on the east and north, and not exposed to damp air rising in the vicinity. The walls and floor are to be coated with marble cement, or at least with clay mixed with grain-chaff and amurca, as this both keeps out mice and worms and makes the grain more solid and firm. Some farmers sprinkle the wheat, too, with amurca, using a quadrantal to about a thousand modii. Different farmers use different powders or sprays, such as Chalcidian or Carian chalk, or wormwood, and other things of this kind. Some use underground caves as granaries, the so-called *sirus,* such as occur in Cappadocia and Thrace; and still others use wells, as in the Carthaginian and Oscensian districts in Hither Spain.[1] They cover the bottom of these with straw, and are careful not to let moisture or air touch them, except when the grain is removed for use; for the weevil does not breed where air does not reach. Wheat stored in this way keeps as long as fifty years, and millet more than a hundred. Some people, as in Hither Spain and in Apulia, build granaries in the field, above ground, so constructed that the wind can cool them not only from the sides, through windows, but also beneath from the ground. Beans and legumes

solo ventus refrigerare possit. Faba et legumina in oleariis vasis oblita cinere perdiu incolumia servantur.

LVIII. Cato ait uvam Aminneam minusculam et maiorem et Apiciam in ollis commodissime condi; eadem in sapa et musto recte; quas suspendas opportunissimas esse duracinas et Aminneas.

LIX. De pomis conditiva, mala struthea, cotonea, Scantiana, Scaudiana, orbiculata et quae antea mustea vocabant, nunc melimela appellant, haec omnia in loco arido et frigido supra paleas posita servari recte putant. Et ideo oporothecas qui faciunt, ad aquilonem ut fenestras habeant atque ut eae perflentur curant, neque tamen sine foriculis, ne, cum umorem 2 amiserint, pertinaci vento vieta fiant; ideoque in iis camaras marmorato et parietes pavimentaque faciunt, quo frigidius sit. In quo etiam quidam triclinium sternere solent cenandi causa. Etenim in quibus luxuria concesserit ut in pinacothece faciant, quod spectaculum datur ab arte, cur non quod natura datum utantur in venustate disposita pomorum? Praesertim cum id non sit faciendum, quod quidam fecerunt, ut Romae coempta poma rus intulerint in 3 oporothecen instruendam convivi causa. In oporotheca mala manere putant satis commode alii in tabulis in opere marmorato, alii substrata palea vel etiam floccis; mala punica demissis suis surculis in dolio harenae, mala cotonea struthea in pensilibus

[1] Cato, 7.

[2] *Cotonea* (*mala*) are the κυδώνια of the Greeks, our quinces; the *struthea* were a smaller variety of the same. In the absence

are kept fresh for a very long time in olive jars sealed with ashes.

LVIII. "Cato says[1] that the smaller and larger Aminnian grape, and the Apician, are best stored in jars, and that the same grapes keep well also in boiled or plain must; and that the best ones for drying are the hard grapes and the Aminnian.

LIX. "The varieties of apples for preserving are the smaller and larger quinces,[2] the Scantian, the Scaudian, the small round, and those formerly called must-apples, but now called honey-apples. It is thought that all these keep well in a dry and cool place, laid on straw. For this reason those who build fruit-houses are careful to let them have windows facing the north and open to the wind; but they have shutters, to keep the fruit from shrivelling after losing its juice, when the wind blows steadily. And it is for this reason, too—to make it cooler—that they coat the ceilings, walls, and floors with marble cement. Some people even spread a dining-table in it to dine there; and, in fact, if luxury allows people to do this in a picture gallery, where the scene is set by art, why should they not enjoy a scene set by nature, in a charming arrangement of fruit? Provided always that you do not follow the example set by some, of buying fruit in Rome and carrying it to the country to pile it up in the fruit-gallery for a dinner-party. Some think that apples keep quite well in the gallery if placed on boards on the cement, but others lay them on straw, or even on wool; that pomegranates are preserved by burying their stems in a jar of sand, and

of scientific descriptions, it is impossible to identify varieties which have only local names.

iunctis; contra in sapa condita manere pira Aniciana sementiva; sorba quidam dissecta et in sole macerata, ut pira; et sorba per se, ubicumque sint posita in arido, facile durare; servare rapa consecta in sinape, nuces iugulandis in harena. Punica mala et in harena iam decerpta ac matura et etiam immatura, cum haereant in sua virga et demiseris in ollam sine fundo, eaque si coieceris in terram et obleris circum ramum, ne extrinsecus spiritus adflet, ea non modo integra eximi, sed etiam maiora, quam in arbore umquam pependerint.

LX. De olivitate oleas esui optime condi scribit Cato orcites et poseas vel virides in muria vel in lentisco contusas. Orcites nigras aridas, sale si sint confriatae dies quinque et tum sale excusso biduum si in sole positae fuerint, manere idoneas solere; easdem sine sale in defrutum condi recte.

LXI. Amurcam periti agricolae tam in doleis condunt quam oleum aut vinum. Eius conditio: cum expressa effluxit, statim de ea decoquuntur duae partes et refrigeratum conditur in vasa. Sunt item aliae conditiones, ut ea in qua adicitur mustum.

LXII. Quod nemo fructus condit, nisi ut promat, de eo quoque vel sexto gradu animadvertenda pauca. Promunt condita aut propterea quod sunt tuenda, aut quod utenda, aut quod vendunda. Ea quod

---

[1] The text seems hopelessly corrupt, and *iunctis* is translated as if it were *iuncis*. Pliny (*N.H.*, XV, 60) expressly warns against the practice here recommended: no breath of air should touch them.

[2] Literally: ripening to seed time. [3] Cato, 7. 4.

large and small quinces in hanging baskets;[1] while, on the other hand, late[2] Anician pears keep best when put down in boiled must. Some hold that sorbs keep best when cut up and dried in the sun, like pears; and that sorbs are easily kept just as they are, wherever they are put, if the place is dry; that rape should be cut up and preserved in mustard, walnuts in sand. Pomegranates are also kept in sand if they are stored freshly gathered and ripe; green ones also, if you keep them on the branch, put them in a pot with no bottom, bury them in the ground, and seal the ends of the branches so that no outside air can reach them; such fruits will be taken out not only sound but even larger than they would ever be if they had hung on the tree.

LX. "Of the olives, Cato writes[3] that the table olives, the orcites, and the posea are best preserved either green in brine, or, when bruised, in mastic oil. The black orcites, if they are covered with salt for five days after being dried, and then, after the salt has been shaken off, are exposed to the sun for two days, usually keep sound; and that the same varieties may be satisfactorily preserved unsalted in boiled must.

LXI. "Experienced farmers store their amurca in jars, just as they do oil and wine. The method of preserving is: as soon as it flows out from the press, two-thirds of it is boiled away, and when it is cool it is stored in vessels. There are also other methods, such as that in which must is added.

LXII. "As no one stores the products of the farm except to bring them forth later, a few remarks must be made about this, the sixth step. Preserved things are brought out of storage because they are to be either protected or consumed or sold. As the

dissimilia sunt inter se, aliut alio tempore tuendum et utendum.

LXIII. Tuendi causa promendum id frumentum, quod curculiones exesse incipiunt. Id enim cum promptum est, in sole ponere oportet aquae catinos, quod eo conveniunt, ut ipsi se necent, curculiones. Sub terra qui habent frumentum in iis quos vocant sirus, quod cum periculo introitur recenti apertione, ita ut quibusdam sit interclusa anima, aliquanto post promere, quam aperueris, oportet. Far, quod in spicis condideris per messem et ad usus cibatus expedire velis, promendum hieme, ut in pistrino pisetur ac torreatur.

LXIV. Amurca cum ex olea expressa, qui est umor aquatilis, ac retrimentum conditum in vas fictile. Id quidam sic solent tueri; diebus XV in eo quod est levissimum ac summum deflatum ut traiciant in alia vasa, et hoc isdem intervallis duodeciens sex mensibus proximis item faciant; cum id novissime, potissimum traiciant, cum senescit luna. Tum decocunt in ahenis leni igni, ad duas partes quaad redegerunt. Tum denique ad usum recte promitur.

LXV. Quod mustum conditur in dolium, ut habeamus vinum, non promendum dum fervet, neque etiam cum processit ita, ut sit vinum factum. Si vetus bibere velis, quod non fit, antequam accesserit annus;[1] anniculum prodit. Si est vero ex eo genere uvae,

[1] *accesserit annus* Iucundus: *accesserunt.*

three operations are for different purposes, the protecting and the consuming take place at different times.

LXIII. "Grain which the weevil has begun to infest should be brought out for protection. When it is brought out, bowls of water should be placed around in the sun; the weevils will congregate at these and drown themselves. Those who keep their grain under ground in the pits which they call *sirus* should remove the grain some time after the pits are opened, as it is dangerous to enter them immediately, some people having been suffocated while doing so. Spelt which you have stored in the ear at harvest-time and wish to prepare for food should be brought out in winter, so that it may be ground in the mill and parched.

LXIV. " Amurca, which is a watery fluid, after it is pressed from the olives is stored along with the dregs in an earthenware vessel. Some farmers use the following method for preserving it: After fifteen days the dregs which, being lighter, have risen to the top are blown off, and the fluid is turned into other vessels; this operation is repeated at the same intervals twelve times during the next six months, the last cleansing being done preferably when the moon is waning. Then they boil in copper vessels over a slow fire until it is reduced to two-thirds its volume. It is then fit to be drawn off for use.

LXV. " Must which is stored in jars to make wine should not be brought out while it is fermenting, and not even after the fermentation has gone far enough to make wine. If you wish to drink old wine (and wine is not old enough until a year has been added to its age), it should be brought out when it is a year old. But if it is of the variety of grapes that sours

quod mature cocescat, ante vindemiam consumi aut venire oportet. Genera sunt vini, in quo Falerna, quae quanto pluris annos condita habuerunt, tanto, cum prompta, sunt fructuosiora.

LXVI. Oleas albas quas condideris, novas si celeriter promas, propter amaritudinem respuit palatum; item nigras, nisi prius eas sale maceraris, ut libenter in os recipiantur.

LXVII. Nucem iuglandem et palmulam et ficum Sabinam quanto citius promas, iucundiore utare, quod vetustate ficus fit pallidior, palmula cariosior, nux aridior.

LXVIII. Pensilia, ut uvae, mala et sorba, ipsa ostendunt, quando ad usum oporteat promi, quod colore mutato et contractu acinorum, si non dempseris ad edendum, ad abiciendum descensurum se minitantur. Sorbum maturum mite conditum citius promi oportet; acerbum enim suspensum lentius est, quod prius domi maturitatem adsequi vult, quam nequit in arbore, quam mitescat.

LXIX. Messum far promendum hieme in pistrino ad torrendum, quod ad cibatum expeditum esse velis; quod ad sationem, tum promendum, cum segetes maturae sunt ad accipiendum. Item quae pertinent ad sationem, suo quoque tempore promenda. Quae vendenda videndum, quae quoque tempore oporteat promi; alia enim, quae manere non possunt, antequam se commutent, ut celeriter promas ac vendas;

quickly, it must be used up or sold before the next vintage. There are brands of wine, the Falernian for instance, which are the more valuable when brought out the more years you have kept them in store.

LXVI. "If you take out the preserved white olives soon, while they are fresh, the palate will reject them because of the bitter taste; and likewise the black olives, unless you first steep them in salt so that they may be taken into the mouth without distaste.

LXVII. "As for the walnut, the date, and the Sabine fig, the sooner you use them the better the flavour; for with time the fig gets too pale, the date too soft, and the nut too dry.

LXVIII. "Fruits that are hung, such as grapes, apples, and sorbs, themselves indicate the proper time for consumption; for by the change of colour and the shrivelling of the skin they put you on notice that if you do not take them down to eat they will come down to be thrown away. If you store sorbs which are ripe and soft, you must use them quickly; those hung up when sour may wait, for they mean, before ripening, to reach in the house a degree of maturity which they cannot reach on the tree.

LXIX. "The part of the spelt harvest which you wish to have ready for food should be taken out in winter to be roasted at the mill; while the part reserved for seed should be taken out when the land is ready to receive it. With regard to seed in general, each kind should be taken out at its proper time. As to the crops intended for market, care must be used as to the proper time for taking out each; thus you should take out and sell at once those which do not

alia, quae servari possunt, ut tum vendas, cum caritas est. Saepe enim diutius servata non modo usuram adiciunt, sed etiam fructum duplicant, si tempore promas.

2 Cum haec diceret, venit libertus aeditumi ad nos flens et rogat ut ignoscamus, quod simus retenti, et ut ei in funus postridie prodeamus. Omnes consurgimus ac simul exclamamus, "Quid? in funus? quod funus? quid est factum?" Ille flens narrat ab nescio quo percussum cultello concidisse, quem qui esset animadvertere in turba non potuisse, sed tan-3 tum modo exaudisse vocem, perperam fecisse. Ipse cum patronum domum sustulisset et pueros dimisisset, ut medicum requirerent ac mature adducerent, quod potius illut administrasset, quam ad nos venisset, aecum esse sibi ignosci. Nec si eum servare non potuisset, quin non multo post animam efflaret, tamen putare se fecisse recte. Non moleste ferentes descendimus de aede et de casu humano magis querentes, quam admirantes id Romae factum, discedimus omnes.

---

[1] The text deliberately expresses the excitement of the freedman. Of course, the funeral was that of his master.

[2] The time of the narrative is placed by Varro at a period of unusual turbulence even in Roman history—the closing years of the Republic.

stand storage before they spoil, while you should sell those which keep well when the price is high. For often products which have been stored quite a long time will not only pay interest on the storage, but even double the profit if they are marketed at the right time."

While he was speaking the sacristan's freedman runs up to us with tears in his eyes and begs us to pardon him for keeping us so long, and asks us to go to a funeral for him [1] the next day. We spring to our feet and cry out in chorus: "What? To a funeral? What funeral? What has happened?" Bursting into tears, he tells us that his master had been stabbed with a knife by someone, and had fallen to the ground; that in the crowd he could not tell who it was, but had only heard a voice saying that a mistake had been made. As he had carried his old master home and sent the servants to find a surgeon and bring him with all speed, he hoped he might be pardoned for attending to his duty rather than coming to us; and though he had not been able to keep him from breathing his last a few moments later, he thought that he had acted rightly. We had no fault to find with him, and walking down from the temple we went our several ways, rather lamenting the mischances of life than being surprised that such a thing had occurred in Rome.[2]

BOOK II

LIBER SECUNDUS

VIRI magni nostri maiores non sine causa praeponebant rusticos Romanos urbanis. Ut ruri enim qui in villa vivunt ignaviores, quam qui in agro versantur in aliquo opere faciendo, sic qui in oppido sederent, quam qui rura colerent, desidiosiores putabant. Itaque annum ita diviserunt, ut nonis modo diebus urbanas res usurparent, reliquis septem ut 2 rura colerent. Quod dum servaverunt institutum, utrumque sunt consecuti, ut et cultura agros fecundissimos haberent et ipsi valetudine firmiores essent, ac ne Graecorum urbana desiderarent gymnasia. Quae nunc vix satis singula sunt, nec putant se habere villam, si non multis vocabulis retinniat[1] Graecis, quom vocent particulatim loca, procoetona, palaestram, apodyterion, peristylon, ornithona, perip- 3 teron, oporothecen. Igitur quod nunc intra murum fere patres familiae correpserunt relictis falce et aratro et manus movere maluerunt in theatro ac circo, quam in segetibus ac vinetis, frumentum locamus qui nobis advehat, qui saturi fiamus ex Africa et Sardinia, et navibus vindemiam condimus ex insula Coa et Chia.

[1] *retinniat* Gesner; *retineant.*

---

[1] According to the Roman method of counting, which includes both ends of the series. He is alluding to the *nundina* or market-day, the last day of the eight-day week.

BOOK II

It was not without reason that those great men, our ancestors, put the Romans who lived in the country ahead of those who lived in the city. For as in the country those who live in the villa are lazier than those who are engaged in carrying out work on the land, so they thought that those who settled in town were more indolent than those who dwelt in the country. Hence they so divided the year that they attended to their town affairs only on the ninth days, and dwelt in the country on the remaining seven.[1] So long as they kept up this practice they attained both objects—keeping their lands most productive by cultivation, and themselves enjoying better health and not requiring the citified gymnasia of the Greeks. In these days one such gymnasium is hardly enough, and they do not think they have a real villa unless it rings with many resounding Greek names—places severally called *procoetion* (ante-room), *palaestra* (exercise-room), *apodyterion* (dressing-room), *peristylon* (colonnade), *ornithon* (aviary), *peripteros* (pergola), *oporotheca* (fruit-room). As therefore in these days practically all the heads of families have sneaked within the walls, abandoning the sickle and the plough, and would rather busy their hands in the theatre and in the circus than in the grain-fields and the vineyards, we hire a man to bring us from Africa and Sardinia the grain with which to fill our stomachs, and the vintage we store comes in ships from the islands of Cos and Chios.

4 Itaque in qua terra culturam agri docuerunt pastores progeniem suam, qui condiderunt urbem, ibi contra progenies eorum propter avaritiam contra leges ex segetibus fecit prata, ignorantes non idem esse agri culturam et pastionem. Alius enim opilio et arator, nec, si possunt in agro pasci armenta, armentarius non aliut ac bubulcus. Armentum enim id quod in agro natum non creat, sed tollit dentibus; contra bos domitus causa fit ut commodius nascatur
5 frumentum in segete et pabulum in novali. Alia, inquam, ratio ac scientia coloni, alia pastoris: coloni ea quae agri cultura factum ut nascerentur e terra, contra pastoris ea quae nata ex pecore. Quarum quoniam societas inter se magna, propterea quod pabulum in fundo compascere quam vendere plerumque magis expedit domino fundi et stercoratio ad fructus terrestres aptissima et maxume ad id pecus appositum, qui habet praedium, habere utramque debet disciplinam, et agri culturae et pecoris pascendi, et etiam villaticae pastionis. Ex ea enim quoque fructus tolli possunt non mediocres ex orni-
6 thonibus ac leporariis et piscinis. E quis quoniam de agri cultura librum Fundaniae uxori propter eius fundum feci, tibi, Niger Turrani noster, qui vehe-

---

[1] A tendency already noted in Cato, Chapter 2, had increased; and by the time of Columella (I, 3) had grown into a peril.

[2] Practically all ploughing and farm work was done by oxen, which had to be broken to the yoke (*boves domiti*). These were kept shut up, and it is their manure (as may be seen on the Continent to-day) which enriches the land which produces both field-crops and forage. What the *armentum* (which seems to include sheep) does is to produce young animals.

[3] *i.e.* Book I; but there he promised all three books to his wife.

And so, in a land where the shepherds who founded the city taught their offspring the cultivation of the earth, there, on the contrary, their descendants, from greed and in the face of the laws, have made pastures out of grain lands [1]—not knowing that agriculture and grazing are not the same thing. For the shepherd is one thing and the ploughman another; and it does not follow that because cattle can graze in a field the herdsman is the same as the ploughman. For grazing cattle do not produce what grows on the land, but tear it off with their teeth; while on the other hand the domestic ox becomes the cause why the grain grows more easily in the ploughed land, and the fodder in the fallow land.[2] The skill and knowledge of the farmer, I repeat, are one thing, and those of the herdsman another; in the province of the farmer are those things which are made to spring from the earth by cultivation of the land; in that of the herdsman, however, those that spring from the herd. As the association between them is very close, inasmuch as it is frequently more profitable to the owner of the farm to feed the fodder on the place than to sell it, and inasmuch as manure is admirably adapted to the fruits of the earth, and cattle especially fitted to produce it, one who owns a farm ought to have a knowledge of both pursuits, agriculture and cattle-raising, and also of the husbandry of the steading. For from it, too, no little revenue can be derived—from the poultry-yards, the rabbit-hutches, and the fishponds. And since I have written a book [3] for my wife, Fundania, on one of these subjects, that of agriculture, on account of her owning a farm, for you, my dear Turranius Niger, who take keen delight

menter delectaris pecore, propterea quod te emptu- rientem in campos Macros ad mercatum adducunt crebro pedes, quo facilius sumptibus multa poscen- tibus ministres, quod eo facilius faciam, quod et ipse pecuarias habui grandes, in Apulia oviarias et in Reatino equarias, de re pecuaria breviter ac summatim percurram ex sermonibus nostris col- latis cum iis qui pecuarias habuerunt in Epiro magnas, tum cum piratico bello inter Delum et Siciliam Graeciae classibus praeessem. Incipiam hinc. . . .[1]

I. Cum Menates discessisset, Cossinius mihi, Nos te non dimittemus, inquit, antequam illa tria expli- caris, quae coeperas nuper dicere, cum sumus inter- pellati. Quae tria? inquit Murrius, an ea quae mihi here dixisti de pastoricia re? Ista, inquit ille, quae coeperat hic disserere, quae esset origo, quae 2 dignitas, quae ars . . .[2] Ego vero, inquam, dicam dumtaxat quod est historicon, de duabus rebus primis quae accepi, de origine et dignitate, de tertia parte, ubi est de arte, Scrofa suscipiet, ut semigraecis pastoribus dicam graece, *ὅς πέρ μου πολλὸν ἀμείνων*. Nam is magister C. Lucili Hirri, generi tui, cuius nobiles pecuariae in Bruttiis haben- tur. Sed haec ita a nobis accipietis, inquit Scrofa,

[1] Goetz, following the MSS., reads *incipiam hinc* [*hic intermisimus*]. The bracketed words, in capitals in the MSS., were perhaps inserted by some copyist, as Storr-Best suggests, in attempted facetiousness or in despair of a corrupt text.

[2] *Locus desperatus*, despite the efforts of many editors to restore the reading of the archetype. Goetz gives, with Keil, an emended reading omitted above, but marks with a dagger: *cum poetam sesum visere venissemus, ni medici adventus nos inpedisset.*

---

[1] A district in Cisalpine Gaul.

[2] In 67 B.C. Varro gained the *corona rostrata* for his services.

in cattle, inasmuch as your feet often carry you, on buying bent, to market at Campi Macri,[1] that you may more easily meet the outlay incurred by the many demands made upon you, I shall run over briefly and summarily the subject of cattle-raising; and I shall the more readily do this because I have myself owned large stocks of cattle, sheep in Apulia and horses in the district of Reate. I shall take as the foundation the conversations I had with extensive cattle-owners in Epirus, at the time when, during the war with the pirates,[2] I was in command of the Greek fleets operating between Delos and Sicily. At this point I shall begin. . . .

I. When Menates had left, Cossinius remarked to me: "We shall not let you go until you have set forth those three topics of which you had begun to speak when we were interrupted." "Which three?" inquired Murrius; "do you mean what you were saying to me yesterday about animal husbandry?" "The points our friend here had begun to discuss," said he, "the origin, the dignity, and the science. . ."[3] "Well," said I, "I shall speak at least of the historical side and tell what I have learned of the two topics first mentioned—the origin and the dignity. Scrofa will take up the tale at the third division, when it becomes a science. He is, if I may quote Greek to half-Greek shepherds, 'a much better man than I am.' For he is the man who taught your son-in-law, Gaius Lucilius Hirrus, whose flocks in the country of the Bruttii are renowned." "But you shall have this discussion by us," said Scrofa, "only on condition that you, who

[3] It is generally agreed that there is at this point a considerable lacuna, providing the setting for the new scene and introducing the new speakers. See also critical note 2.

ut vos, qui estis Epirotici pecuariae athletae, remuneremini nos ac quae scitis proferatis in medium;
3 nemo enim omnia potest scire. Cum accepissem condicionem et meae partes essent primae, non quo non ego pecuarias in Italia habeam, sed non omnes qui habent citharam sunt citharoedi: Igitur, inquam, et homines et pecudes cum semper fuisse sit necesse natura — sive enim aliquod fuit principium generandi animalium, ut putavit Thales Milesius et Zeno Citieus, sive contra principium horum extitit nullum, ut credidit Pythagoras Samius et Aristoteles Stagerites — necesse est humanae vitae ab summa memoria gradatim descendisse ad hanc aetatem, ut scribit Dicaearchus, et summum gradum fuisse naturalem,
4 cum viverent homines ex his rebus, quae inviolata ultro ferret terra, ex hac vita in secundam descendisse pastoriciam, e feris atque agrestibus ut arboribus ac virgultis decarpendo glandem, arbutum, mora, poma colligerent ad usum, sic ex animalibus cum propter eandem utilitatem, quae possent, silvestria deprenderent ac concluderent et mansuescerent. In quis primum non sine causa putant oves assumptas et propter utilitatem et propter placiditatem; maxime enim hae natura quietae et aptissimae ad vitam hominum. Ad cibum enim lacte et caseum adhibi-
5 tum, ad corpus vestitum et pelles adtulerunt. Tertio denique gradu a vita pastorali ad agri culturam descenderunt, in qua ex duobus gradibus superioribus retinuerunt multa, et quo descenderant, ibi processerunt longe, dum ad nos perveniret. Etiam nunc

---

[1] Thales made this principle water, and Zeno made it fire. Pythagoras taught an unending transmigration of souls, and Aristotle that they exist by φύσις, nature, as a generating cause.

are the cattle-raising champions of Epirus, shall repay us by disclosing what you know of the subject; for no man can know everything." "When I had accepted the proposal and was to open the play—not that I do not own flocks myself in Italy, but not all who own a harp are harpers"—I began: "As it is a necessity of nature that people and flocks have always existed (whether there was an original generating principle of animals, as Thales of Miletus and Zeno of Citium thought, or, on the contrary, as was the view of Pythagoras of Samos and of Aristotle of Stagira, there was no point of beginning for them),[1] it is a necessity that from the remotest antiquity of human life they have come down, as Dicaearchus teaches, step by step to our age, and that the most distant stage was that state of nature in which man lived on those products which the virgin earth brought forth of her own accord; they descended from this stage into the second, the pastoral, in which they gathered for their use acorns, arbutus berries, mulberries, and other fruits by plucking them from wild and uncultivated trees and bushes, and likewise caught, shut up, and tamed such wild animals as they could for the like advantage. There is good reason to suppose that, of these, sheep were first taken, both because they are useful and because they are tractable; for these are naturally most placid and most adapted to the life of man. For to his food they brought milk and cheese, and to his body wool and skins for clothing. Then by a third stage man came from the pastoral life to that of the tiller of the soil; in this they retained much of the former two stages, and after reaching it they went far before reaching

in locis multis genera pecudum ferarum sunt aliquot, ab ovibus, ut in Phrygia, ubi greges videntur complures, in Samothrace caprarum, quas Latine rotas appellant. Sunt enim in Italia circum Fiscellum et Tetricam montes multae. De subus nemini ignotum, nisi qui apros non putat sues vocari. Boves perferi etiam nunc sunt multi in Dardanica et Maedica et Thracia, asini feri in Phrygia et Lycaonia, equi feri in Hispania citeriore regionibus aliquot.

6 Origo, quam dixi; dignitas, quam dicam. De antiquis illustrissimus quisque pastor erat, ut ostendit et Graeca et Latina lingua et veteres poetae, qui alios vocant polyarnas, alios polymelos, alios polybutas; qui ipsas pecudes propter caritatem aureas habuisse pelles tradiderunt, ut Argis, Atreus quam sibi Thyesten subduxe queritur; ut in Colchide ad Aeetam, ad cuius arietis pellem profecti regio genere dicuntur Argonautae; ut in Libya ad Hesperidas, unde aurea mala, id est secundum antiquam consuetudinem capras et oves, Hercules ex Africa in Graeciam exportavit. Ea enim a sua voce Graeci

7 appellarunt mela. Nec multo secus nostri ab eadem voce, sed ab alia littera (vox earum non me, sed be sonare videtur), oves baelare [1] vocem efferentes, e quo post balare dicunt extrita littera, ut in multis. Quod si apud antiquos non magnae dignitatis pecus

[1] *baelare* Friedrich: *balare*.

---

[1] The text is clearly corrupt. Scaliger conjectures *platycerotae* (with spreading horns); Schneider suggests *strepsicerotae* (with twisted horns). But Keil notes that some Latin term would seem to be indicated because of the following clause.

[2] In Macedonia, on the western bank of the Strymon river; cf. Livy, XXVI, 25, 8.

[3] *i.e.* πολυάρνους; πολυμήλους; πολυβούτας.

[4] *i.e.* μῆλα.

our stage. Even now there are several species of wild animals in various places: as of sheep in Phrygia, where numerous flocks are seen, and in Samothrace those goats which are called in Latin *rotae*[1]; for there are many wild goats in Italy in the vicinity of Mount Fiscellum and Mount Tetrica. As to swine, everybody knows—except those who think that wild boars ought not to be called swine. There are even now many quite wild cattle in Dardania, Maedica,[2] and Thrace; wild asses in Phrygia and Lycaonia, and wild horses at several points in Hither Spain.

"The origin is as I have given it; the dignity, as I shall now show. Of the ancients the most illustrious were all shepherds, as appears in both Greek and Latin literature, and in the ancient poets, who call some men 'rich in flocks,' others 'rich in sheep,' others 'rich in herds';[3] and they have related that on account of their costliness some sheep actually had fleeces of gold—as at Argos the one which Atreus complains that Thyestes stole from him; or as in the realm of Aeetes in Colchis, the ram in search of whose golden fleece the Argonauts of royal blood are said to have fared forth; or as among the Hesperides in Libya, from which Hercules brought from Africa to Greece golden *mala*,[4] which is the ancient manner of naming goats and sheep. For the Greeks called these *mela* from the sound of their bleating; and in fact our people give them a similar name from the same bleating, but with a different consonant (for the bleating seems to give the sound *be*, and not *me*), and they call the bleating of sheep *baelare*, and hence later, by the excision of a letter, as occurs in many words, *balare*. But if the flock had not been held in high honour among the ancients, the

esset, in caelo describendo astrologi non appellassent eorum vocabulis signa, quae non modo non dubitarunt ponere, sed etiam ab iis principibus duodecim signa multi numerant, ab ariete et tauro, cum ea praeponerent Apollini et Herculi. Ii enim dei ea
8 secuntur, sed appellantur Gemini. Nec satis putarunt de duodecim signis sextam partem obtinere pecudum nomina, nisi adiecissent, ut quartam tenerent, capricornum. Praeterea a pecuariis addiderunt capram, haedos, canes. An non etiam item in mari terraque ab his regionibus notae, in mari, quod nominaverunt a capris Aegaeum pelagus, ad Syriam montem Taurum, in Sabinis Cantherium montem, Bosporum unum Thracium, alterum Cimmerium?
9 Nonne in terris multa, ut oppidum in Graecia Hippion Argos? Denique non Italia a vitulis, ut scribit Piso? Romanorum vero populum a pastoribus esse ortum quis non dicit? quis Faustulum nescit pastorem fuisse nutricium, qui Romulum et Remum educavit? Non ipsos quoque fuisse pastores obtinebit, quod Parilibus potissimum condidere urbem? Non idem, quod multa etiam nunc ex vetere instituto bubus et ovibus dicitur, et quod aes antiquissimum quod est flatum pecore est notatum, et quod, urbs cum condita est,
10 tauro et vacca qua essent muri et portae definitum,

---

[1] The Egyptians represented the constellation by two kids. The Greeks altered the symbol to two children, variously said to be Castor and Pollux, Apollo and Hercules, or Triptolemus and Jason. [2] *i.e.* *αἶγες*. [3] "Bull."

[4] = *κανθήλιος*, a beast of burden. [5] *i.e.* *βόσπορος*.

[6] Another explanation is given in Chapter 5, 3, below.

[7] The shepherds' festival, held on April 21. See Ovid, *Fasti*, IV, 721–782.

[8] The *as*, first coined and stamped with figures of cattle by Servius Tullius; cf. Pliny, *N.H.*, XVIII, 12.

astronomers, in laying out the heavens, would not have called by their names the signs of the zodiac; they not only did not hesitate to give such names, but many of them begin their enumeration of the twelve signs with the names of the Ram and the Bull, placing them ahead of Apollo and Hercules. For those gods follow them, but are called the Twins.[1] And they were not content to have one-sixth of the twelve signs bear the names of domestic animals, but added Capricornus, so that one-fourth might have them. And besides this, they added of the domestic animals the she-goat, the kid, and the dog. Or are not tracts on land and sea also known by the names of animals? For instance, they named a sea Aegean from the word for goats,[2] a mountain on the border of Syria, Taurus,[3] a mountain in the Sabine country, Cantherius,[4] and two straits Bosphorus (ox-ford)[5]—the Thracian and the Cimmerian. Did they not give such names to many places on land, as, for instance, the city in Greece called 'Hippion (horse-rearing) Argos'? And, finally, is not Italy named from *vituli* (bullocks), as Piso states?[6] Further, does not everyone agree that the Roman people is sprung from shepherds? Is there anyone who does not know that Faustulus, the foster-father who reared Romulus and Remus, was a shepherd? Will not the fact that they chose exactly the Parilia[7] as the time to found a city demonstrate that they were themselves shepherds? Is not the same thing proved by the following facts: that up to this day a fine is assessed after the ancient fashion in oxen and sheep; that the oldest copper coins are marked with cattle;[8] that when the city was founded the position of walls and gates was marked out by a bull and a

et quod, populus Romanus cum lustratur suovitaurilibus, circumaguntur verres aries taurus, et quod nomina multa habemus ab utroque pecore, a maiore et a minore — a minore Porcius, Ovinius, Caprilius; sic a maiore Equitius, Taurius, Asinius — et idem cognomina adsignificare quod dicuntur, ut Anni Caprae, Statili Tauri, Pomponi Vituli, sic a pecudibus alia multa?

11 Relicum est de scientia pastorali, de qua est dicendum, quod Scrofa noster, cui haec aetas defert rerum rusticarum omnium palmam, quo melius potest, dicet. Cum convertissent in eum ora omnes, Scrofa, Igitur, inquit, est scientia pecoris parandi ac pascendi, ut fructus quam possint maximi capiantur ex eo, a quibus ipsa pecunia nominata est; nam 12 omnis pecuniae pecus fundamentum. Ea partes habet novem, discretas ter ternas, ut sit una de minoribus pecudibus, cuius genera tria, oves capra sus, altera de pecore maiore, in quo sunt item ad tres species natura discreti, boves asini equi. Tertia pars est in pecuaria quae non parantur, ut ex iis capiatur fructus, sed propter eam aut ex ea sunt, muli canes pastores. Harum una quaeque in se generalis partis habet minimum novenas, quarum in pecore parando necessariae quattuor, alterae in pascendo totidem; praeterea communis una. Ita fiunt omnium partes minimum octoginta et una, et

[1] Cato, Chapter 141, describes the ceremony, and gives the formula used.

[2] The names are from the words for pig, sheep, goat, horse, bull, ass, respectively.

[3] Varro's usual pun on a proper name; see Chapter 4 of this book, where Scrofa explains the origin of the name.

[4] *pecus*, from which comes *pecunia* as here used, not of a "flock," but of all sorts of domestic animals.

cow; that when the Roman people is purified by the *suovetaurilia*,[1] a boar, a ram, and a bull are driven around; that many of our family names are derived from both classes, the larger and the smaller, such as Porcius, Ovinius, Caprilius from the smaller, and Equitius, Taurius, Asinius from the larger;[2] and that the so-called *cognomina* [surnames] prove the same thing, as, for instance, the Annii Caprae, the Statilii Tauri, the Pomponii Vituli, and many others, derived from the names of domestic animals?

"It remains to speak of the science of animal husbandry, and our friend, Scrofa,[3] to whom this generation presents the palm in all agricultural matters, and who is therefore better fitted, will discuss it." When the eyes of all were turned on him Scrofa began:—"Well, there is a science of assembling and feeding cattle in such fashion as to secure the greatest returns from them; the very word for money is derived from them, for cattle[4] are the basis of all wealth. The science embraces nine divisions under three topics of three divisions each: the topic of the smaller animals, with its three divisions, sheep, goats, swine; the second topic, that of the larger animals, with likewise its three classes naturally separate, oxen, asses, horses. The third topic comprises animals which are kept not for the profit derived from them, but for the purpose of the above groups, or as a result of them, mules, dogs, and herdsmen. Each one of these divisions includes at least nine general subdivisions; and of these four are necessarily involved in assembling and an equal number in feeding; while one is common to both. There are, then, at the lowest 81 subdivisions, all of them important and not one insignificant.

13 quidem necessariae nec parvae. Primum ut bonum pares pecus, unum scire oportet, qua aetate quamque pecudem parare habereque expediat. Itaque in bubulo pecore minoris emitur annicula et supra decem annorum, quod a bima aut trima fructum ferre incipit neque longius post decimum annum
14 procedit. Nam prima aetas omnis pecoris et extrema sterilis. E quattuor primis altera pars est cognitio formae unius cuiusque pecudis, qualis sit. Magni enim interest, cuius modi quaeque sit ad fructum. Ita potius bovem emunt cornibus nigrantibus quam albis, capram amplam quam parvam, sues procero corpore, capitibus ut sint parvis. Tertia pars est, quo sit seminio quaerendum. Hoc nomine enim asini Arcadici in Graecia nobilitati, in Italia Reatini, usque eo ut mea memoria asinus venierit sestertiis milibus sexaginta et unae quadrigae Romae constiterint
15 quadringentis milibus. Quarta pars est de iure in parando, quem ad modum quamque pecudem emi oporteat civili iure. Quod enim alterius fuit, id ut fiat meum, necesse est aliquid intercedere, neque in omnibus satis est stipulatio aut solutio nummorum ad mutationem domini. In emptione alias stipulandum sanum esse, alias e sano pecore, alias neu-
16 trum.

Alterae partes quattuor sunt, cum iam emeris, observandae, de pastione, de fetura, de nutricatu, de sanitate. Pascendi primus locus qui est, eius

First: in order to assemble a sound flock one must know one item—at what age it is profitable to get and keep each several kind. Thus, in the matter of cattle, they can be purchased at a lower price below the age of one year and beyond that of ten years, for they begin to yield a profit after the age of two or three, and do not continue to do so much beyond the age of ten years—the earliest youth and extreme age of all animals being barren. The second of the first four heads is a knowledge of the proper characteristics of each species of animal, as this has a very important bearing on the profit. Thus, one buys an ox with dark rather than with white horns, a full-bodied she-goat rather than a thin one, and swine with long bodies provided the head be small. The third point of inquiry is as to the breed; it is for this reason that in Greece the asses of Arcadia are noted, and in Italy those of Reate—so much so that within my recollection an ass fetched 60,000 sesterces, and one team of four at Rome sold for 400,000. The fourth topic is the law of purchase—the proper legal form to be followed in the purchase of each separate species. In order that the property of another may become mine an intermediate step is necessary, and not in all purchases is an agreement or the payment of money sufficient to effect a change of ownership; and in a purchase it is sometimes to be stipulated that the animal is sound, sometimes that it is from a sound flock, while at other times neither stipulation is made.

"After the purchase has been made we come to the second group of four points which are to be observed: they are those concerned with pasturage, breeding, feeding, and health. Of pasturage, which

ratio triplex, in qua regione quamque potissimum pascas et quando et qui, ut capras in montuosis potius locis fruticibus quam in herbidis campis, equas contra. Neque eadem loca aestiva et hiberna idonea omnibus ad pascendum. Itaque greges ovium longe abiguntur ex Apulia in Samnium aestivatum atque ad publicanum profitentur, ne, si inscriptum pecus 17 paverint, lege censoria committant. Muli e Rosea campestri aestate exiguntur in Burbures[1] altos montes. Qui potissimum quaeque pecudum pascatur, habenda ratio, nec solum quod faeno fit satura equa aut bos, cum sues hoc vitent et quaerant glandem, sed quod hordeum et faba interdum sit quibusdam obiciendum et dandum bubus lupinum et lactariis medica et cytisum; praeterea quod ante admissuram diebus triginta arietibus ac tauris datur plus cibi, ut vires habeant, feminis bubus demitur, quod macescentes melius concipere dicuntur. 18 Secunda pars est de fetura. Nunc appello feturam a conceptu ad partum; hi enim praegnationis primi et extremi fines. Quare primum videndum de admissione, quo quaeque tempore ut ineant facere oporteat. Nam ut suillo pecori a favonio ad aequinoctium vernum putant aptum, sic ovillo ab arcturi occasu usque ad aquilae occasum. Praeterea ha-

[1] No such mountains are known, and the word is certainly corrupt.

---

[1] Flocks so registered could be pastured on public lands on payment of the *scriptum* (registration tax).

[2] A very fertile plain near Reate. See Book I, 7, 10.

[3] See critical note.

[4] For the dates see Book I, Chapter 28.

is the first point, there are three divisions: the preferable locality for the pasturage of the several species, the time, and the manner; thus, it is better to pasture goats on a bushy hillside than on a grassy plain, while the opposite is true of mares. Again, the same localities are not equally suited in summer and winter to the pasturing of all species. Hence, flocks of sheep are driven all the way from Apulia into Samnium for summering, and are reported to the tax-collectors, for fear of offending against the censorial regulation forbidding the pasturing of unregistered[1] flocks; and mules are driven in summer from the level Rosea[2] into the high mountains of Burbur.[3] One must also consider the preferable method of pasturing each species—by which I do not mean merely that a horse or an ox is content with hay, while swine will have none of it and seek mast, but that barley and beans should be fed at intervals to some of them, and that lupines should be fed to oxen, and alfalfa and clover to milch cows; and besides that for thirty days before mating more food should be given rams and bulls to increase their vigour, and food should be lessened for the females, because it is claimed that they conceive more readily when they are thin. The second topic is that of breeding—by which I mean the process from conception to birth, these being the limits of pregnancy. The first point to be observed, therefore, is that of mating—the time at which opportunity for coition should be allowed each species; thus, the period from the beginning of the west wind to the vernal equinox[4] is deemed best suited to swine, while that from the setting of Arcturus to the setting of Aquila is considered best for sheep. Consideration should

benda ratio, quanto antequam incipiat admissura fieri mares a feminis secretos habeant, quod fere in omnibus binis mensibus ante faciunt et armentarii et
19 opiliones. Altera pars est, in fetura quae sint observanda, quod alia alio tempore parere solet. Equa enim ventrem fert duodecim menses, vacca decem, ovis et capra quinos, sus quattuor. In fetura res incredibilis est in Hispania, sed est vera, quod in Lusitania ad oceanum in ea regione, ubi est oppidum Olisipo, monte Tagro quaedam e vento concipiunt certo tempore equae, ut hic gallinae quoque solent, quarum ova hypenemia appellant. Sed ex his equis qui nati pulli, non plus triennium vivunt. Quae nata sunt matura et corda, ut pure et molliter stent videndum et ne obterantur. Dicuntur agni cordi, qui post tempus nascuntur ac remanserunt in volvis intimis . . . .[1] vocant chorion, a quo cordi appellati.
20 Tertia res est, in nutricatu quae observari oporteat, in quo quot diebus matris sugant mammam et id quo tempore et ubi; et si parum habet lactis mater, ut subiciat sub alterius mammam, qui appellantur subrumi, id est sub mamma. Antiquo enim vocabulo mamma rumis, ut opinor. Fere ad quattuor menses

[1] Keil thinks words are lost; but the sentence seems quite Varronian.

---

[1] The same "wondrous tale" is told by Virgil, *Georg.*, III, 273 ff., and by many other writers, including Aristotle.

[2] *ὑπηνέμια*, "wind-eggs," *i.e.* unfertilized eggs.

[3] The membrane surrounding the foetus in the womb. Cf. Arist., *H. A.*, VI, 10, 58: *χόριον δὲ καὶ ὕμενες ἴδιοι περὶ ἕκαστον γίγνονται τῶν ἐμβρύων*. The term is still in use in physiology.

[4] From the name was derived that of the goddess, Rumina, "she who gives suck," who had a temple near the fig-tree

also be given to the proper period before breeding begins, during which the males should be kept away from the females—a period which both stockmen and shepherds usually fix at two months for all animals. The second division comprises the points to be watched in breeding, arising from the difference among species in the period of gestation; thus, the mare carries her young twelve months, the cow ten, the sheep and the goat five, the sow four. (Speaking of breeding, there is a story from Spain which, though incredible,[1] is quite authentic, that on the shore of the ocean in Lusitania, in the district in which is situated the town of Olisipo, certain mares on Mount Tagrus, at a particular time of year, are impregnated by the wind; just as in this country frequently occurs in the case of those hens the eggs of which are called *hypenemia*.[2] But the foals of these mares do not live beyond three years.) Care must be taken that the young which have gone the full time or longer have a clean, soft place to stand in, and that they be not trampled. Lambs which are born after the full period are called *cordi*, the name being derived from the fact that they have remained in those deep-lying folds which are called *chorion*.[3] There is a third item—the practice to be observed in the matter of feeding, including the number of days on which the young may have the teat, at what times, and where; and if the mother is deficient in milk, that the young be allowed to suckle the udder of another mother. Such animals are called *subrumi*, 'under the udder,' the udder being called *rumis*, as I suppose, in old Latin.[4] As a rule, lambs are not

(*ficus Ruminalis*) under which Romulus and Remus were suckled.

a mamma non diiunguntur agni, haedi tres, porci duo. E quis qui iam puri sunt ad sacrificium, ut immolentur, olim appellati sacres, quos appellat Plautus cum ait "quanti sunt porci sacres?" Sic boves altiles ad sacrificia publica saginati dicuntur 21 opimi. Quarta pars est de sanitate, res multiplex ac necessaria, quod morbosum pecus est vitiosum, et quoniam non valet, saepe magna adficiuntur calamitate. Cuius scientiae genera duo, ut in homine, unum ad quae adhibendi medici, alterum quae ipse etiam pastor diligens mederi possit. Eius partes sunt tres. Nam animadvertendum, quae cuiusque morbi sit causa, quaeque signa earum causarum sint, et quae quemque morbum ratio curandi sequi debeat. 22 Fere morborum causae erunt, quod laborant propter aestus aut propter frigora, nec non etiam propter nimium laborem aut contra nullam exercitationem, aut si, cum exercueris, statim sine intervallo cibum aut potionem dederis. Signa autem sunt, ut eorum qui e labore febrem habent adapertum os umido spiritu crebro et corpore calido. Curatio autem, 23 cum hic est morbus, haec: Perfunditur aqua et perunguitur oleo et vino tepefacto, et item cibo sustinetur, et inicitur aliquid, ne frigus laedat; sitienti aqua tepida datur. Si hoc genus rebus non proficitur, demittitur sanguis, maxime e capite. Item ad alios morbos aliae causae et alia signa, in

---

[1] *Menaechmi*, 290.

weaned under four months, kids under three, and pigs under two. Those of the last named which are pure for sacrifice and may be offered up, used to be called *sacres;* Plautus uses the term in his sentence[1]: 'What's the price of sacred pigs?' Similarly, oxen, fattened for public offerings, are called *opimi*, fatlings. The fourth division is that of health—a complicated but extremely important matter, inasmuch as a sickly herd is a losing investment, and men frequently come to grief because it is not strong. There are two divisions of such knowledge, as there are in the treatment of human beings: in the one case the physician should be called in, while in the other even an attentive herdsman is competent to give the treatment. The topic has three heads: we must observe the cause of the several diseases, the symptoms displayed by such causes, and the proper method of treatment to be followed for each disease. In general, sickness is caused by the fact that the animals are suffering from heat or from cold, or else from excessive work, or, on the other hand, from lack of exercise; or in case food or drink has been given them immediately after working, without a period of rest. The symptoms are that those which have fever from overwork keep the mouth open, pant fast with moist breath, and have hot bodies. The following is the treatment in such cases: The animal is drenched with water, rubbed down with oil and warm wine, and, further, is sustained with food, and a covering is thrown over it to prevent a chill; in case of thirst tepid water is administered. If improvement is not obtained by such treatment, blood is let, usually from the head. Other diseases have other causes and other symptoms, and the man

omni pecore quae scripta habere oportet magistrum
24 pecoris.

Relinquitur nonum quod dixi, de numero, utriusque partis commune. Nam et qui parat pecus necesse est constituat numerum, quot greges et quantos sit pasturus, ne aut saltus desint aut supersint et ideo fructus dispereant. Praeterea scire oportet, in grege quot feminas habeat, quae parere possint, quot arietes, quot utriusque generis suboles, quot reiculae sint alienandae. In alimoniis, si sunt plures nati, ut quidam faciunt, sequendum ut quosdam subducas, quae res facere solet ut reliqui melius crescant.

25 Vide, inquit Atticus, ne te fallat et novenae istae partes non exeant extra pecoris minoris ac maioris nomen. Quo pacto enim erunt in mulis et pastoribus novenae partes, ubi nec admissurae nec feturae ob-
26 servantur? In canibus enim video posse dici. Sed do etiam in hominibus posse novenarium retineri numerum, quod in hibernis habent in villis mulieres, quidam etiam in aestivis, et id pertinere putant, quo facilius ad greges pastores retineant, et puerperio familiam faciunt maiorem et rem pecuariam fructuosiorem. Sic, inquam, numerus non est ut sit ad amussim, ut non est, cum dicimus mille naves isse ad Troiam, centumvirale esse iudicium Romae.

---

[1] See Cato, 2, 7.

[2] Varro is the speaker here, and in support of Scrofa's contention.

in charge of the herd should keep them all in written form.

"There remains the ninth point I have mentioned, common to both divisions—the proper number. For the man who is founding a herd must decide on the size, determining how many herds and how large he is going to graze, so that his pasturage will not run short, and so that he will not have idle pasturage and hence lose his profit. He must also decide how many females to have in the flock for breeding, how many males, how many young of each sex, and how many culls[1] are to be cut out. In the matter of feeding, if too many young are born you should follow the practice of some breeders, and wean some of them; the result usually being that the rest grow better."

"Don't get confused," said Atticus, "and let your ninefold division get away from the matter of smaller and larger animals. How will you get a ninefold division in the case of mules and herdsmen, where there is neither breeding nor bearing? I see how you can use it in the case of dogs. I grant you also that even in the case of the humans the ninefold division can be retained, as they keep women in their huts in the winter ranches, and some have them even in the summer, thinking that this is worth while in order the more easily to keep the herdsmen with their herds; and by the natural increase they enlarge their slave gangs and make the cattle-raising more profitable." "The number," I remark,[2] "is not to be taken as precisely accurate, just as we do not mean to be taken exactly when we say that a thousand ships set forth against Troy, or speak of the centumviral court at Rome. So, if you wish, subtract two of the topics,

Quare deme, si vis, duas res de mulis, admissuram
27 et parturam. Vaccius, Parturam? inquit, proinde ut non aliquotiens dicatur Romae peperisse mulam. Cui ego ut succinerem, subicio Magonem et Dionysium scribere, mula et equa cum conceperint, duodecimo mense parere. Quare non, si hic in Italia cum peperit mula sit portentum, adsentiri omnes terras. Neque enim hirundines et ciconiae, quae in Italia pariunt, in omnibus terris pariunt. Non scitis palmulas careotas Syrias parere in Iudaea, in Italia
28 non posse? Sed Scrofa, Si exigere mavis sine mulorum fetura et nutricatu numerum octoginta et unum, est qui expleas duplicem istam lacunam, quod extraordinariae fructum species duae accedunt magnae, quarum una est tonsura, quod oves ac capras detondunt aut vellunt; altera, quae latius patet, de lacte et caseo, quam scriptores Graeci separatim *τυροποιίαν* appellaverunt ac scripserunt de ea re permulta.

II. Sed quoniam nos nostrum pensum absolvimus ac limitata est pecuaria quaestio, nunc rursus vos reddite nobis, o Epirotae, de una quaque re, ut videamus, quid pastores a Pergamide Maledove potis
2 sint. Atticus, qui tunc Titus Pomponius, nunc Quintus Caecilius cognomine eodem, Ego opinor, inquit, incipiam primus, quoniam in me videre coniecisse

---

[1] Aristotle (*H. A.*, VI, 24) states that there are no authenticated instances. Columella (VI, 37, 3) cites this statement of Varro, but in a different form: "that in parts of Africa the offspring of mules is so far from being a prodigy, that it is as familiar as the offspring of mares among us." Livy gives several cases of such prodigies (XXVI, 23; XXXVII, 3)—both, incidentally, at Varro's Reate!

[2] The places are not mentioned elsewhere.

[3] Well known as the intimate friend and correspondent of Cicero. In 58 B.C. he was adopted by his uncle, Quintus

coition and foaling, when you speak of mules." "Foaling?" asked Vaccius; "why, don't you know that it has several times been asserted that a mule has borne a colt at Rome?"[1] To back up his statement, I add that both Mago and Dionysius remark that the mule and the mare bring forth in the twelfth month after conception. Hence we must not expect all lands to agree, even if it is considered a portent when a mule bears young here in Italy. Swallows and storks, for instance, which bear in Italy, do not bear in all lands. Surely you are aware that the date-palms of Syria bear fruit in Judea but cannot in Italy. Scrofa, however, remarked: "If you insist on having 81 sub-heads, omitting the breeding and feeding of mules, you may easily fill that double gap; two very important sources of revenue fall outside the enumeration. There is the shearing—the clipping or pulling of wool and goat hair—and another, which is even more important, the matter of milk and cheese. The Greek authorities treat this as a separate topic, calling it *tyropoiia* (cheese-making), and have had a great deal to say about it.

II. "But since I have completed my task, and the subject of stock-raising has been sketched in outline, you gentlemen of Epirus should take up the tale in your turn and let us see what the shepherds from Pergamis and Maledos[2] can tell us under each head." Then Atticus,[3] who at the time bore the name Titus Pomponius, but is now called Quintus Caecilius though he retains the same cognomen, began: "I suppose I should start the discus-

Caecilius, and his name became, according to Cicero (*ad Att.*, III, 20) and Nepos (*Att.*, 5), Quintus Caecilius Pomponianus Atticus.

oculos, et dicam de primigenia pecuaria. E feris enim pecudibus primum dicis oves comprehensas ab hominibus ac mansuefactas. Has primum oportet bonas emere, quae ita ab aetate, si neque vetulae sunt neque merae agnae, quod alterae nondum, alterae iam non possunt dare fructum. Sed ea melior aetas, quam sequitur spes, quam ea quam mors.
3 De forma ovem esse oportet corpore amplo, quae lana multa sit et molli, villis altis et densis toto corpore, maxime circum cervicem et collum, ventrem quoque ut habeat pilosum. Itaque quae id non habent, maiores nostri apicas appellabant ac reiciebant. Esse oportet cruribus humilibus; caudis observate ut sint in Italia prolixis, in Syria brevibus. In primis videndum ut boni seminis pecus habeas.
4 Id fere ex duabus rebus potest animadverti, ex forma et progenie: ex forma, si arietes sint fronte lana vestiti bene, tortis cornibus pronis ad rostrum, ravis oculis, lana opertis auribus, ampli, pectore et scapulis et clunibus latis, cauda lata et longa. Animadvertendum quoque lingua ne nigra aut varia sit, quod fere qui eam habent nigros aut varios procreant agnos. Ex progenie autem animadvertitur, si agnos
5 procreant formosos. In emptionibus iure utimur eo, quo lex praescripsit. In ea enim alii plura, alii pauciora excipiunt; quidam enim pretio facto in singulas oves, ut agni cordi duo pro una ove adnumerentur, et si quoi vetustate dentes absunt, item binae pro singulis ut procedant. De reliquo antiqua

[1] But the word includes more than "form," as can be seen from Section 4 below. "Type" would perhaps be a better word.
[2] The word is possibly from ἄποκος, "fleece-less."
[3] So the accepted text; but Schneider says it is clear that it should read *arietem*, "sire," because of the next sentence.
[4] So Virgil, *Georg.*, III, 387–395, but it is no longer believed.

sion, as you all seem to be looking to me, and I shall speak of the earliest branch of animal husbandry, as you claim that sheep were the first of the wild animals to be caught and tamed by man. The first consideration is that these be in good condition when purchased; with respect to age that they be neither too old nor mere lambs, the latter being not yet, and the former no longer profitable—though the age which is followed by hope is better than the one which is followed by death. As to form,[1] sheep should be full-bodied, with abundant soft fleece, with fibres long and thick over the whole body, especially about the shoulders and neck, and should have a shaggy belly also. In fact, sheep which did not have this our ancestors called 'bald'[2] (*apicas*), and would have none of them. The legs should be short; and observe that the tail should be long in Italy but short in Syria. The most important point to watch is to have a flock[3] from good stock. This can usually be judged by two points—the form and the progeny; by the form if the rams have a full coating of fleece on the forehead, have flat horns curving towards the muzzle, grey eyes, and ears overgrown with wool; if they are full-bodied, with wide chest, shoulders, and hindquarters, and a wide, long tail. A black or spotted tongue is also to be avoided, for rams with such a tongue usually beget black or spotted lambs.[4] The stock is determined by the progeny if they beget handsome lambs. In purchasing we take advantage of the variation which the law allows, some making more and others fewer exceptions; thus, some purchasers, when the price is fixed by the head, stipulate that two late-born lambs count as one sheep, and in the case of those which have lost their teeth from age,

fere formula utuntur. Cum emptor dixit "tanti sunt mi emptae?" Et ille respondit "sunt" et expromisit nummos, emptor stipulatur prisca formula sic,
6 "illasce oves, qua de re agitur, sanas recte esse, uti pecus ovillum, quod recte sanum est extra luscam surdam minam, id est ventre glabro, neque de pecore morboso esse habereque recte licere, haec sic recte fieri spondesne?" Cum id factum est, tamen grex dominum non mutavit, nisi si est adnumeratum; nec non emptor pote ex empto vendito illum damnare, si non tradet, quamvis non solverit nummos, ut ille emptorem simili iudicio, si non reddit pretium.
7 De alteris quattuor rebus deinceps dicam, de pastione, fetura, nutricatu, sanitate. Primum providendum ut totum annum recte pascantur intus et foris; stabula idoneo loco ut sint, ne ventosa, quae spectent magis ad orientem quam ad meridianum tempus. Ubi stent, solum oportet esse eruderatum et proclivum, ut everri facile possit ac fieri purum. Non enim solum ea uligo lanam corrumpit ovium,
8 sed etiam ungulas, ac scabras fieri cogit. Cum aliquot dies steterunt, subicere oportet virgulta alia, quo mollius requiescant purioresque sint; libentius

---

[1] If the words *id est ventre glabro* are not an interpolation, *mina* and *apica* (Sec. 3) mean the same thing. The seller could not know whether they were *luscae* or *surdae*; and the buyer could see if they were *minae*.

that they also be reckoned two for one. With this exception, the ancient formula is generally employed: when the purchaser has said, 'They are sold at such a price?' and the seller has replied, 'Yes,' and the money has passed, the purchaser, using the old formula, says: 'You guarantee that the sheep in question are perfectly sound, up to the standard of a flock which is perfectly sound, excepting those blind of one eye, deaf, or *minae* (that is, with belly bare of wool),[1] that they do not come from a diseased flock, and that title may legally pass—that all this may be properly done?' Even after this has been agreed on, the flock does not change owners unless the money has been counted; and the purchaser still has the right to obtain a judgment against the vendor under the law of purchase and sale if he does not make delivery, even though no money has passed; just as the vendor may obtain a judgment against the purchaser under the same law if he does not make payment.

"I shall discuss next the remaining four points—pasturage, breeding, feeding, health. It is first to be arranged that they feed properly the year round, indoors and out. The fold should be placed in a suitable situation, protected from the wind, and facing the east rather than the south. The ground on which they are to stand should be clear of undergrowth and sloping, so that it can easily be swept and kept clean; for the moisture of the ground injures not only the fleece of the sheep but their hoofs as well, and causes them to become scabby. When they have been standing for some days, fresh brush should be spread for them, to give them a softer bed and keep them cleaner; for this increases

enim ita pascuntur. Faciendum quoque saepta secreta ab aliis, quo incientes secludere possis, item quo corpore aegro. Haec magis ad villaticos greges 9 animadvertenda. Contra illae in saltibus quae pascuntur et a tectis absunt longe, portant secum crates aut retia, quibus cohortes in solitudine faciant, ceteraque utensilia. Longe enim et late in diversis locis pasci solent, ut multa milia absint saepe hibernae pastiones ab aestivis. Ego vero scio, inquam; nam mihi greges in Apulia hibernabant, qui in Reatinis montibus aestivabant, cum inter haec bina loca, ut iugum continet sirpiculos, sic calles publicae distantes 10 pastiones. Eaeque ibi, ubi pascuntur in eadem regione, tamen temporibus distinguntur, aestate quod cum prima luce exeunt pastum, propterea quod tunc herba ruscida meridianam, quae est aridior, iucunditate praestat. Sole exorto potum propellunt, ut redintegrantes rursus ad pastum alacriores faciant. 11 Circiter meridianos aestus, dum defervescant, sub umbriferas rupes et arbores patulas subigunt, quaad refrigeratur. Aere vespertino rursus pascunt ad solis occasum. Ita pascere pecus oportet, ut averso sole agat; caput enim maxime ovis molle est. Ab occasu parvo intervallo interposito ad bibendum appellunt et rursus pascunt, quaad contenebravit; iterum enim tum iucunditas in herba redintegrabit. Haec a vergiliarum exortu ad aequinoctium autumnale maxime 12 observant. Quibus in locis messes sunt factae, ini-

[1] Varro himself speaks, cf. II, Introd. 6.

[2] Atticus resumes.

[3] Compare Virgil, *Georg.*, III, 324 ff. for a beautiful paraphrase of the whole passage.

[4] Varro regularly uses words which imply that sheep were driven, instead of led. This word is used of cattle in Chapter 5, 15.

[5] See I, 28, 2.

their appetite. Separate enclosures should also be built, so that you may take the pregnant ones away from the flock, and also those that are sick. These measures concern most the flocks which are folded at the steading. On the other hand, in the case of those that feed on the ranges and are far from cover, hurdles or nets are carried with which to make enclosures in a desolate district, as well as other necessary things; for they usually graze far and wide in all sorts of places, so that frequently the winter grazing grounds are many miles away from the summer." "I am well aware of that," said I,[1] "for I had flocks that wintered in Apulia and summered in the mountains around Reate, these two widely separated ranges being connected by public cattle-trails, as a pair of buckets by their yoke." "Such[2] flocks, even when they feed in the same locality, are treated differently at different seasons; thus, in summer they begin feeding at daybreak, because at that time the grass, filled with dew,[3] is superior to the grass of midday, which is drier. At sunrise they are driven[4] to water, to make them more eager to graze when they come back. During the midday heat they are driven under shady cliffs and widespreading trees to cool off until the day grows cooler; and they feed again in the evening until sunset. Sheep should be headed in grazing in such a way as to have the sun behind them, as the head of the sheep is its weakest part. A short time after sunset they are driven to water, and then again they graze until it becomes quite dark; for at this time the succulence comes again to the grass. This practice is usually kept up from the rising of the Pleiades until the autumnal equinox.[5] It is profitable to drive

gere est utile duplici de causa: quod et caduca spica saturantur et obtritis stramentis et stercoratione faciunt in annum segetes meliores. Reliquae pastiones hiberno ac verno tempore hoc mutant, quod pruina iam exalata propellunt in pabulum et pascunt diem totum ac meridiano tempore semel agere potum satis habent.

13 Quod ad pastiones attinet, haec fere sunt; quod ad feturam, quae dicam. Arietes, quibus sis usurus ad feturam, bimestri tempore ante secernendum et largius pabulo explendum. Cum redierunt ad stabula e pastu, hordeum si est datum, firmiores fiunt ad laborem sustinendum. Tempus optimum ad admittendum ab arcturi occasu ad aquilae occasum, quod quae postea concipiuntur, fiunt vegrandes atque

14 imbecillae. Ovis praegnas est diebus CL. Itaque fit partus exitu autumnali, cum aer est modice temperatus et primitus oritur herba imbribus primoribus evocata. Quam diu admissura fit, eadem aqua uti oportet, quod commutatio et lanam facit variam et corrumpit uterum. Cum omnes conceperunt, rursus arietes secernendi, iam factis praegnatibus quod sunt molesti, obsunt. Neque pati oportet minores quam bimas saliri, quod neque natum ex his idoneum est, neque non ipsae fiunt deteriores; et non meliores quam trimae admissae. Deterrent ab saliendo, et

[1] Pliny, *N.H.*, VIII, 187, gives precise dates—from May 13 to July 23.

them into stubble fields for two reasons: they get their fill of the ears that have fallen, and make the crop better the next year by trampling the straw and by their dung. The feeding during the rest of the year, winter and spring, varies from this, in that when the frost has melted they are driven out to feed and range the whole day, and it is considered sufficient for them to be driven to water only once, at midday.

"With regard to pasturage the foregoing remarks will suffice; the following apply to breeding. The rams which are to be used for breeding are to be removed from the flock two months ahead, and fed more generously. If barley is fed them on their return to the pens from the pasture, they are strengthened for the work before them. The best time for mating is from the setting of Arcturus to the setting of Aquila;[1] as lambs which are conceived after that time grow undersized and weak. As the period of pregnancy of the sheep is 150 days, the birth thus occurs at the close of autumn, when the air is fairly temperate, and the grass which is called forth by the early rains is just growing. During the whole time of breeding they should drink the same water, as a change of water causes the wool to spot and is injurious to the womb. When all the ewes have conceived, the rams should again be removed, as they are troublesome in worrying the ewes which have now become pregnant. Ewes less than two years old should not be allowed to breed, for the offspring of these is not sturdy and the ewes themselves are injured; and no others are better than the three-year-olds for breeding. They may be protected from the male by binding behind them

fiscellas e iunco aliave qua re quod alligant ad naturam; commodius servantur, si secretas pascunt.

15 In nutricatu, cum parere coeperunt, inigunt in stabula, eaque habent ad eam rem seclusa, ibique nata recentia ad ignem prope ponunt, quaad convaluerunt. Biduum aut triduum retinent, dum adcognoscant matrem agni et pabulo se saturent. Deinde matres cum grege pastum prodeunt, retinent agnos, ad quos cum reductae ad vesperum, aluntur lacte et rursus discernuntur, ne noctu a matribus conculcentur. Hoc item faciunt mane, antequam matres in 16 pabulum exeant, ut agni satulli fiant lacte. Circiter decem dies cum praeterierunt, palos offigunt et ad eos alligant libro aut qua alia re levi distantes, ne toto die cursantes inter se teneri delibent aliquid membrorum. Si ad matris mammam non accedet, admovere oportet et labra agni unguere buturo aut adipe suilla et olfacere labra lacte. Diebus post paucis obicere iis viciam molitam aut herbam teneram, antequam exeunt pastum et cum reverterunt. 17 Et sic nutricantur, quaad facti sunt quadrimestres. Interea matres eorum iis temporibus non mulgent quidam; qui id melius, omnino perpetuo, quod et lanae plus ferunt et agnos plures. Cum depulsi sunt agni a matribus, diligentia adhibenda est, ne desiderio senescant. Itaque deliniendum in nutricatu pabuli bonitate et a frigore et aestu ne quid laboret

[1] Cf. Columella, VII, 3, 17: "When a lamb is born, it should be set on its feet and put to the udder; then, too, its mouth should be opened and moistened, by pressing the teats, so that it may learn to draw nourishment from its mother."

baskets made of rushes or other material; but they are protected more easily if they feed apart.

"As to feeding: when they begin to bear they are driven into the pens which are kept separate for that purpose; and there the new-born lambs are placed near a fire until they get their strength. They are kept penned for two or three days, while they are learning to recognise their dams and are getting their fill of nourishment. Then the dams go to pasture with the flock, and the lambs are kept penned; when the dams are brought back to them toward evening, the lambs are suckled by them and are again separated to keep them from being trampled by the dams during the night. The same thing takes place in the morning, before the dams go out to pasture, so that the lambs may be filled with milk. When about ten days have passed, stakes are set to which the lambs are fastened at intervals by bark or other smooth ropes, so that the tender young things may not knock the skin off any of their legs while frisking about together during the whole day. If the lamb will not come to its dam's udder, it should be held close and its lips smeared with butter or hog's lard and the lips be given the savour of milk.[1] A few days later ground vetch or tender grass is thrown out to them before they go out to pasture and when they come back; and this feeding is continued until they are four months old. During this time some breeders do not milk the dams; and it is even better not to milk them at all, as they both yield more wool and bear more lambs. When the lambs are removed from the dams, care must be taken that they do not sicken from the separation; and so in feeding they must be coaxed by the

18 curandum. Cum oblivione iam lactis non desiderat matrem, tum denique compellendum in gregem ovium. Castrare oportet agnum non minorem quinque mensum, neque antequam calores aut frigora se fregerunt. Quos arietes summittere volunt, potissimum eligunt ex matribus, quae geminos parere solent. Pleraque similiter faciendum in ovibus pellitis, quae propter lanae bonitatem, ut sunt Tarentinae et Atticae, pellibus integuntur, ne lana inquinetur, quo minus vel infici recte possit vel lavari ac

19 putari. Harum praesepia ac stabula ut sint pura maiorem adhibent[1] diligentiam, quam hirtis. Itaque faciunt lapide strata, urina necubi in stabulo consistat. His quaecumque lubenter vescuntur, ut folia ficulnea et palea, vinacea, furfures, obiciuntur modice, ne parum aut nimium saturentur. Utrumque enim ad corpus alendum inimicum, ut maxime amicum cytisum et medica. Nam et pingues facit facillime et genit lacte.

20 De sanitate sunt multa; sed ea, ut dixi, in libro scripta magister pecoris habet, et quae opus ad medendum, portat secum. Relinquitur de numero, quem faciunt alii maiorem, alii minorem. Nulli enim huius moduli naturales. Illut fere omnes in Epiro facimus, ne minus habeamus in centenas oves hirtas singulos homines, in pellitas binos.

[1] *adhibent* Schneider: *adhibeant.*

---

[1] Cf. Horace, *Odes*, II, 6, 10, and Pliny, *N.H.*, VIII, 189–190.

daintiness of the food and guarded from being harmed by cold and heat. They must be driven into the flock only after they no longer miss the dam, because they have forgotten the taste of milk. Lambs should be castrated not earlier than the fifth month, and then not until the heat or the cold has broken. Those they wish to rear for rams are chosen preferably from the young of dams which usually bear twins. The treatment is, in general, the same in the case of jacketed sheep—those which, on account of the excellence of the wool, are jacketed with skins, as is the practice at Tarentum[1] and in Attica, to prevent the fleece from being soiled, in which case it cannot be so well dyed, or washed and bleached. More care is employed in the case of these than in the case of rough-fleeced sheep, to keep the folds and stalls clean; and so they are covered with a stone pavement so that the urine may not stand anywhere in the stalls. To these the food which they prefer, such as fig leaves, straw, grape dregs, and bran, is fed in moderate quantities, to avoid under-feeding or over-feeding; either of which is harmful to their fattening, while alfalfa and snail-clover are both beneficial, as these fatten them very easily and produce milk.

"In the matter of health there are many rules; but, as I said, the head shepherd keeps these written down in a book, and carries with him the remedies he may need. The only remaining division is that of number, and some make this larger, others smaller; for there are no natural limits in this respect. Our almost universal practice in Epirus is not to have less than one shepherd to the hundred rough-fleeced sheep, and two to the hundred jacketed sheep."

III. Cui Cossinius, Quoniam satis balasti, inquit, o Faustule noster, accipe a me cum Homerico Melanthio cordo de capellis, et quem ad modum breviter oporteat dicere, disce. Qui caprinum gregem constituere vult, in eligendo animadvertat oportet primum aetatem, ut eam paret, quae iam ferre possit fructum, et de iis eam potius, quae diutius; novella
2 enim quam vetus utilior. De forma videndum ut sint firmae, magnae, corpus leve ut habeant, crebro pilo, nisi si glabrae sunt (duo enim genera earum); sub rostro duas ut mammulas pensiles habeant, quod eae fecundiores; ubere sint grandiore, ut et lac multum et pingue habeant pro portione. Hircus molliori et potissimum pilo albo ac cervice et collo brevi, gurgulione longiore. Melior fit grex, si non est ex collectis comparatus, sed ex consuetis una.
3 De seminio dico eadem, quae Atticus in ovibus; hoc aliter, ovium semen tardius esse, quo eae sint placidiores; contra caprile mobilius esse, de quarum velocitate in Originum libro Cato scribit haec: "in Sauracti et Fiscello caprae ferae sunt, quae saliunt e saxo pedes plus sexagenos." Oves enim, quas pascimus, ortae sunt ab ovibus feris, sic quas alimus caprae a capris feris ortae, a quis propter Italiam

---

[1] The name of the shepherd who brought up Romulus and Remus; cf. II, 1, 9.

[2] A goatherd of Odysseus (*Odyss.*, XVII, 217) who supplied the suitors with the best of the flock, and was killed by Telemachus. The term *cordus* was explained in I, 19.

[3] In sections 2, 3, 4 of Chapter 2.

III. Cossinius, addressing him, said: "As you have bleated long enough, my dear Faustulus,[1] now hear from me, as from Homer's Melanthius [2] born out of due season, with regard to goats; and learn how one ought to speak, briefly and to the point. One who decides to assemble a flock of goats should, in choosing his animals, have regard first to age, picking those of the age which is already capable of bringing in a profit, and this age preferably the one which can bring it longer; for the young goat is more profitable than the old. As to conformation, see that they be strong and large, and have a smooth coat with thick hair, unless, to be sure, they belong to the hairless breed, for there are two breeds of goats. They should have two teat-like growths hanging under the chin, as such goats are more fertile; they should have rather large udders, so that they may give a greater quantity of milk and of richer quality in proportion. The buck should have hair which is rather soft and by preference white; short shoulders and neck; and a somewhat long throat. The flock is better if it is not formed of animals bought here and there, but of those which are accustomed to run together. As to the breed, I make the remark which Atticus made [3] with regard to sheep; with this exception, that the race of sheep is more quiet, inasmuch as they are gentler, while on the other hand that of goats is more nimble. As to their activity, Cato says in his *Origines:* 'On Soracte and Fiscellum there are wild goats which make leaps of more than sixty feet from the cliffs.' For just as the domesticated sheep is sprung from the wild sheep, so the domesticated goat is sprung from the wild goat; and the island Caprasia, off the

4 Caprasia insula est nominata. De capris quod meliore semine eae quae bis pariant, ex his potissimum mares solent summitti ad admissuras. Quidam etiam dant operam ut ex insula Melia capras habeant, quod ibi maximi ac pulcherrimi existimantur fieri haedi.

5 De emptione aliter dico atque fit, quod capras sanas sanus nemo promittit ; numquam enim sine febri sunt. Itaque stipulantur paucis exceptis verbis, ac Manilius scriptum reliquit sic: " illasce capras hodie recte esse et bibere posse habereque recte licere, haec spondesne ? " de quibus admirandum illut, quod etiam Archelaus scribit: non ut reliqua animalia naribus, sed auribus spiritum ducere solere pastores curiosiores aliquot dicunt.

6 De alteris quattuor quod est de pastu, hoc dico: Stabulatur pecus melius, ad hibernos exortos si spectat, quod est alsiosum. Id, ut pleraque, lapide aut testa substerni oportet, caprile quo minus sit uliginosum ac lutulentum. Foris cum est pernoctandum, item in eandem partem caeli quae spectent saepta oportet substerni virgultis, ne oblinantur. Non multo aliter tuendum hoc pecus in pastu atque ovillum, tamen habent sua propria quaedam, quod potius

7 silvestribus saltibus delectantur quam pratis ; studiose enim de agrestibus fruticibus pascuntur atque in locis cultis virgulta carpunt. Itaque a carpendo caprae nominatae. Ab hoc in lege locationis fundi excipi solet, ne colonus capra natum in fundo pascat.

---

[1] Cf. the selection of young rams, II, 2, 18.

[2] Scaliger thinks that Melos is meant.

[3] Manius Manilius, consul 149 B.C., author of laws concerning purchase and sale.

coast of Italy, derives its name from these. As she-goats which bear twins are of better stock, it is from these, preferably, that the males are usually chosen for service.[1] Some owners are even careful to import she-goats from the island Melia,[2] because it is thought that the largest and finest kids are produced there.

"With regard to purchase, my rule is different from the usual practice, as no man sound of mind guarantees that goats (which are never free of fever) are sound of body. And so the bargain is struck with only a few exceptions made, after a formula derived from the code of Manilius:[3] 'Do you guarantee that the said goats are to-day in good condition and able to drink, and that the title is in proper form?' There is a remarkable thing about these animals, and even Archelaus is authority for the statement: some shepherds who have watched quite closely claim that goats do not breathe, as other animals do, through the nostrils, but through the ears.

"Of the other four points, I have this to say with regard to feeding: It is better to have the goat stalls face the sunrise in winter, as the animals feel the cold acutely. Such stalls, and in fact all stalls, should be floored with stone or tile, to prevent the goat-house from being wet and muddy. When they have to spend the night outdoors, their pens should face in the same direction, and they should be bedded down with twigs so that they may not be muddied. The care of this animal in the matter of feeding is about the same as that of the sheep, though each has certain peculiarities; thus, the goat prefers wooded glades to meadows, as it eats eagerly the field bushes and crops the undergrowth on cultivated land. Indeed, their name *capra* is derived from *car-*

Harum enim dentes inimici sationi, quas etiam astrologi ita receperunt in caelum, ut extra lembum duodecim signorum excluserint; sunt duo haedi at capra
8 non longe a tauro. Quod ad feturam pertinet, desistente autumno exigunt a grege in campo hircos in caprilia, item ut in arietibus dictum. Quae concepit, post quartum mensem reddit tempore verno. In nutricatu haedi, trimestres cum sunt facti, tum submittuntur et in grege incipiunt esse. Quid dicam de earum sanitate, quae numquam sunt sanae? Nisi tamen illud unum: quaedam scripta habere magistros pecoris, quibus remediis utantur ad morbos quosdam earum ac vulneratum corpus, quod usu venit iis saepe, quod inter se cornibus pugnant atque in
9 spinosis locis pascuntur. Relinquitur de numero, qui in gregibus est minor caprino quam in ovillo, quod caprae lascivae et quae dispargant se; contra oves quae se congregent ac condensent in locum unum. Itaque in agro Gallico greges plures potius faciunt quam magnos, quod in magnis cito existat pestilentia,
10 quae ad perniciem eum perducat. Satis magnum gregem putant esse circiter quinquagenas. Quibus adsentiri putant id quod usu venit Gaberio, equiti Romano. Is enim cum in suburbano mille iugerum haberet et a caprario quodam, qui adduxit capellas ad urbem decem, sibi in dies singulos denarios

---

[1] Cf. II, 1, 8. [2] See p. 217, note 5.

pere, to crop. It is because of this fact that in a contract for the lease of a farm the exception is usually made that the renter may not pasture the offspring of a goat on the place. For their teeth are injurious to all forms of growth; and though the astronomers have placed them in the sky, they have put them outside the circle of the twelve signs—there are two kids and a she-goat not far from Taurus.[1] As to breeding, at the close of autumn, while the herd is at pasture, the bucks are driven from it into the goat-houses, as was directed with regard to rams. The female which has conceived drops her kid four months later, during the spring. As to rearing, when the kids reach the age of three months they are turned out and begin to form part of the flock. What can I say of the health of animals which are never healthy? I can only make one remark: that the head goatherds keep written directions as to the remedies to be used for some of their diseases and for flesh wounds which they frequently receive, as they are always fighting one another with their horns, and as they crop in thorny places. One topic remains—that of number. This is smaller in the goat herd than in the flock, as goats are wanton and scatter widely, while sheep, on the contrary, huddle together and crowd into the same space. Hence in the Ager Gallicus[2] breeders keep numerous herds rather than large ones, because in large herds an epidemic quickly spreads, and this may ruin the owner. A flock of about fifty is considered quite large enough. The experience of the Roman knight, Gaberius, is thought to prove this: He had a place containing 1000 iugera near the city, and hearing from a certain goatherd who drove ten

singulos dare audisset, coegit mille caprarum, sperans se capturum de praedio in dies singulos denarium mille. Tantum enim fefellit, ut brevi omnes amiserit morbo. Contra in Sallentinis et in Casinati ad centenas pascunt. De maribus et feminis idem fere discrimen, ut alii ad denas capras singulos parent hircos, ut ego; alii etiam ad quindecim, ut Menas; non nulli etiam, ut Murrius, ad viginti.

IV. Sed quis e portu potius Italico prodit ac de suillo pecore expedit? Tametsi Scrofam potissimum de ea re dicere oportere cognomen eius significat. Cui Tremelius, Ignorare, inquit, videre, cur appeller Scrofa. Itaque ut etiam hi propter te sciant, cognosce meam gentem suillum cognomen non habere, nec me esse ab Eumaeo ortum. Avus meus primum appellatus est Scrofa, qui quaestor cum esset Licinio Nervae praetori in Macedonia provincia relictus, qui praeesset exercitui, dum praetor rediret, hostes, arbitrati occasionem se habere victoriae, 2 impressionem facere coeperunt in castra. Avos, cum cohortaretur milites ut caperent arma atque exirent contra, dixit celeriter se illos, ut scrofa porcos, disiecturum, id quod fecit. Nam eo proelio hostes

---

[1] The other speakers are "half-Greek" (II, 1, 2). Now a genuine Italian is to speak of swine. And who more fittingly than one who bears a name Scrofa, which also means "brood-sow"?

[2] The swineherd of Odysseus (*Odyss.*, XIV, 22) who received and fed his master on his return.

[3] This cannot refer to the year 167 B.C., in which Nerva was

goats to the city that they yielded him a denarius a day per head, he bought 1000 goats, hoping that he would make 1000 denarii a day profit. In which he was sadly mistaken, for within a short time he lost the whole flock by disease. Among the Sallentini, however, and around Casinum, they have herds running as high as 100. As to the proportion of males to females, there is about the same difference of opinion, some (and this is my own practice) keeping one buck to every ten does; others, such as Menas, one to fifteen; and still others, such as Murrius, one to twenty.

IV. "But who sails forth from harbour, and preferably from an Italian harbour, to discourse about swine?[1] I need hardly ask, for that Scrofa should be chosen to speak on that subject this surname of his indicates." "You seem," said Tremelius in reply, "not to know why I have the nickname Scrofa. That these gentlemen, too, may learn the reason while you are being enlightened, you must know that my family does not bear a swinish surname, and that I am no descendant of Eumaeus.[2] My grandfather was the first to be called Scrofa. He was quaestor to the praetor Licinius Nerva, in the province of Macedonia, and was left in command of the army until the return of the praetor.[3] The enemy, thinking that they had an opportunity to win a victory, began a vigorous assault on the camp. In the course of his plea to the soldiers to seize arms and go to meet them, my grandfather said that he would scatter those people as a sow scatters her pigs; and he was as good as his word. For he so scattered and

praetor (Livy, XLV, 44); and it is possible that it occurred in 142 B.C., during a revolt in Macedonia.

ita fudit ac fugavit, ut eo Nerva praetor imperator sit appellatus, avus cognomen invenerit ut diceretur Scrofa. Itaque proavos ac superiores de Tremeliis nemo appellatus Scrofa, nec minus septimus sum deinceps praetorius in gente nostra. Nec tamen 3 defugio quin dicam quae scio de suillo pecore. Agri enim culturae ab initio fui studiosus, nec de pecore suillo mihi et vobis, magnis pecuariis, ea res non est communis. Quis enim fundum colit nostrum, quin sues habeat, et qui non audierit patres nostros dicere ignavum et sumptuosum esse, qui succidiam in carnario suspenderit potius ab laniario quam e domestico fundo?

Ergo qui suum gregem vult habere idoneum, eligere oportet primum bona aetate, secundo bona forma (ea est cum amplitudine membrorum, praeterquam pedibus capite), unicoloris potius quam varias. Cum haec eadem ut habeant verres videndum, 4 tum utique sint cervicibus amplis. Boni seminis sues animadvertuntur a facie et progenie et regione caeli: a facie, si formosi sunt verris et scrofa; a progenie, si porcos multos pariunt; a regione, si potius ex his locis, ubi nascuntur amplae quam exiles, 5 pararis. Emi solent sic: "illasce sues sanas esse habereque recte licere noxisque praestari neque de

[1] But Macrobius (*Saturn.*, I, 6) gives a different story: His slaves had stolen and killed a neighbour's sow, and had hidden it under his wife's bed. When the house was searched, he swore that he had no other sow in the house than the one under the bed-clothes, where his wife was lying.

routed the enemy in that battle that because of it the praetor Nerva received the title of Imperator, and my grandfather earned the surname of Scrofa.[1] Hence neither my great-grandfather nor any of the Tremelii who preceded him was called by this surname of Scrofa; and I am no less than the seventh man of praetorian rank in succession in our family. Still, I will not shrink from the task of telling what I know about swine. For I have been a close student of agriculture since my earliest days, and this matter of swine is of equal interest to me and to you who are large cattle-owners. For who of our people cultivates a farm without keeping swine? and who has not heard that our fathers called him lazy and extravagant who hung in his larder a flitch of bacon which he had purchased from the butcher rather than got from his own farm?

"A man, then, who wishes to keep his herd in good condition should select, first, animals of the proper age, secondly, of good conformation (that is, with heavy members, except in the case of feet and head), of uniform colour rather than spotted. You should see that the boars have not only these same qualities, but especially that their shoulders are well developed. The breed of swine is determined by their appearance, their litter, and the locality from which they come: from their appearance if both boar and sow are handsome; from their litter if they produce numerous pigs; from the locality if you get them from places where fat rather than thin swine are produced. The formula of purchase usually runs as follows: 'Do you guarantee that the said swine are sound, and that the title is good, and that I am protected from suits for damage, and

pecore morboso esse spondesne?" Quidam adiciunt perfunctas esse a febri et a foria.

In pastu locus huic pecori aptus uliginosus, quod delectatur non solum aqua sed etiam luto. Itaque ob eam rem aiunt lupos, cum sint nancti sues, trahere usque ad aquam, quod dentes fervorem carnis ferre

6 nequeant. Hoc pecus alitur maxime glande, deinde faba et hordeo et cetero frumento, quae res non modo pinguitudinem efficiunt, sed etiam carnis iucundum saporem. Pastum exigunt aestate mane et, antequam aestus incipiat, subigunt in umbrosum locum, maxime ubi aqua sit; post meridiem rursus lenito fervore pascunt. Hiberno tempore non prius exigunt pastum, quam pruina evanuit ac colliquefacta

7 est glacies. Ad feturam verres duobus mensibus ante secernendi. Optimum ad admissuram tempus a favonio ad aequinoctium vernum; ita enim contingit ut aestate pariat. Quattuor enim menses est praegnas et tunc parit, cum pabulo abundat terra. Neque minores admittendae quam anniculae; melius viginti menses expectare, ut bimae pariant. Cum coeperunt, id facere dicuntur usque ad septimum

8 annum recte. Admissuras cum faciunt, prodigunt in lutosos limites ac lustra, ut volutentur in luto, quae enim illorum requies, ut lavatio hominis. Cum omnes conceperunt, rursus segregant verres. Verris

---

[1] See I, 28, 2.

[2] Schneider suggests *lamas*, "swamps," for *limites*; cf. Hor. *Epist.*, I, 13, 10, *Viribus uteris per clivos flumina lamas*, on which the scholiast quotes a verse from Ennius containing the words *lamasque lutosas*.

that they are not from a diseased herd?' Some buyers add the stipulation that they have got through with fever and diarrhoea.

"In the matter of feeding, ground proper for this animal is wet, as it likes not only water but even mud. It is for this reason, they say, that wolves, when they catch swine, always drag them to water, because their teeth cannot endure the heat of the flesh. As this animal feeds chiefly on mast, and next on beans, barley, and other grains, this food produces not only fat but a pleasant flavour in the flesh. In summer they are driven to pasture early in the day, and before the heat grows intense they are driven into a shady spot, preferably where there is water; then in the afternoon, when the heat has diminished, they are again turned out to pasture. In winter they are not turned out until the frost has disappeared and the ice has melted. In the matter of breeding, the boars are to be separated out two months ahead. The best time for service is from the beginning of the west wind to the spring equinox,[1] as in this case the litter is produced in summer. For the sow is pregnant for four months and will thus bear her young when the land is rich in food. Sows should not be bred when less than a year old, and it is better to wait until they are twenty months old, so that they will be two years old when they bear. When they once begin bearing it is said that they continue to do so satisfactorily up to the seventh year. At the time of breeding they are driven into muddy lanes[2] and pools, so that they may wallow in the mud; for this is their form of refreshment, as bathing is to human beings. After all the sows have conceived, the boars are again separated. The

octo mensum incipit salire, permanet ut id recte facere possit ad trimum, deinde it retro, quaad pervenit ad lanium. Hic enim conciliator suillae carnis datus populo.

9 Sus graece dicitur ὗς, olim θῦς dictus ab illo verbo quod dicunt θύειν, quod est immolare. Ab suillo enim pecore immolandi initium primum sumptum videtur, cuius vestigia, quod initiis Cereris porci immolantur, et quod initiis pacis, foedus cum feritur, porcus occiditur, et quod nuptiarum initio antiqui reges ac sublimes viri in Etruria in coniunctione nuptiali nova nupta et novus maritus primum porcum

10 immolant. Prisci quoque Latini, etiam Graeci in Italia idem factitasse videntur. Nam et nostrae mulieres, maxime nutrices, naturam qua feminae sunt in virginibus appellant porcum, et Graecae choeron, significantes esse dignum insigne nuptiarum. Suillum pecus donatum ab natura dicunt ad epulandum; itaque iis animam datam esse proinde ac salem, quae servaret carnem. E quis succidias Galli optimas et maximas facere consuerunt. Optimarum signum, quod etiam nunc quotannis e Gallia adportantur Romam pernae Comacinae et Cavarae et

11 petasiones. De magnitudine Gallicarum succidiarum Cato scribit his verbis: "in Italia Insubres terna atque quaterna milia aulia[1] succidia uere sus usque

[1] Jordan and Keil would omit *aulia*, which is meaningless.

---

[1] Cf. Virgil, *Aeneid*, VIII, 641: *caesa iungebant foedera porca.*

[2] Ursinus would read *esse id insigne nuptiarum*. A scene in the *Acharnians* of Aristophanes is based on the pun, especially 758-759.

[3] Pliny (*N.H.*, VIII, 207), commenting on the great stupidity of the pig, passes on the humorous saying (attributed by Cicero (*Nat. Deor.*, II, 64; *De Fin.*, V, 13) to Chrysippus the Stoic) that this animal was given life instead of salt (wit) as a preservative.

boar begins to cover at eight months and keeps his vigour up to three years; after which time he begins to deteriorate until he reaches the butcher, the appointed go-between of pork and the populace.

"The Greek name for the pig is ὗς, once called θῦς from the verb θύειν, that is, 'to sacrifice'; for it seems that at the beginning of making sacrifices they first took the victim from the swine family. There are traces of this in these facts: that pigs are sacrificed at the initial rites of Ceres; that at the rites that initiate peace, when a treaty is made, a pig is killed;[1] and that at the beginning of the marriage rites of ancient kings and eminent personages in Etruria, the bride and groom, in the ceremonies which united them, first sacrificed a pig. The ancient Latins, too, as well as the Greeks living in Italy, seem to have had the same custom; for our women, and especially nurses, call that part which in girls is the mark of their sex *porcus*, as Greek women call it *choeros*, meaning thereby that it is a distinctive part mature enough for marriage.[2] There is a saying that the race of pigs is expressly given by nature to set forth a banquet; and that accordingly life was given them just like salt, to preserve the flesh.[3] The Gauls usually make the best and largest flitches of them; it is a sign of their excellence that annually Comacine and Cavarine[4] hams and shoulders are still imported from Gaul to Rome. With regard to the size of the Gallic flitches, Cato[5] uses this language: 'The Insubrians in Italy salt down three and four thousand

[4] Both tribes probably lived in Gallia Narbonensis.

[5] The passage does not occur verbatim in the works of Cato as we have them; but cf. Jordan, *Catonis Frag.*, p. 11.

adeo pinguitudine crescere solet, ut se ipsa stans sustinere non possit neque progredi usquam. Itaque eas siquis quo traicere volt, in plaustrum imponit." In Hispania ulteriore in Lusitania sus cum esset occisus, Atilius Hispaniensis, minime mendax et multarum rerum peritus in doctrina, dicebat L. Volumnio senatori missam esse offulam cum duabus costis, quae penderet tres et viginti pondo, eiusque

12 suis a cute ad os pedem et tres digitos fuisse. Cui ego, Non minus res admiranda cum mi esset dicta in Arcadia, scio me isse spectatum suem, quae prae pinguitudine carnis non modo surgere non posset, sed etiam ut in eius corpore sorex exesa carne nidum fecisset et peperisset mures. Hoc etiam in Venetia [1] factum accepi.

13 Sus ad feturam quae sit fecunda, animadvertunt fere ex primo partu, quod non multum in reliquis mutat. In nutricatu, quam porculationem appellant, binis mensibus porcos sinunt cum matribus; secundo, cum iam pasci possunt, secernunt. Porci, qui nati hieme, fiunt exiles propter frigora et quod matres aspernantur propter exiguitatem lactis, quod dentibus sauciantur propterea mammae. Scrofa in sua quaeque hara suos alat oportet porcos, quod alienos non aspernatur et ideo, si conturbati sunt, in fetura

14 fit deterior. Natura divisus earum annus bifariam, quod bis parit in anno: quaternis mensibus fert

[1] *Venetia* Victorius: *vineta.*

---

[1] Columella (VII, 9, 11): "For pigs, if they get out of the sty, very easily become mixed, and the sow, when she lies down, offers her teats to a strange pig as readily as to her own." Hence Keil has here inserted *non* before *aspernatur*. Compare Section 19 below.

flitches; in spring the sow grows so fat that she cannot stand on her own feet, and cannot take a step; and so when one is to be taken anywhere it is placed in a wagon.' Atilius of Spain, a thoroughly truthful man and one widely versed in a variety of subjects, used to tell the story that when a sow was killed in Lusitania, a district of Farther Spain, there was sent to the senator Lucius Volumnius a piece of the meat with two ribs attached which weighed three-and-twenty pounds; and that the meat of that sow was one foot three fingers thick from skin to bone." "No less remarkable a thing was told me in Arcadia," I remarked; "I recall that I went to look at a sow which was so fat that not only could she not rise to her feet, but actually a shrew-mouse had eaten a hole in her flesh, built her nest, and borne her young. I have heard that the same thing occurred in Venetia."

"It may usually be determined from the first litter which sow is prolific in breeding, as there is not much difference in the number of pigs in the succeeding litters. As to rearing, which is called *porculatio*, the pigs are allowed to remain with their mothers for two months; in the second month, after they are able to feed, they are removed. Pigs born in winter are apt to grow thin on account of the cold and because the mothers drive them off on account of the scantiness of the milk, and the consequent bruising of their teats by the teeth of the pigs. Each sow should have her separate sty in which to feed her pigs; because she does not drive away the pigs of a strange litter, and so, if they become mixed, she deteriorates in breeding.[1] Her year is naturally divided into two parts, as she bears twice a year,

ventrem, binis nutricat. Haram facere oportet circiter trium pedum altam et latam amplius paulo, ea altitudine abs terra, ne, dum exilire velit praegnas, abortet. Altitudinis modus sit, ut subulcus facile circumspicere possit nequi porcellus a matre opprimatur, et ut facile purigare possit cubile. In haris ostium esse oportet et limen inferius altum palmipedale, ne porci, ex hara cum mater prodit, transilire
15 possint. Quotienscumque haras subulcus purgat, totiens harenam inicere oportet, aut quid item quod exsugat umorem in singulas haras inicere debet; et cum peperit, largiore cibatu sustentare, quo facilius lac suppeditare possit. Quibus hordei circiter binas libras aqua madefactas dare solent, quod quidam duplicant, ut sit mane et vesperi, si alia quae obiciant
16 non habuerint. Cum porci depulsi sunt a mamma, a quibusdam delici appellantur neque iam lactantes dicuntur, qui a partu decimo die habentur puri, et ab eo appellantur ab antiquis sacres, quod tum ad sacrificium idonei dicuntur primum. Itaque aput Plautum in Menaechimis, cum insanum quem putat, ut pietur, in oppido Epidamno interrogat " quanti hic porci sunt sacres? " Si fundus ministrat, dari
17 solent vinacea ac scopi ex uvis. Amisso nomine lactantes dicuntur nefrendes ab eo, quod nondum fabam

[1] Literally, "weaned." [2] See note 1, page 326.

being with young for four months and giving suck for two. The sty should be constructed about three feet high, and a little more than that across, at such a height from the ground that if the sow when pregnant should try to jump out, she will not cast her young. The height of the pen should be such that the swineherd can easily look around it, to prevent the little pigs from being crushed by the mother, and be able to clean the bottom without trouble. The sty should have a door with the lower sill one and a third feet high, so that the pigs cannot jump over it when the mother leaves the sty. Whenever the swineherd cleans the sty he should always cover the floor with sand, or throw into each sty something to soak up the moisture; and when a sow has young he should feed her more bountifully so that she may more easily supply milk. They are usually fed about two pounds of barley soaked in water; some double this amount, feeding both morning and evening, if they have no other food to give. Pigs when weaned are by some people no longer called 'sucking-pigs,' but *delici*[1] or shoats. On the tenth day after birth they are considered 'pure,' and for that reason the ancients called them *sacres*, because they are said to be fit for sacrifice first at that age. Hence in Plautus's play, the *Menaechmi*,[2] the scene of which is laid in Epidamnus, a character who, thinking that another is mad, wants him to make sacrifice and be cured, asks: 'What's the price of *porci sacres* in this town?' Wine dregs and grape refuse are usually fed them if the farm produces these. When they have outgrown the name of sucking-pigs they are called *nefrendes*, from the fact that they are not able to

frendere possunt, id est frangere. Porcus Graecum est nomen antiquum, sed obscuratum, quod nunc eum vocant choeron. In eorum partu scrofae bis die ut bibant curant lactis causa. Parere dicunt oportere porcos, quot mammas habeat; si minus pariat, fructuariam idoneam non esse; si plures

18 pariat, esse portentum. In quo illud antiquissimum fuisse scribitur, quod sus Aeneae Lavini triginta porcos peperit albos. Itaque quod portenderit factum, post tricesimum annum ut Lavinienses condiderint oppidum Albam. Huius suis ac porcorum etiam nunc vestigia apparent, quod et simulacra eorum ahenea etiam nunc in publico posita, et corpus matris ab sacerdotibus, quod in salsura fuerit, de-

19 monstratur. Nutricare octonos porcos parvulos primo possunt; incremento facto a peritis dimidia pars removeri solet, quod neque mater potest sufferre lac, neque congenerati alescendo roborari. A partu decem diebus proximis non producunt ex haris matrem, praeterquam potum. Praeteritis decem diebus sinunt exire pastum in propincum locum villae, ut crebro

20 reditu lacte alere possint porcos. Cum creverunt, patiuntur sequi matrem pastum domique secernunt a matribus ac seorsum pascunt, ut desiderium ferre possint parentis nutricis, quod decem diebus assecuntur. Subulcus debet consuefacere, omnia ut faciant ad bucinam. Primo cum incluserunt, cum

---

[1] Varro, *de Ling. Lat.*, V, 19: *Porcus nisi si a Graecis quod Athenis in libris sacrorum scriptum est porcae, porco.*

[2] Cf. Virgil, *Aeneid*, III, 390–393.

[3] *i.e.* "white."

'crunch' (*frendere*), that is crush, beans. *Porcus* is an old Greek[1] word, but it is obsolete, as they now use the word *choeros*. At the time of bearing, care is taken to see that the sows drink twice a day for the sake of the milk. The saying is that a sow should bear as many pigs as she has teats; if she bear less she will not pay for herself, and if she bear more it is a portent. It is recorded that the most ancient portent of this kind is the sow of Aeneas [2] at Lavinium, which bore thirty white pigs; and the portent was fulfilled in that thirty years later the people of Lavinium founded the town of Alba.[3] Traces of this sow and her pigs are to be seen even to this day; there are bronze images of them standing in public places even now, and the body of the sow is exhibited by the priests, having been kept in brine, according to their account. A sow can feed eight little pigs at first; but when they have taken on weight it is the practice of experienced breeders to remove half of them, as the mother cannot supply enough milk and the whole of the litter cannot grow fat. The mother is not driven out of the sty except for water during the first ten days after delivery; but after this time they are allowed to range for food in near-by parts of the steading, so that they may come back often and feed their pigs. When these are grown they are allowed to follow the mother to pasture; but when they come home they are separated from the mothers and fed apart, so as to grow accustomed to the lack of the mother's nourishment, a point which they reach in ten days. The swineherd should accustom them to do everything to the sound of the horn. At first they are penned in; and then, when the

bucinatum est, aperiunt, ut exire possint in eum locum, ubi hordeum fusum in longitudine. Sic enim minus disperit, quam si in acervo positum, et plures facilius accedunt. Ideo ad bucinam convenire di- 21 cuntur, ut silvestri loco dispersi ne dispereant. Castrantur verres commodissime anniculi, utique ne minores quam semestres; quo facto nomen mutant atque e verribus dicuntur maiales. De sanitate suum unum modo exempli causa dicam: porcis lactentibus si scrofa lac non potest suppeditare, triticum frictum dari oportet (crudum enim solvit alvom) vel 22 hordeum obici ex aqua, quaad fiant trimestres. De numero in centum sues decem verres satis esse putant; quidam etiam hinc demunt. Greges inaequabiles habent; sed ego modicum puto centenarium; aliquot maiores faciunt, ita ut ter quinquagenos habeant. Porcorum gregem alii duplicant, alii etiam maiorem faciunt. Minor grex quam maior minus sumptuosus, quod comites subulcus pauciores quaerit. Itaque gregis numerum pastor ab sua utilitate constituit, non ut quot verres habeat; id enim ab natura sumendum.

V. Haec hic. At Quintus Lucienus senator, homo quamvis humanus ac iocosus, introiens, familiaris omnium nostrum, Synepirotae, inquit, χαίρετε; Scrofam enim et Varronem nostrum, ποιμένα λαῶν, mane salutavi. Cum alius eum salutasset, alius conviciatus

---

[1] The word *maialis* was derived from Maia, to whom this sacrifice was made.

[2] Cf. II, 4, 1, where Atticus and Cossinius are called *semi-Graeci*. This explains the use of the greeting, χαίρετε.

[3] The usual Homeric address to kings and generals. Varro was the commanding officer at the time; cf. the Introduction to this book, Sec. 6. But the entire passage is jocular.

horn sounds, the sty is opened so that they can come out into the place where barley is spread out. This is spread in a row because in that way less is wasted than if it is heaped up, and more of them can reach it easily. The idea in having them gather at the sound of the horn is that they may not become lost when scattered in wooded country. The best time for castrating the boars is when one year old, and certainly not less than six months; when this is done their name is changed, and they are called 'barrows'[1] instead of boars. As to the health of swine, I shall give but one illustration: if the sow cannot furnish enough milk for the sucking-pigs, toasted wheat should be fed (for raw wheat loosens the bowels), or barley soaked in water, until they are three months old. As to numbers, ten boars are considered enough for 100 sows, and some breeders even lessen this number. The number in a herd varies; for myself I consider a herd of 100 a reasonable number, but some breeders have larger ones, the number sometimes going as high as 150. Some double the size of the herd, and others have even a larger herd. A rather small herd is less expensive than one too large, as the herdsman requires fewer helpers; and so the breeder determines the size of the herd by his own advantage, and not as he determines the number of boars to keep, as this latter point is derived from nature."

V. So far he. At this point Quintus Lucienus, the senator, a thoroughly kindly and jovial person, and a friend to all the company, entered and said: "Greetings, fellow-citizens of Epirus;[2] for to Scrofa and to our friend Varro, shepherd of the people,[3] I paid my greetings this morning." One returned his greeting

esset, qui tam sero venisset ad constitutum, Videbo iam vos, inquit, balatrones, et hoc adferam meum corium et flagra. Tu vero, Murri, veni mi advocatus, dum asses solvo Laribus,[1] si postea a me repetant, 2 ut testimonium perhibere possis. Atticus Murrio, Narra isti, inquit, eadem, qui sermones sint habiti et quid reliqui sit, ut ad partes paratus veniat; nos interea secundum actum de maioribus adtexamus. In quo quidem, inquit Vaccius, meae partes, quoniam boves ibi. Quare dicam, de bubulo pecore quam acceperim scientiam, ut, siquis quid ignorat, discat; siquis scit, nuncubi labar observet. Vide quid agas, 3 inquam, Vacci. Nam bos in pecuaria maxima debet esse auctoritate, praesertim in Italia, quae a bubus nomen habere sit existimata. Graecia enim antiqua, ut scribit Timaeus, tauros vocabat italos, a quorum multitudine et pulchritudine et fetu vitulorum Italiam dixerunt. Alii scripserunt, quod ex Sicilia Hercules persecutus sit eo nobilem taurum, qui diceretur italus. Hic socius hominum in rustico opere 4 et Cereris minister, ab hoc antiqui manus ita abstineri voluerunt, ut capite sanxerint, siquis occidisset. Qua

[1] The MSS. have *Palibus*. Keil suggests *Laribus*, quoting a fragment (Nonius 538) of Varro which states that "asses" were paid to the Lares.

---

[1] For the whipping he has earned for being late; but no satisfactory explanation of the next sentence has been given.

[2] The inevitable pun on *vacca*, "cow."

[3] A Greek historian of whose history of Sicily down to 264 B.C. only a few fragments remain. The etymology is accepted.

[4] Doubtless one of the cattle of Geryon, perhaps one stolen by Cacus, to whom Columella refers in I, 3, 7.

[5] Cf. Columella, Book VI, Sec. 7 of Introduction: "At Athens in Attica the servant of Ceres and Triptolemus,

and another chid him for coming late to his appointment; whereupon he remarked: "I'll see you again presently, my merry men, and bring my skin and whips back with me.[1] But as for you, Murrius, come along as my backer while I am paying my pence to the Lares, so that if they demand them from me later you can bear me witness." "Tell him while you are going," said Atticus to Murrius, "how far our conversation has gone and what has not been discussed, so that he may come back ready for his part; and let us meanwhile tack on the second act, on the larger animals." "That is where my part comes in," said Vaccius,[2] "since there are cows in it. So I shall give you the advantage of the knowledge I have acquired on the subject of the cattle herd, so that he who is ignorant may learn, and he who knows may see where I go wrong." "Watch your step, Vaccius," said I; "for the cow should be in the highest esteem among cattle, and especially in Italy, which is supposed to have derived its name from the word for oxen. For the ancient Greeks, according to Timaeus,[3] called bulls *itali*, and the name Italy was bestowed because of the number and beauty of its cattle, and the great number of calves. Others say it is so named from the fact that Hercules chased hither from Sicily a noble bull which was called *italus*.[4] This is man's partner in his rustic labours and is the servant of Ceres; and hence the ancients so wished his life to be safe that they made it a capital offence to kill one.[5] In this matter Attica is witness

shares the skies with the brightest constellations, is the most hard-working comrade of man, and such was the veneration for this animal among our ancestors, that it was as much an offence to kill an ox as to kill a citizen." Cf. also Aratus, *Phaen.*, 134 (p. 390 of L. C. L.); Virgil, *Georg.*, II, 537.

in re testis Attice, testis Peloponnesos. Nam ab hoc pecore Athenis Buzuges nobilitatus, Argis Bomagiros.[1] Novi, inquit ille, maiestatem boum et ab his dici pleraque magna, ut busycon bupaeda bulimon 5 boopin, uvam quoque bumammam. Praeterea scio hunc esse, in quem potissimum Iuppiter se convertit, cum exportavit per mare e Phoenice amans Europam; hunc esse, qui filios Neptuni a Menalippa servarit, ne in stabulo infantes grex boum obtereret; denique ex hoc putrefacto nasci dulcissimas apes, mellis matres, a quo eas Graeci bugenes appellant; et hunc[2] Plautium locutum esse Latine quam Hirrium praetorem renuntiatum Romam in senatum scriptum habemus. Sed bono animo es, non minus satisfaciam tibi, quam qui Bugoniam scripsit.

6 Primum in bubulo genere aetatis gradus dicuntur quattuor, prima vitulorum, secunda iuvencorum, tertia bovum novellorum, quarta vetulorum. Discernuntur in prima vitulus et vitula, in secunda iuvencus et iuvenca, in tertia et quarta taurus et

[1] *Bomagiros* Wilamowitz: *Homogyros*.
[2] So the accepted text; but it is evidently corrupt.

---

[1] Buzuges, "he who yoked oxen," is either Triptolemus or Epimenides; later it was the title of the keeper of the sacred cattle at Athens or Eleusis. Because *Homogyros*, the reading of the older editions, does not contain the prefix *Bo*, Wilamowitz suggested *Bomagiros*, which Keil accepts into his text.

[2] *i.e.* with the prefix *bu*=*βου*.

[3] The text is hopelessly corrupt, and various efforts have been made either to emend or to explain. The speaking of an ox is frequently mentioned as a prodigy (Livy, XXXV, 21, 4–5; Pliny, VIII, 183), but we have no anecdote of this nature. The translation follows the explanation of Schöll.

as well as Peloponnesus; for it is to this animal that Buzuges owes his fame at Athens, and Bomagiros at Argos."[1] "I am acquainted," replied Vaccius, "with the high esteem in which oxen are held, and the fact that many large things are named from them,[2] such as *busycos* (bull fig), *bupais* (bull-boy), *bulimos* (bull hunger), *boopis* (cow-eyed), and that a grape also has the name *bumamma* (cow's udder). I know, further, that it was this animal into which Jupiter chose to change himself when he carried his beloved Europa over the sea from Phoenicia; that it was this animal which saved the sons of Neptune by Menalippa from being trampled in the stall, when they were infants, by a herd of cattle; further, that it is from the putrefied body of this animal that there spring the sweetest bees, those honey-mothers from which the Greeks therefore call bees 'the ox-sprung' (*βουγενεῖς*); and we have the official record that the praetor reported to the Senate at Rome that it was this animal which said, in Latin, 'Plautius rather than Hirrius.'[3] So be of good cheer; I shall give you as much satisfaction as the author of the Bugonia[4] could.

"First: in the race of cattle four stages of life are distinguished, the first that of calf, the second that of yearling, the third that of prime, the fourth that of old; and a distinction of sex is indicated in each stage, in the first by bull-calf and heifer-calf, in the second by bullock and heifer, and in the

[4] The work is unknown. Various editors have taken this to be a poem in praise of bees because of the connection with *βουγενεῖς* above. There seems no evidence for this and *βουγονία* must refer to the birth of oxen, who form the main subject of the whole paragraph.

vacca. Quae sterilis est vacca, taura appellata; quae praegnas, horda. Ab eo in fastis dies hordicidia
7 nominatur, quod tum hordae boves immolantur. Qui gregem armentorum emere vult, observare debet primum, ut sint eae pecudes aetate potius ad fructos ferendos integrae quam iam expartae; ut sint bene compositae, ut integris membris, oblongae, amplae, nigrantibus cornibus, latis frontibus, oculis magnis et nigris, pilosis auribus, compressis malis subsimae, ne gibberae, spina leviter remissa, apertis naribus,
8 labris subnigris, cervicibus crassis ac longis, a collo palea demissa, corpore bene costato, latis umeris, bonis clunibus, codam profusam usque ad calces ut habeant, inferiorem partem frequentibus pilis subcrispam, cruribus potius minoribus rectis, genibus eminulis distantibus inter se, pedibus non latis, neque ingredientibus qui displudantur, nec cuius ungulae divarent, et cuius ungues sint leves et pares, corium tactu non asperum ac durum, colore potissimum nigro, deinde robeo, tertio helvo, quarto albo; mol-
9 lissimus enim hic, ut durissimus primus. De mediis duobus prior quam posterior in eo prior, utrique plures quam nigri et albi. Neque non praeterea ut mares seminis boni sint, quorum et forma est spectanda, et qui ex his orti sunt respondent ad parentum speciem. Et praeterea quibus regionibus nati sint refert; boni enim generis in Italia plerique Gallici
10 ad opus, contra nugatorii Ligusci; transmarini Epiro-

[1] A festival in honour of Tellus, held at Rome, 15th April.

third and fourth by bull and cow. A sterile cow is called *taura*, and a pregnant cow is called *horda*. It is from the fact that at that time pregnant cows are sacrificed that one of the days in the calendar is called *hordicidia*.[1] One who wishes to buy a herd of cattle should be careful to have animals of such an age that they are sound for bearing calves rather than those which have already reached the age of barrenness. They should be well formed, that is, clean-limbed, square-built, large, with blackish horns, wide foreheads, large black eyes, hairy ears, narrow jaws, somewhat snub-nosed, not humpbacked, but with a slight depression of the spine, spreading nostrils, blackish lips, thick, long neck, with dewlap hanging from it, body well ribbed, broad shoulders, sturdy rump, a long tail hanging down to the ankles, curling somewhat at the end with thick hair, with legs rather short and straight, knees prominent and a good distance apart, feet not wide and not splaying as they walk, the hoofs not widely cloven but with the two toes smooth and of equal size, the skin not hard and rough to the touch. The best colour is black, next red, then dun, and then white; for those of the last mentioned colour are most delicate, and those of the first most hardy. Of the other two colours the first is preferable to the second, while both are more common than the black and the white. He should furthermore see that the males be of good breed, and their conformation should be looked to, for their young reproduce the characteristics of the parents. It is also a matter of importance where they are born; thus in Italy many Gallic oxen of good breed are good workers, while the Ligurian are of small account; of foreign cattle those of

tici non solum meliores totius Graeciae, sed etiam quam Italiae. Tametsi quidam de Italicis, quos propter amplitudinem praestare dicunt, victimas faciunt atque ad deorum servant supplicia, qui sine dubio ad res divinas propter dignitatem amplitudinis et coloris praeponendi. Quod eo magis fit, quod albi in Italia non tam frequentes quam in Thracia ad μέλανα κόλπον, ubi alio colore pauci. Eos cum emimus domitos, stipulamur sic: " illosce boves sanos

11 esse noxisque praestari "; cum emimus indomitos, sic: " illosce iuvencos sanos recte deque pecore sano esse noxisque praestari spondesne? " Paulo verbosius haec, qui Manili actiones secuntur lanii, qui ad cultrum bovem emunt; qui ad altaria, hostiae sanitatem non solent stipulari.

Pascuntur armenta commodissime in nemoribus, ubi virgulta et frons multa; hieme cum hibernant secundum mare, aestu abiguntur in montes frondosos.

12 Propter feturam haec servare soleo. Ante admissuram mensem unum ne cibo et potione se impleant, quod existimantur facilius macrae concipere. Contra tauros duobus mensibus ante admissuram herba et palea ac faeno facio pleniores et a feminis secerno. Habeo tauros totidem, quot Atticus, ad matrices LXX duo, unum anniculum, alterum bimum. Haec

---

[1] Now the Gulf of Samos [= Saros].

[2] For the white oxen of Umbria, used especially at triumphs, cf. Virgil, *Georg.*, II, 146.

[3] Cf. Chap. 2, 5-6; Chap. 3, 5; Chap. 4, 5 of this book.

Epirus are not only the best in all Greece, but are even better than the Italian. Yet some people use cattle of Italian breeds, which they claim excel in size, as offerings, and these they reserve for solemn offerings to the gods. These are doubtless to be preferred for sacrificial purposes because of the splendour of their size and colour; and this is done all the more because white cattle are not so common in Italy as they are in Thrace on the shores of the Black Gulf[1] where there are few of any other colour.[2] In the purchase of oxen which have been broken in, the bargain is in these terms: 'Do you guarantee that the said oxen are sound, and that I am protected from suits for damage?' In buying them unbroken, the formula runs: 'Do you guarantee that the said bullocks are quite sound and of a sound herd, and that I am protected from suits for damage?' Butchers use a somewhat fuller form, following the rule of Manilius,[3] in buying for slaughter; those who buy for sacrifice do not usually demand a guarantee of soundness in the victim.

"Large cattle are most conveniently pastured on wooded land where there is much undergrowth and foliage; and those that spend the winter along the coast are driven in summer into the leafy hills. In the matter of breeding I usually follow these principles: for one month before they are mated, cows should not have their fill of food and drink, because it is thought that when thin they are in better condition to conceive. On the other hand, I keep the bulls filled with grass, straw, and hay for two months before mating; and I keep them away from the females. I keep the same number of bulls as Atticus—two to every 70 brood cows—one a

secundum astri exortum facio, quod Graeci vocant
13 lyran, fidem nostri. Tum denique tauros in gregem redigo. Mas an femina sit concepta, significat descensu taurus, cum init, quod, si mas est, in dexteriorem partem abit; si femina, in sinisteriorem. Cur hoc fiat, vos videritis, inquit mihi, qui Aristotelem legitis. Non minores oportet inire bimas, ut trimae pariant, eo melius, si quadrimae. Pleraeque pariunt in decem annos, quaedam etiam plures. Maxime idoneum tempus ad concipiendum a delphini exortu usque ad dies quadraginta aut paulo plus. Quae enim ita conceperunt, temperatissimo anni tempore pariunt; vaccae enim mensibus decem
14 sunt praegnates. De quibus admirandum scriptum inveni, exemptis testiculis si statim admiseris taurum, concipere. Eas pasci oportet locis viridibus et aquosis. Cavere oportet ne aut angustius stent aut feriantur aut concurrant. Itaque quod eas aestate tabani concitare solent et bestiolae quaedam minutae sub cauda ali, ne concitentur, aliqui solent includere saeptis. Iis substerni oportet frondem aliudve quid in cubilia, quo mollius conquiescant. Aestate ad
15 aquam appellendum bis, hieme semel. Cum parere

---

[1] Columella, XI, 2, 40, gives the date 13th May.

[2] *De Gen. An.*, IV, 1. states that sex is distinguishable in the embryonic stage, but whether such distinction is made before we can observe it, is debated. "For some say that this distinction is in the very seeds; thus Anaxagoras and some of the physiologists: that the seed is produced from the male and that the female *vagina* supplies the female; that the male comes from the right and the female from the left, and that the male is on the right side of the womb and the female on the left." Columella (VI, 24, 3) repeats this statement of Varro's, which is absurd.

[3] *Tabanus* was the popular name for the Greek οἶστρος.

yearling, the other a two-year-old. I attend to this matter following the rising of the constellation which the Greeks call Lyra, and which our people call Fides[1]—it is only then that I turn the bulls into the herd. The bull shows by the way he dismounts whether a male or female has been conceived by his act: if it is a male he comes down on the right side, and if a female on the left. Why this is true," he remarked to me, "you who read Aristotle[2] will have to find out. Cows should not be covered which are less than two years old, so that they may be three years old when they bear; and it will be all the better if they are four years old. Most of them continue bearing up to ten years, and some of them even longer. The best time for mating is from the rising of the Dolphin up to forty days or a little more; for cows which conceive at that time drop their calves at the most temperate season, as cows carry their calves for ten months. On this subject I have seen a remarkable statement—that if you turn in a bull immediately after he has been castrated, he can get a calf. The cows should be pastured in grassy and watered ground, and care should be taken not to let them crowd, be struck, or run against one another. As cattle-flies[3] have a way of tormenting them in summer and certain minute insects grow under their tails, some breeders keep them shut up in pens, to keep them from being worried. Their pens should be strewn with a bedding of leaves or some such thing, so that they may rest in greater comfort. In summer they should be driven to water twice, in winter once. When they come to the

Virgil, *Georg.*, III, 146, calls it *asilus* and says Juno sent it to plague Io.

coeperunt, secundum stabula pabulum servari oportet integrum, quod egredientes degustare possint; fastidiosae enim fiunt. Et providendum, quo recipiunt se, ne frigidus locus sit; algor enim eas et famis
16 macescere cogit. In alimoniis armenticium pecus sic contuendum. Lactantes cum matribus ne cubent; obteruntur enim. Ad eas mane adigi oportet, et cum redierunt e pastu. Cum creverunt vituli, levandae matres pabulo viridi obiciendo in praesepiis. Item his, ut fere in omnibus stabulis, lapides sub-sternendi aut quid item, ne ungulae putrescant. Ab aequinoctio autumnali una pascuntur cum matribus.
17 Castrare non oportet ante bimum, quod difficulter, si aliter feceris, se recipiunt; qui autem postea castrantur, duri et inutiles fiunt. Item ut in reliquis gregibus pecuariis dilectus quotannis habendus et reiculae reiciundae, quod locum occupant earum quae ferre possunt fructus. Siquae amisit vitulum, ei supponere oportet eos, quibus non satis praebent matres. Semestribus vitulis obiciunt furfures triticios et farinam hordeaceam et teneram herbam et
18 ut bibant mane et vesperi curant. De sanitate sunt complura, quae exscripta de Magonis libris armentarium meum crebro ut aliquid legat curo. Numerus de tauris et vaccis sic habendus, ut in sexaginta

time of calving, fresh fodder should be kept near the stalls for them to nibble at as they go out, for they become dainty. Care should also be taken that the place into which they are turned shall not be chilly, for chill and hunger make them grow thin. In the matter of rearing, the following rules should be observed with this kind of animal: Sucklings must not sleep with their dams, as they will be trampled; they should be admitted to their dams in the morning and when they come back from pasture. When the calves have made some growth, the dams should be relieved by throwing green food before the calves in the pens. These stalls (and this holds good for practically all stalls) should be paved with stones or something of the sort, so that their hoofs may not rot. After the autumnal equinox calves pasture along with their dams. They should not be castrated until they are two years old, because it is hard for them to recover otherwise; while those which are castrated later become tough and worthless. Just as in the case of other herds, there should be a culling once a year, and the culls should be cut out of the herd, as they take up the room of those which can bring in a profit. If a cow has lost her calf she should be given some whose dams do not give enough milk. Calves six months old are fed wheat bran and barley-meal and tender grass, and care is taken that they drink morning and evening. On the subject of health there are many rules; these have been copied down from Mago's treatise, and I see to it that my head herdsman is reading some of them repeatedly. As to the number of bulls and cows, the rule is that there be, to every sixty cows, one yearling bull and one two-year-old. Some breeders

unus sit anniculus, alter bimus. Quidam habent aut minorem aut maiorem numerum; nam apud Atticum duo tauri in septuaginta matribus sunt. Numerum gregum alius facit alium, quidam centenarium modicum putant esse, ut ego. Atticus centum viginti habet, ut Lucienus.

VI. Haec ille. At Murrius, qui, dum loquitur Vaccius, cum Lucieno redisset, Ego, inquit, de asinis potissimum dicam, quod sum Reatinus, ubi optimi et maximi fiunt, e quo seminio ego hic procreavi

2 pullos et ipsis Arcadibus vendidi aliquotiens. Igitur asinorum gregem qui facere vult bonum, primum videndum ut mares feminasque bona aetate sumat, utrique ut quam diutissime fructum ferre possint; firmos, omnibus partibus honestos, corpore amplo, seminio bono, ex his locis, unde optumi exeunt, quod faciunt Peloponnesi cum potissimum eos ex Arcadia emant, in Italia ex agro Reatino. Non enim, si murenae optimae flutae sunt in Sicilia et helops ad Rhodon, continuo hi pisces in omni mari similes nascuntur. Horum genera duo: unum ferum, quos

3 vocant onagros, ut in Phrygia et Lycaonia sunt greges multi; alterum mansuetum, ut sunt in Italia omnes. Ad seminationem onagrus idoneus, quod et e fero fit mansuetus facile et e mansueto ferus

---

[1] Cf. II, 1, 14, and II, 8, 3.

[2] *Murenae flutae* were so designated for the reason that they *floated* because of their fatness. Columella, VIII, 17, 8, calls them the most highly valued of the lampreys. Pliny says (IX, 60) that the *helops* was the same as the *acipenser* (perhaps

make the number smaller or larger; as, for instance, in Atticus's herd there are two bulls to seventy breeding cows. The number of animals in a herd varies with the owner, some breeders (and I am one of them) considering a hundred a reasonable number. But Atticus has 120, as does Lucienus."

VI. So far Vaccius. Then Murrius, who had returned with Lucienus while Vaccius was speaking, said: "I shall speak by preference on the subject of asses, as I am from Reate, where the best and largest are grown; out of this stock I have bred colts here and several times sold them even to Arcadians.[1] One who wishes, then, to start a good herd of asses should first be careful to get males and females of the proper age, so that they both may continue to bring in a profit as long as possible. They should be sturdy, sound in all parts, full-bodied, of good stock, and from those districts from which the best come; this is a point considered by those breeders in Peloponnesus who, by preference, buy in Arcadia, and those in Italy who buy in the neighbourhood of Reate.[1] For it does not at all follow that, because the best 'floating' lampreys grow in Sicily and the *helops*[2] off Rhodes, these fish grow of the same excellence in all seas. There are two species of these animals: the wild ass, called *onagrus*, of which there are many herds, as, for instance, in Phrygia and Lycaonia; and the domesticated, such as are all those in Italy. The wild ass is well suited for breeding, because he is easily changed from wild to tame and never changes

the sturgeon) and was of no esteem in his day. Columella, VIII, 16, 9, says that it fed only in the depths of the Pamphylian Sea (Gulf of Adalia), in Asia Minor.

numquam. Quod similes parentum genuntur, eligendi et mas et femina cum dignitate ut sit. In mercando item ut ceterae pecudes emptionibus et traditionibus dominum mutant, et de sanitate ac

4 noxa solet caveri. Commode pascuntur farre et furfuribus hordeaceis. Admittuntur ante solstitium, ut eodem tempore alterius anni pariant; duodecimo enim mense conceptum semen reddunt. Praegnates opere levant; venter enim labore nationem reddit deteriorem. Marem non deiungunt ab opere, quod remissione laboris fit deterior. In partu eadem fere observant, quae in equis. Secundum partum pullos anno non removent a matre. Proximo anno noctibus patiuntur esse cum his et leniter capistris aliave qua re habent vinctos. Tertio anno domare incipiunt ad eas res, ad quas quisque eos vult habere in usu.

5 Relinquitur de numero, quorum greges non sane fiunt, nisi ex eis[1] qui onera portant, ideo quod plerique diducuntur ad molas aut ad agri culturam, ubi quid vehendum est, aut etiam ad arandum, ubi levis est terra, ut in Campania. Greges fiunt fere mercatorum, ut eorum qui e Brundisino aut Apulia asellis dossuariis comportant ad mare oleum aut vinum itemque frumentum aut quid aliut.

VII. Lucienus: Ego quoque adveniens aperiam carceres, inquit, et equos emittere incipiam, nec solum

[1] *nisi ex eis* Keil: *niestei.*

---

[1] But this statement seems to be contradicted by Job, xxxix, 4 ff.; and Columella, VI, 37, 4, seems also to contradict it indirectly, saying that the offspring of an *onager* and a mare is unbroken and stubborn. He advises that the male offspring be put to a mare, so that the wildness of the *onager* may be mitigated in the third generation.

[2] Undoubtedly the summer solstice.

back from tame to wild.[1] As the young reproduce the qualities of their parents, both sire and dam should be chosen with an eye to their worth. In the matter of transfer of title, they change owners, as do other animals, by purchase and delivery; and there is the usual guarantee of soundness and against liability for damage. The best food for them is spelt and barley bran. They are bred before the solstice,[2] so that they may drop their colts at the same season the next year; for they foal in the twelfth month after conception. Pregnant jennies are relieved of work, as work makes the womb bear a poorer offspring. The male is not kept from work, as he loses vigour from lack of labour. In the matter of foaling about the same rules are followed as in the case of mares. The young are not separated from their dams for a year after birth; but during the next year they are allowed to be with them at night, and are kept loosely tied with a leather halter or the like. In the third year they begin their training for the work for which their owners wish to keep them. There remains the question of number; but there really are no herds of these animals except of those which form pack trains, for the reason that they are usually separated and sent to the mills, or to the fields for hauling, or even for ploughing where the land is porous, as it is in Campania. The trains are usually formed by the traders, as, for instance, those who pack oil or wine and grain or other products from the region of Brundisium or Apulia to the sea in donkey panniers."

VII. "I too," broke in Lucienus, "shall open the barriers[3] as I come, and begin to let out the steeds,

[3] In the race, the horses started from stalls which were closed by barriers.

mares, quos admissarios habeo, ut Atticus, singulos in feminas denas. E quis feminas Q. Modius Equiculus, vir fortissimus, etiam patre militari, iuxta ac mares habere solebat. Horum equorum et equarum greges qui habere voluerunt, ut habent aliqui in Peloponneso et in Apulia, primum spectare oportet aetatem, quam praecipiunt sic. Videmus ne sint 2 minores trimae, maiores decem annorum. Aetas cognoscitur et equorum et fere omnium qui ungulas indivisas habent et etiam cornutarum, quod equus triginta mensibus primum dentes medios dicitur amittere, duo superiores, totidem inferiores. Incipientes quartum agere annum itidem eiciunt et totidem eiciunt proxumos eorum quos amiserunt, et 3 incipiunt nasci, quos vocant columellares. Quinto anno incipiente item eodem modo amittere binos, cum cavos habeat tum renascentes, ei sexto anno impleri, septumo omnes habere solet renatos et completos. Hoc maiores qui sunt, intellegi negant posse, praeterquam cum dentes sint facti brocchi et supercilia cana et sub ea lacunae, ex observatu 4 dicunt eum equom habere annos sedecim. De forma esse oportet magnitudine modica, quod nec vastos nec minutos decet esse, equas clunibus ac ventribus latis. Equos, ad admissuram quos velis habere, legere oportet amplo corpore, formosos, nulla parte 5 corporis inter se non congruenti. Qualis futurus sit equus, e pullo coniectari potest: si caput habet non

---

[1] Lit. "little columns." The four sharp-pointed tearing-teeth, between the incisors and the molars.

[2] Varro here follows Aristotle (*H. A.*, VI, 22), and is quite accurate. Xenophon, in his treatise *On Horsemanship*, Chapter I, goes into more detail.

[3] *brocchi.* Nonius (Bk. I, *s.v.*) says that the word was used of horses *producto ore dentibus prominentibus.*

and not the stallions only, which I keep for breeding, as Atticus does, one to every ten mares. The females of these Quintus Modius Equiculus, a very gallant gentleman whose father was also a soldier, used to value as highly as the males. Those who wish to establish a herd of horses and mares, as some do in the Peloponnesus and in Apulia, should first have an eye to age; and the following rules are laid down: We are careful to have them not less than three nor more than ten years old. The age of horses and of almost all animals with solid hoof, and in fact of those with horns, is determined by the teeth, the horse being said to drop, at thirty months, first the middle teeth, two upper and as many lower; at the beginning of the fourth year they again cast, this time dropping the same number of those coming next to those which they have lost; and the so-called canine teeth [1] begin to grow. At the beginning of the fifth year they again shed two in each jaw in the same way, as at that time the animal has hollow fresh teeth which fill out in the sixth year, so that in the seventh it usually has a full set of permanent teeth.[2] It is said that there is no way of determining those which are older than this, except that when the teeth become prominent [3] and the brows grey with hollows under them, they determine by looking at him that such a horse is sixteen years old. As to conformation, they should be of moderate size, neither over nor under size, and the mares should have broad quarters and bellies. Stallions kept for breeding should be chosen of broad body, handsome, with no part of the body breaking the harmony. What sort of a horse is going to turn out can be determined from the colt: if it has a head not over

magnum nec membris confusis si est, oculis nigris, naribus non angustis, auribus applicatis, iuba crebra, fusca, subcrispa subtenuibus saetis, inplicata in dexteriorem partem cervicis, pectus latum at plenum, umeris latis, ventre modico, lumbis deorsum versus pressis, scapulis latis, spina maxime duplici, si minus, non extanti, coda ampla subcrispa, cruribus rectis aequalibus intro versus potius figuratis, genibus rutundis ne magnis, ungulis duris; toto corpore ut habeat venas, quae animadverti possint, quod qui huiusce modi sit, cum est aeger, ad medendum appositus.

6 De stirpe magni interest qua sint, quod genera sunt multa. Itaque ab hoc nobiles a regionibus dicuntur, in Graecia Thessalici equi a Thessalia, in Italia ab Apulia Apuli, ab Rosea Roseani.[1] Equi boni futuri signa, si cum gregalibus in pabulo contendit in currendo aliave qua re, quo potior sit; si, cum flumen travehundum est gregi, in primis progreditur ac non respectat alios. Emptio equina similis fere ac boum et asinorum, quod eisdem rebus in emptione dominum mutant, ut in Manili actionibus sunt perscripta.

7 Equinum pecus pascendum in pratis potissimum herba, in stabulis ac praesepibus arido faeno; cum pepererunt, hordeo adiecto, bis die data aqua. Horum feturae initium admissionis facere oportet ab aequinoctio verno ad solstitium, ut partus idoneo tempore fiat; duodecimo enim mense die decimo

[1] *a Thessalia . . . Roseani* restored by Keil, following Pontedera: *a terra apulia brosea roseani.*

---

[1] Meaning with the spine lying in a groove between two ridges of muscles; cf. Virgil, *Georg.*, III, 75–88, and Columella, VI, 29, 2.

[2] By means of letting blood.

[3] Cf. II, 3, 5.

size and well-proportioned limbs, dark eyes, full nostrils, close-lying ears; mane abundant, dark, slightly curling, with very fine hair falling on the right side of the neck; broad, full chest, broad shoulders, fair-sized barrel, flanks converging downward, broad shoulder-blades, preferably with a double spine[1] or at least with the backbone not prominent, full, somewhat curly tail, legs straight and sloping symmetrically rather inward than outward, the knees round but not large, and hard hoofs. The veins should be visible over the whole body, as a horse of this kind is capable of easy treatment when it is sick.[2] The stock from which they come is of great importance, as there are a number of breeds; hence noted breeds are named from the districts from which they come, as in Greece the Thessalian from Thessaly, in Italy the Apulian from Apulia, and the Rosean from Rosea. It is a sign that the horse will be a good one if, when in pasture with its mates, it vies with them in racing or in other ways to show its superiority; if, when a river is to be crossed by the herd, it runs with the leaders and does not look back at the rest of the herd. The terms of purchase for horses are practically the same as those for cattle and asses, as they change owners by purchase on the same terms, as laid down in the decisions of Manilius.[3]

"The breeding stud of horses is best fed in meadows on grass, and in stalls and enclosures on dry hay; and when they have foaled, with an additional ration of barley, and with water twice a day. In the matter of breeding, the beginning of mating should be at the vernal equinox and it should continue until the solstice, so that the foal may come at a seasonable time; for it is said that they are born on the tenth

aiunt nasci. Quae post tempus nascuntur, fere
8 vitiosa atque inutilia existunt. Admittere oportet,
cum tempus anni venerit, bis die, mane et vespere,
per origam ; sic appellatur quiqui admittit. Eo enim
adiutante, equa alligata, celerius admittuntur, neque
equi frustra cupiditate impulsi semen eiciunt. Quaad
satis sit admitti, ipsae significant, quod se defendunt.
Si fastıdium saliendi est, scillae medium conterunt
cum aqua ad mellis crassitudinem ; tum ea re naturam
equae, cum menses ferunt, tangunt ; contra ab
9 locis equae nares equi tangunt. Tametsi incredibile,
quod usu venit, memoriae mandandum. Equus
matrem salire cum adduci non posset, cum eum
capite obvoluto auriga adduxisset et coegisset matrem
inire, cum descendenti dempsisset ab oculis,
ille impetum fecit in eum ac mordicus interfecit.
10 Cum conceperunt equae, videndum ne aut laborent
plusculum aut ne frigidis locis sint, quod algor
maxime praegnatibus obest. Itaque in stabulis et
umore prohibere oportet humum, clausa habere ostia
ac fenestras, et inter singulas a praesepibus intericere
longorios, qui eas discernant, ne inter se pugnare
possint. Praegnatem neque implere cibo neque esurire
11 oportet. Alternis qui admittant, diuturniores
equas, meliores pullos fieri dicunt, itaque ut restibiles

[1] *origa*, an old (and popular) form of *auriga*.

day of the twelfth month after conception. Foals which are born after this time are usually defective and unprofitable. When the proper season arrives, the stallion should be admitted twice daily, morning and evening, with the help of the groom[1]—as they call the man who attends to the mating. For with his help, when the mare is tied, the coition takes place more quickly, and the stallions do not, in their eagerness, eject the seed to no purpose. The mares show when they have conceived by defending themselves. If the horse will not cover the mare, the centre of a squill is crushed in water and reduced to the consistency of honey; with this the natural parts of the mare are touched when she is in heat, and on the other hand the nostrils of the horse are touched with what comes from the natural parts of the mare. (Though it is incredible, as it actually happened the following story should be recorded: when a horse could not be induced to mount his dam, the groom covered his head, led him up, and forced him to do so; but when he took the cloth from the horse's eyes after he had come down, the horse dashed at him and killed him with his teeth.) Care must be taken that the mares, after conceiving, are not worked over hard or kept in cold places, as chill is extremely injurious to those with foal. So in their stalls the ground should be kept free of dampness, the doors and windows should be kept shut, and poles should be placed in the pen to separate each mare, so that they cannot fight one another. A mare with foal should not be over-fed or under-fed. Those who mate their mares every other year claim that they breed for a longer time and that the colts are better; and that those which become pregnant every

segetes esse exuctiores, sic quotannis quae praegnates fiant.

In decem diebus secundum partum cum matribus in pabulum prodigendum, ne ungulas comburat stercus tenellas. Quinquemestribus pullis factis, cum redacti sunt in stabulum, obiciendum farinam hordeaciam molitam cum furfuribus, et siquid 12 aliud terra natum libenter edent. Anniculis iam factis dandum hordeum et furfures, usque quaad erunt lactantes. Neque prius biennio confecto a lacte removendum; eosque, cum stent cum matribus, interdum tractandum, ne, cum sint deiuncti, exterreantur; eademque causa ibi frenos suspendendum, ut eculi consuescant et videre eorum faciem et e 13 motu audire crepitus. Cum iam ad manus accedere consuerint, interdum imponere iis puerum bis aut ter pronum in ventrem, postea iam sedentem. Haec facere, cum sit trimus; tum enim maxime crescere ac lacertosum fieri. Sunt qui dicant post annum et sex menses eculum domari posse, sed melius post trimum, a quo tempore farrago dari solet. Haec enim purgatio maxime necessaria equino pecori. Quod diebus decem facere oportet, nec pati alium 14 ullum cibum gustare. Ab undecimo die usque ad quartum decimum dandum hordeum, cottidie adicientem minutatim; quod quarto die feceris, in eo decem diebus proximis manendum. Ab eo tempore

[1] Cf. I, 44, 2 and 3.

year are sooner exhausted, just as are fields which are planted every year.[1]

"Within ten days after birth colts should be driven to pasture with their dams, so that the dung may not burn their tender hoofs. When the colts are five months old, on being driven back to the stable they should have spread before them barley-meal ground with bran, and whatever other product of the soil they relish. When they become yearlings they should be fed with barley and bran so long as they suckle. And they should not be removed before the end of two years; and while they are still with the dam they should be handled from time to time, to prevent them from being frightened when they are separated. For the same purpose harness should be hung in the stall, so that the young horses may become accustomed both to the sight of it and to its jingling when it moves. As soon as they have become accustomed to coming up and being handled it is well to let a boy mount them two or three times, first lying flat on his stomach and then seated. This should be done when the colt is a three-year-old, for at that age it is growing most rapidly, and putting on muscle. Some breeders claim that a young horse can be broken at eighteen months; but it is better to wait until they are three-year-olds; from which time it is customary to feed mixed forage, for this is a most necessary form of purging for horses. It should be fed for ten days, and the horse should be allowed to taste no other food. From the eleventh to the fourteenth day barley should be fed, the amount being increased gradually from day to day; and the amount fed on the fourth day should be continued for the next ten days. After that time

mediocriter exercendum et, cum consudarit, perunguendum oleo. Si frigus erit, in equili faciendus ignis.

15 Equi quod alii sunt ad rem militarem idonei, alii ad vecturam, alii ad admissuram, alii ad cursuram, non item sunt spectandi atque habendi. Itaque peritus belli alios eligit atque alit ac docet; aliter quadrigarius ac desultor; neque idem qui vectorios facere vult ad ephippium aut ad raedam, quod qui ad rem militarem, quod ut ibi ad castra habere volunt acres, sic contra in viis habere malunt placidos. Propter quod discrimen maxime institutum ut castrentur equi. Demptis enim testiculis fiunt quietiores, ideo quod semine carent. Ii cantherii appellati, ut

16 in subus maiales, gallis gallinaceis capi. De medicina vel plurima sunt in equis et signa morborum et genera curationum, quae pastorem scripta habere oportet. Itaque ab hoc in Graecia potissimum medici pecorum ἱππίατροι appellati.

VIII. Cum haec loqueremur, venit a Menate libertus, qui dicat liba absoluta esse et rem divinam paratam; si vellent, venirent illuc et ipsi pro se sacrificarentur. Ego vero, inquam, vos ante ire non patiar,

---

[1] The word *cantherius* is perhaps the same as κανθήλιος, pack-ass.

[2] Cf. II, 4, 21.

[3] Cf. II, 1, 1.

[4] Cf. II, 1, 12. Varro is the speaker.

he should have gentle exercise, and be rubbed down with oil after he has sweated. If the weather is chilly, a fire should be built in the stall.

"As some horses are fitted for military service, others for hauling, others for breeding, and others for racing, all are not to be judged and valued by the same standards. Thus the experienced soldier chooses his horses by one standard and feeds and trains them in one way, and the charioteer and circus-rider in another; and the trainer who is breaking horses for riding under the saddle or for the carriage does not use the same system as the man who has military service in view; for as on the one hand, in the army, they want spirited horses, so on the other hand they prefer more docile ones for road service. It is for this reason that there has grown up the greatest difference in the matter of castrating horses; for when the testicles have been removed they become more steady for the reason that they no longer have seed. Such horses are called geldings,[1] just as castrated boars are called barrows,[2] and castrated cocks are called capons. In the matter of treatment there are, in the case of horses, a great many symptoms of disease and methods of treatment, and the head groom should have these written out. It is for this reason that in Greece those who treat cattle in general are called by the special name *ἱππίατροι*, 'horse-doctors.'"

VIII. While we were thus speaking a freedman comes from Menates[3] to tell us that the cakes had been offered and the sacrifice made ready; if the gentlemen wished they might come there and perform their sacrifices for themselves. "But," I said, "I shan't let you go until you have played out the third act[4]—on

antequam mihi reddideritis tertium actum de mulis, de canibus, de pastoribus. Brevis oratio de istis, inquit Murrius. Nam muli et item hinni[1] bigeneri atque insiticii, non suopte genere ab radicibus. Ex equa enim et
2 asino fit mulus, contra ex equo et asina hinnus. Uterque eorum ad usum utilis, partu fructus neuter. Pullum asininum a partu recentem subiciunt equae, cuius lacte ampliores fiunt, quod id lacte quam asininum ad alimonia dicunt esse melius. Praeterea educant eum paleis, faeno, hordeo. Matri suppositiciae quoque inserviunt, quo equa ministerium lactis cibum pullo praebere possit. Hic ita eductus a trimo potest admitti; neque enim aspernatur propter consuetu-
3 dinem equinam. Hunc minorem si admiseris, et ipse citius senescit, et quae ex eo concipiuntur fiunt deteriora. Qui non habent eum asinum, quem supposuerunt equae, et asinum admissarium habere volunt, de asinis quem amplissimum formosissimumque possunt eligunt, quique seminio natus sit bono, Arcadico, ut antiqui dicebant, ut nos experti sumus, Reatino, ubi tricenis ac quadragenis milibus admissarii aliquot venierunt. Quos emimus item ut equos stipulamurque in emendo ac facimus in accipiendo idem, quod
4 dictum est in equis. Hos pascimus praecipue faeno atque hordeo, et id ante admissuram et largius facimus, ut cibo suffundamus vires ad feturam, eodem tempore quo equos adducentes, itemque ut ineat

[1] *hinni* added by Iucundus.

---

[1] *i.e.* with a mare. [2] Cf. II, 7, 6.

mules, dogs, and herdsmen." "It will take only a short time to discuss them," said Murrius; "for mules and hinnies are hybrids and grafts, not from roots after their own kind; for the mule is the offspring of a mare and an ass, while the hinny is the offspring of a horse and a jenny; each is useful for work, but neither brings any return from young. When an ass colt is newly born it is placed under a mare and becomes fatter on her milk, as they claim that such nourishment is more nutritious than the ass's milk. They are reared, in addition, on straw, hay, or barley. Special care is also taken of the foster-mother, so that the mare may furnish the colt with an abundant supply of milk. A jack so reared may be used for breeding after three years, and because it is accustomed to horses it will not refuse to mate.[1] If you use him at an earlier age, he himself tires sooner, and his offspring will be of poorer quality. Those who do not have such a jack, reared on mare's milk, but want a breeding jack, pick one as heavy and handsome as they can find and of good breed—of the Arcadian breed, our ancestors used to say, but of Reatine breed, as we have found by experience; in that district several breeding asses have sold for three hundred and even four hundred thousand sesterces. In purchasing we observe the same rules as in the case of horses, and make the same stipulations in the matter of purchase and acceptance as were named in the case of horses.[2] We feed these chiefly on hay and barley and increase the amount before breeding, so that we may furnish strength from the food for begetting; and we mate them at the same season in which we mate horses, and we are careful also to have them cover the mares with the help of a groom.

equas per origas curamus. Cum peperit equa mulum 5 aut mulam, nutricantes educamus. Hi si in palustribus locis atque uliginosis nati, habent ungulas molles; idem si exacti sunt aestivo tempore in montes, quod fit in agro Reatino, durissimis ungulis fiunt. In grege mulorum parando spectanda aetas et forma, alterum, ut in vecturis sufferre labores possint; alterum, ut oculos aspectu delectare queant. Hisce enim binis coniunctis omnia vehicula in viis 6 ducuntur. Haec me Reatino auctore probares, mihi inquit, nisi tu ipse domi equarum greges haberes ac mulorum greges vendidisses. Hinnus qui appellatur, est ex equo et asina, minor quam mulus corpore, plerumque rubicundior, auribus ut equinis, iubam et caudam habet[1] similem asini. Item in ventre est, ut equus, menses duodecim. Hosce item ut eculos et educant et alunt et aetatem eorum ex dentibus cognoscunt.

IX. Relinquitur, inquit Atticus, de quadripedibus quod ad canes attinet, quod pertinet[2] maxime ad nos, qui pecus pascimus lanare. Canes enim ita custos pecoris eius quod eo comite indiget ad se defendendum. In quo genere sunt maxime oves, deinde caprae. Has enim lupus captare solet, cui opponimus canes defensores. In suillo pecore tamen sunt quae se vindicent, verres, maiales, scrofae. Prope enim haec apris, qui in silvis saepe dentibus 2 canes occiderunt. Quid dicam de pecore maiore?

[1] *habet* added by Keil.
[2] *quod pertinet* added by Keil.

When a mare drops a horse-mule or a mare-mule we rear it at the teat. If these are born in swampy or damp ground they have soft hoofs; but if they are driven into the mountains in summer, as is done in the district of Reate, their hoofs grow quite hard. In assembling a herd of mules both age and build must be watched—the former that they may be strong enough to bear the labour of hauling, and the latter that they may please the eyes with their appearance; for it is by pairs of these animals that all vehicles are drawn on the roads. You would take my word for this as being an expert from Reate," he remarked to me, "if you did not keep herds of mares at home yourself, and had not sold herds of mules. The so-called hinny is the offspring of a horse and a jenny; smaller than the mule, usually rather redder, with ears like a horse's, but with mane and tail like those of the ass. It also, like the horse, is carried for twelve months. These are reared and fed just as young horses are, and their age is determined by the teeth."

IX. "There is left," said Atticus, "of the discussion of quadrupeds only the topic of dogs; but it is of great interest to those of us who keep fleece-bearing flocks, the dog being the guardian of the flock, which needs such a champion to defend it. Under this head come especially sheep but also goats, as these are the common prey of the wolf, and we use dogs to protect them. In a herd of swine, however, there are some members which can defend themselves, namely, boars, barrows, and sows; for they are very much like wild boars, which have often killed dogs in the forest with their tusks. And why speak about the larger animals? For I know that while

cum sciam mulorum gregem, cum pasceretur et eo venisset lupus, ultro mulos circumfluxisse et ungulis caedendo eum occidisse, et tauros solere diversos adsistere clunibus continuatos et cornibus facile propulsare lupos. Quare de canibus quoniam genera duo, unum venaticum et pertinet ad feras bestias silvestres, alterum quod custodiae causa paratur et pertinet ad pastorem, dicam de eo ad formam artis expositam in novem partes.

3 Primum aetate idonea parandi, quod catuli et vetuli neque sibi neque ovibus sunt praesidio et feris bestiis non numquam praedae. Facie debent esse formosi, magnitudine ampla, oculis nigrantibus aut ravis, naribus congruentibus, labris subnigris aut rubicundis neque resimis superioribus nec pendulis subtus, mento suppresso et ex eo enatis duobus dentibus dextra et sinistra paulo eminulis, superiori-
4 bus directis potius quam brocchis, acutos quos habeant labro tectos, capitibus et auriculis magnis ac flaccis, crassis cervicibus ac collo, internodis articulorum longis, cruribus rectis et potius varis quam vatiis, pedibus magnis et latis, qui ingredienti ei displodantur, digitis discretis, unguibus duris ac curvis, solo ne ut corneo ne nimium duro, sed ut fermentato ac molli; a feminibus summis corpore suppresso, spina neque eminula neque curva, cauda crassa; latrato gravi, hiatu magno, colore potissimum albo, quod in tenebris facilius agnoscuntur, specie
5 leonina. Praeterea feminas volunt esse mammosas

---

[1] Cf. II, 1, 12.

[2] Columella, VII, 12, discusses the same subject with his usual eloquence.

a herd of mules was feeding and a wolf came upon them, the animals actually whirled about and kicked him to death; that bulls often stand facing different ways, with their hind-quarters touching, and easily drive off wolves with their horns. As there are, then, two sorts of dogs—the hunting-dog suited to chase the beasts of the forest, and the other which is procured as a watch-dog and is of importance to the shepherd—I shall speak of the latter under nine divisions, according to the scientific division which has been set forth.[1]

" In the first place, they should be procured of the proper age, as puppies and dogs over age are of no value for guarding either themselves or sheep, and sometimes fall a prey to wild beasts. They should be comely in face, of good size, with eyes either darkish or yellowish, symmetrical nostrils, lips blackish or reddish, the upper lip neither raised too high nor drooping low, stubby jaw with two fangs projecting somewhat from it on the right and left, the upper straight rather than curved, their sharp teeth covered by the lip, large head, large and drooping ears, thick shoulders and neck, the thighs and shanks long, legs straight and rather bowed in than out, large, wide paws which spread as he walks, the toes separated, the claws hard and curving, the sole of the foot not horny or too hard, but rather spongy, as it were, and soft; with the body tapering at the top of the thigh, the backbone neither projecting nor swayed, tail thick; with a deep bark, wide gape, preferably white in colour, so that they may the more readily be distinguished in the dark; and of a leonine appearance.[2] Bitches, in addition, should have well formed dugs with teats of equal size.

aequalibus papillis. Item videndum ut boni seminii sint ; itaque et a regionibus appellantur Lacones, Epirotici, Sallentini. Videndum ne a venatoribus aut laniis canes emas ; alteri quod ad pecus sequendum inertes, alteri, si viderint leporem aut cervum, quod eum potius quam oves sequentur. Quare a pastoribus empta melior, quae oves sequi consuevit, aut sine ulla consuetudine quae fuerit. Canis enim facilius quid adsuescit, eaque consuetudo firmior, 6 quae fit ad pastores, quam quae ad pecudes. Publius Aufidius Pontianus Amiterninus cum greges ovium emisset in Umbria ultima, quibus gregibus sine pastoribus canes accessissent, pastores ut deducerent in Metapontinos saltus et Heracleae emporium, inde cum domum redissent qui ad locum deduxerant, e desiderio hominum diebus paucis postea canes sua sponte, cum dierum multorum via interesset, sibi ex agris cibaria praebuerunt atque in Umbriam ad pastores redierunt. Neque eorum quisquam fecerat, quod in agri cultura Saserna praecepit : qui vellet se a cane sectari, ut ranam obiciat coctam. Magni interest ex semine esse canes eodem, quod cognati 7 maxime inter se sunt praesidio. Sequitur quartum de emptione : fit alterius, cum a priore domino secundo traditus est. De sanitate et noxa stipulationes fiunt eaedem, quae in pecore, nisi quod hic

---

[1] The distance was some 300 miles.

[2] Cf. I, 2, 22-26.

[3] Cf. II, 2, 5-6.

Care should also be taken that they be of good breed; accordingly they receive their names from the districts from which they come: Spartans, Epirotes, Sallentines. You should be careful not to buy dogs from huntsmen or butchers—in the latter case because they are too sluggish to follow the flock, and in the other because if they see a hare or a stag they will follow it rather than the sheep. It is better, therefore, to buy from a shepherd a bitch which has been trained to follow sheep, or one that has had no training at all; for a dog forms a habit for anything very easily, and the attachment he forms for shepherds is more lasting than that he forms for sheep. Publius Aufidius Pontianus, of Amiternum, had bought some herds of sheep in furthest Umbria, the purchase including the dogs but not the shepherds, but providing that the shepherds should take them to the pastures of Metapontum and to market at Heraclea. When the men who had taken them there had returned home, the dogs, without direction and simply from their longing for their masters, returned to the shepherds in Umbria a few days later, though it was a journey of many days,[1] having lived off the country. And yet not one of those shepherds had done what Saserna, in his book on agriculture,[2] directed: that a man who wanted a dog to follow him should throw him a boiled frog! It is very important that the dogs be all of the same family, as those which are related are the greatest protection to one another. The fourth point is that of purchase: possession passes when the dog is delivered by the former owner to the next. With regard to health and liability to damage, the same precautions are taken as in the case of sheep,[3]

utiliter exceptum est: alii pretium faciunt in singula capita canum, alii ut catuli sequantur matrem, alii ut bini catuli unius canis numerum obtineant, ut solent bini agni ovis, plerique ut accedant canes, qui consuerunt esse una.

8 Cibatus canis propior hominis quam ovis. Pascitur enim eduliis et ossibus, non herbis aut fronde. Diligenter ut habeat cibaria providendum. Fames enim hos ad quaerendum cibum ducet, si non prae-9 bebitur, et a pecore abducet; nisi si, ut quidam putant, etiam illuc pervenerint, proverbium ut tollant anticum vel etiam ut μῦθον aperiant de Actaeone 10 atque in dominum adferant dentes. Nec non ita panem hordeacium dandum, ut non potius eum in lacte des intritum, quod eo consueti cibo uti a pecore non cito desciscunt. Morticinae ovis non patiuntur vesci carne, ne ducti sapore minus se abstineant. Dant etiam ius ex ossibus et ea ipsa ossa contusa. Dentes enim facit firmiores et os magis patulum, propterea quod vehementius diducuntur malae, acrioresque fiunt propter medullarum saporem. Cibum capere consuescunt interdiu, ubi pascuntur, vesperi, 11 ubi stabulantur. Feturae principium admittendi faciunt veris principio; tum enim dicuntur catulire, id est ostendere velle se maritari. Quae tum admissae, pariunt circiter solstitium; praegnates enim solent

---

[1] In his *De Ling. Lat.* (VII, 31) Varro quotes the proverb: *canis caninam non est:* "dog doesn't eat dog."

[2] A famous hunter who, having come upon Diana in her bath, was changed into a stag and torn to pieces by his own dogs.

[3] Literally, "to want puppies."

except that it is advisable to make the following stipulation: some people fix the price of dogs per head, others stipulate that pups go with their mother, others that two pups count as one dog just as usually two lambs count as one sheep, and many that dogs be included which have become accustomed to being together.

"The food of dogs is more like that of man than that of sheep: they eat scraps of meat and bones, not grass and leaves. Great care must be taken for their supply of food; for hunger will drive them to hunt for food, if it is not provided, and take them away from the flock—even if they do not, as some think, come to the point of disproving the ancient proverb,[1] or even go so far as to enact the story of Actaeon,[2] and sink their teeth in their master. You should also feed them barley bread, but not without soaking it in milk; for when they have become accustomed to eating that kind of food they will not soon stray from the flock. They are not allowed to feed on the flesh of a dead sheep, for fear that the taste will make them less inclined to spare the flock. They are also fed on bone soup and even broken bones as well; for these make their teeth stronger and their mouths of wider stretch, because their jaws are spread with greater force, and the savour of the marrow makes them more keen. Their habit is to eat during the day when they are out with the flocks, and at evening when these are folded. The beginning of breeding is fixed at the opening of spring, for at that time they are said to be 'in heat,'[3] that is, to show their desire for mating. Those that conceive at that time have a litter about the time of the summer solstice, for they usually

esse ternos menses. In fetura dandum potius hordeacios quam triticios panes; magis enim eo aluntur et lactis praebent maiorem facultatem. In nutricatu secundum partum, si plures sunt, statim eligere oportet quos habere velis, reliquos abicere. Quam paucissimos reliqueris, tam optimi in alendo fiunt propter copiam lactis. Substernitur eis acus aut quid item aliut, quod molliore cubili facilius educentur. Catuli diebus XX videre incipiunt. Duobus mensibus primis a partu non diiunguntur a matre, sed minutatim desuefiunt. Educunt eos plures in unum locum et inritant ad pugnandum, quo fiunt acriores, neque defatigari patiuntur, quo fiunt segniores.

13 Consue quoque faciunt ut alligari possint primum levibus vinclis; quae si abrodere conantur, ne id consuescant facere, verberibus eos deterrere solent. Pluviis diebus cubilia substernenda fronde aut pabulo duabus de causis: ut ne oblinantur aut perfrigescant.

14 Quidam eos castrant, quod eo minus putant relinquere gregem; quidam non faciunt, quod eos credunt minus acres fieri. Quidam nucibus Graecis in aqua tritis perungunt aures et inter digitos, quod muscae et ricini et pulices soleant, si hoc unguine non sis

15 usus, ea exulcerare. Ne vulnerentur a bestiis, imponuntur his collaria, quae vocantur melium, id est cingulum circum collum ex corio firmo cum clavulis

carry their young for three months. During the period of gestation they should be fed barley bread rather than wheat bread, for they are better nourished on the former and yield a larger supply of milk. In the matter of rearing after birth, if the litter is large you should at once pick those that you wish to keep and dispose of the others. The fewer you leave the better they will grow, because of the abundance of milk. Chaff and other like stuff is spread under them, because they are more easily reared on a soft bedding. The pups open their eyes within twenty days; for the first two months after birth they are not taken from the mother, but are weaned by degrees. Several of them are driven into one place and teased to make them fight, so as to make them more keen; but they are not allowed to tire themselves out, as this makes them sluggish. They are also accustomed to being tied, at first with slight leashes; and if they try to gnaw these they are whipped to keep them from forming the habit of doing this. On rainy days the kennels should be bedded with leaves or fodder, and this for two purposes: to keep them from being muddied, and to keep them from getting chilled. Some people castrate them, because they think that by this means they are less likely to leave the flock; others do not, because they think this makes them less keen. Some people crush filberts in water and rub the mixture over their ears and between their toes, as the flies and worms and fleas make ulcers there if you do not use this ointment. To protect them from being wounded by wild beasts, collars are placed on them—the so-called *melium*, that is, a belt around the neck made of stout leather with nails having heads; under

capitatis, quae intra capita insuitur pellis mollis, ne noceat collo duritia ferri; quod, si lupus aliusve quis his vulneratus est, reliquas quoque canes facit, quae
16 id non habent, ut sint in tuto. Numerus canum pro pecoris multitudine solet parari; fere modicum esse putant, ut singuli sequantur singulos opiliones. De quo numero alius alium modum constituit, quod, si sunt regiones, ubi bestiae sint multae, debent esse plures, quod accidit his qui per calles silvestres longinquos solent comitari in aestiva et hiberna. Villatico vero gregi in fundum satis esse duo, et id marem et feminam. Ita enim sunt adsiduiores, quod cum altero item alter fit acrior, et si alteruter aeger est, ne sine cane grex sit.

X. Cum circumspiceret, nequid praeterisset, Hoc silentium, inquam, vocat alium ad partes. Relicum enim in hoc actu, quot et quod genus sint habendi pastores. Cossinius: Ad maiores pecudes aetate superiores, ad minores etiam pueros, utrosque horum firmiores qui in callibus versentur, quam eos qui in fundo cotidie ad villam redeant (itaque in saltibus licet videre iuventutem, et eam fere armatam, cum in fundis non modo pueri sed etiam puellae
2 pascant). Qui pascunt, eos cogere oportet in pastione diem totum esse, pascere communiter, contra pernoctare ad suum quemque gregem, esse omnes sub uno magistro pecoris; eum esse maiorem natu potius quam alios et peritiorem quam reliquos, quod ei qui

---

[1] *i.e.* Varro, cf. II, 8, 1.

the nail heads there is sewed a piece of soft leather, to prevent the hard iron from injuring the neck. The reason for this is that if a wolf or other beast has been wounded by these nails, this makes the other dogs also, which do not have the collar, safe. The number of dogs is usually determined by the size of the flock; it is thought to be about right for one dog to follow each shepherd. But the number varies with the circumstances; thus in countries where wild beasts are plentiful there should be more, as is usually the case with those who escort the flocks to summer and winter pastures through distant woodland trails. On the other hand, for a flock feeding near the steading two dogs to the farm are sufficient. These should be a male and a female, for in this case they are more watchful, as one makes the other more keen, and if one of the two is sick that the flock may not be without a dog."

X. As he glanced around to see if he had overlooked anything, I[1] remarked: "Your silence gives the cue to another actor; for the remaining scene in this act concerns the number and kind of herdsmen to be kept." Whereupon Cossinius: "For herds of larger cattle older men, for the smaller even boys; but in both cases those who range the trails should be sturdier than those on the farm who go back to the steading every day. Thus on the range you may see young men, usually armed, while on the farm not only boys but even girls tend the flocks. The herdsmen should be required to stay on the range the entire day and have the herds feed together; but, on the other hand, to spend the night each with his own herd. They should all be under one herd-master; he should preferably be older than the

aetate et scientia praestat animo aequiore reliqui
3 parent. Ita tamen oportet aetate praestare, ut ne propter senectutem minus sustinere possit labores. Neque enim senes neque pueri callium difficultatem ac montium arduitatem atque asperitatem facile ferunt, quod patiendum illis, qui greges secuntur, praesertim armenticios ac caprinos, quibus rupes ac silvae ad pabulandum cordi. Formae hominum legendae ut sint firmae ac veloces, mobiles, expeditis membris, qui non solum pecus sequi possint, sed etiam a bestiis ac praedonibus defendere, qui onera extollere in iumenta possint, qui excurrere, qui iacu-
4 lari. Non omnis apta natio ad pecuariam, quod neque Bastulus neque Turdulus idonei, Galli appositissimi, maxime ad iumenta. In emptionibus dominum legitimum sex fere res perficiunt: si hereditatem iustam adiit; si, ut debuit, mancipio ab eo accepit, a quo iure civili potuit; aut si in iure cessit, qui potuit cedere, et id ubi oportuit; aut si usu cepit; aut si e praeda sub corona emit; tumve cum
5 in bonis sectioneve cuius publice veniit. In horum

---

[1] Inhabitants of the Baetic Province in Southern Spain, modern Andalusia.

[2] *Mancipium* was the most formal act of purchase. In the presence of six Roman citizens of full age, the purchaser laid his hand on the object purchased (here the slave), asserted his ownership, struck with a piece of money the scale held by one of the witnesses (*per aes et libram*), and gave the coin to the seller. See Gaius, *Inst.*, I, 119.

[3] A legal fiction, in which the owner (*dominus qui cessit*) and the prospective purchaser (*cui cedebatur*) appeared before the magistrate (*qui addixit*). The purchaser claimed the object as his own; the magistrate asked the owner if he had any defence; and when he replied that he had none, the magistrate adjudged the object to the claimant. See Gaius, *Inst.*, I, 2.

rest and also more experienced, as the other herdsmen will be more disposed to take orders from one who surpasses them in both age and knowledge. Still, he should not be so much older that his age will prevent him from being as able to stand hard work; for neither old men nor boys can easily endure the hardships of the trail and the steepness and roughness of the mountains—all of which must be encountered by those who follow the herd, and especially herds of cattle and goats, which like cliffs and woods for pasturage. The men chosen for this work should be of a sturdy sort, swift, nimble, with supple limbs; men who can not only follow the herd but can also protect it from beasts and robbers, who can lift loads to the backs of pack animals, who can dash out, and who can hurl the javelin. It is not every people that is fitted for herding; thus neither a Bastulan nor a Turdulan[1] is suited, while Gauls are admirably adapted, especially for draught cattle. In the matter of purchase there are some six methods of acquiring a legitimate title: by legal inheritance; by receiving, in due form, through mancipation[2] from one who had a legal right to transfer; by legal cession,[3] from one who had the right to cede, and that at the proper time; by right of possession;[4] by purchase at auction from war-booty; and lastly by official sale among other property or in confiscated[5] property. In the purchase of slaves, it is customary

[4] *Usucapio* is unchallenged possession for one year in the case of movable property, for two years in the case of immovable property. See Gaius, *Inst.*, II, 41.

[5] *Sectio* is the official term for the sale at auction of confiscated property, *e.g.* the property of a person who had been proscribed.

emptione solet accedere peculium aut excipi et stipulatio intercedere, sanum esse, furtis noxisque solutum; aut, si mancipio non datur, dupla promitti, aut, si ita pacti, simpla. Cibus eorum debet esse interdius separatim unius cuiusque gregis, vespertinus in cena, qui sunt sub uno magistro, communis. Magistrum providere oportet ut omnia sequantur instrumenta, quae pecori et pastoribus opus sunt, maxime ad victum hominum et ad medicinam pecudum. Ad quam rem habent iumenta dossuaria domini, alii equas, alii pro iis quid aliut, quod onus dosso ferre possit.

6 Quod ad feturam humanam pertinet pastorum, qui in fundo perpetuo manent, facile est, quod habent conservam in villa, nec hac venus pastoralis longius quid quaerit. Qui autem in saltibus et silvestribus locis pascunt et non villa, sed casis repentinis imbres vitant, iis mulieres adiungere, quae sequantur greges ac cibaria pastoribus expediant eosque assiduiores 7 faciant, utile arbitrati multi. Sed eas mulieres esse oportet firmas, non turpes, quae in opere multis regionibus non cedunt viris, ut in Illyrico passim videre licet, quod vel pascere pecus vel ad focum afferre ligna ac cibum coquere vel ad casas instru- 8 mentum servare possunt. De nutricatu hoc dico, easdem fere et nutrices et matres. Simul aspicit

---

[1] In the case of transfer without mancipation, the seller was bound by law in double the value of the property. This guarantee is exacted in case the title prove bad, before the purchaser is secured by *usucapio*.

for the *peculium* to go with the slave, unless it is expressly excepted; and for a guarantee to be given that he is sound and has not committed thefts or damage; or, if the transfer is not by mancipation, double the amount is guaranteed, or merely the purchase price, if this be agreed on.[1] They should eat during the day apart, each with his own herd, but in the evening all those who are under one head-herdsman should eat together. The head-herdsman is to see that all equipment needed for the animals and herdsmen, and especially for the sustenance of the men and the treatment of the cattle, shall accompany them; for which purpose owners keep pack animals, in some cases mares, in others any animal instead, which can carry a load on its back.

"As to the breeding of herdsmen; it is a simple matter in the case of those who stay all the time on the farm, as they have a female fellow-slave in the steading, and the Venus of herdsmen looks no farther than this. But in the case of those who tend the herds in mountain valleys and wooded lands, and keep off the rains not by the roof of the steading but by makeshift huts, many have thought that it was advisable to send along women to follow the herds, prepare food for the herdsmen, and make them more diligent. Such women should, however, be strong and not ill-looking. In many places they are not inferior to the men at work, as may be seen here and there in Illyricum, being able either to tend the herd, or carry firewood and cook the food, or to keep things in order in their huts. As to feeding their young, I merely remark that in most cases they suckle them as well as bear them." At the same time, turning to me, he said: "As I have

ad me et, Ut te audii dicere, inquit, cum in Liburniam venisses, te vidisse matres familias eorum afferre ligna et simul pueros, quos alerent, alias singulos, alias binos, quae ostenderunt fetas nostras, quae in conopiis iacent dies aliquot, esse eiuncidas ac con-
9 temnendas. Cui ego, Certe, inquam; nam in Illyrico hoc amplius, praegnatem saepe, cum venit pariendi tempus, non longe ab opere discedere ibique enixam puerum referre, quem non peperisse, sed invenisse putes; nec non etiam hoc, quas virgines ibi appellant, non numquam annorum viginti, quibus mos eorum non denegavit, ante nuptias ut succumberent quibus vellent et incomitatis ut vagari liceret et
10 filios habere. Quae ad valitudinem pertinent hominum ac pecoris et sine medico curari possunt, magistrum scripta habere oportet. Is enim sine litteris idoneus non est, quod rationes dominicas pecuarias conficere nequiquam recte potest. De numero pastorum alii angustius, alii laxius constituere solent.
11 Ego in octogenas hirtas oves singulos pastores constitui, Atticus in centenas. Greges ovium si magni sunt,[1] quos miliarios faciunt quidam, facilius de summa hominum detrahere possis, quam de minoribus, ut sunt et Attici et mei. Septingenarii enim mei; tu, opinor, octingenarios habuisti, nec tamen non, ut nos, arietum decumam partem. Ad equarum gre-

[1] *ovium si magni sunt* Goetz after Keil: *ovium sed magnum*.

---

[1] Horace (*Epod.*, 9, 16) and Propertius (IV, 11, 45) use the word *canopium* contemptuously of Cleopatra's luxury.

heard you say that you, when you were in Liburnia, saw mothers carrying logs and children at the breast at the same time, sometimes one, sometimes two; showing that our newly-delivered women, who lie for days under their mosquito-nets,[1] are worthless and contemptible." "It is quite true," I replied; "and in Illyricum I have seen something even more remarkable: for it often happens there that a pregnant woman, when her time has come, steps aside a little way from her work, bears her child there, and brings it back so soon that you would say she had not borne it but had found it. They have also another remarkable practice: their custom does not refuse to allow women, often as much as twenty years old (and they call them maidens, too), before marriage to mate with any man they please, to wander around by themselves, and to bear children." (Cossinius resumes), "All directions for caring for the health of human beings and cattle, and all sicknesses which can be treated without the aid of a physician, the head-herdsman should keep in writing. For one who does not know his letters is not fit for the place, because he cannot possibly keep his master's cattle accounts correctly. The number of herdsmen is determined differently, some having a smaller, some a larger number. My own practice is to have a herdsman to every eighty wool-bearing sheep, while Atticus has one to every hundred. If flocks of sheep are very large (and some people have as many as 1000) you can decrease the number of shepherds more easily than you can in smaller flocks, such as those of Atticus and mine. My own flocks contain 700, and yours, I think, had 800; but still you had one tenth of them rams, as I do.

gem quinquagenarium bini homines, utique uterque horum ut secum habeat equas domitas singulas in his regionibus, in quibus in stabula solent equas abigere, ut in Apulia et in Lucanis accidit saepe.

XI. Quoniam promissa absolvimus, inquit, eamus. Si quidem, inquam, adieceritis de extraordinario pecudum fructu, ut praedictum est, de lacte in eo et tonsura. Lacte[1] est omnium rerum, quas cibi causa capimus, liquentium maxime alibile, et id ovillum, dein caprinum. Quod autem maxime perpurget, est equinum, tum asininum, dein bubulum, 2 tum caprinum. Sed horum sunt discrimina quaedam et a pastionibus et a pecudum natura et a mulgendo:[2] a pastionibus, quod ad alendum utile quod[3] fit ab hordeo et stupla et omnino arido et firmo cibo pecude pasta; ad perpurgandum ab ea, quae a viridi pasta, eo magis, si fuerit ex herbis, quae ipsae sumptae perpurgare solent corpora nostra; a pecudum natura, quod lac melius est a valentibus et ab his quae nondum veteres sunt, quam si est contra. A mulgendo atque ortu optimum est id quod neque nimium longe abest a mulso neque a partu continuo est sump-3 tum. Ex hoc lacte casei qui fiunt, maximi cibi sunt bubuli et qui difficillime transeant sumpti, secundo

[1] *lacte* Pontedera after Ursinus: *lanae*.
[2] *mulgendo* suggested by Goetz: *motu*.
[3] *ad alendum utile quod* added by Keil.

Two men are needed for a herd of fifty mares, and each of these should certainly have for his use a mare which has been broken to the saddle, in those districts where it is customary for the mares to be rounded up and driven to stalls, as is frequently true in Apulia and Lucania.

XI. "As we have completed what we promised," he said, "let us leave." "Yes," said I, "but not until you have added, as was promised, something about supplementary profit from the flock, including under it the milk and the shearing." (Cossinius continues) "Of all the liquids which we take for sustenance, milk is the most nourishing—first sheep's milk, and next goat's milk. Mare's milk, however, has the greatest purgative effect, secondly ass's milk, then cow's milk, and lastly goat's milk. But there are certain differences among these which arise from a difference of pasturage, a difference in the nature of the animal, and a difference in the milking. As affected by pasturage, milk is best for nourishment which comes from animals fed on barley and straw, and, in general, on solid dry food; while that from animals fed on green food is best for purging, and especially if the green food be such as purges us when we eat it ourselves. As affected by the nature of the animal, milk from healthy animals and those not yet old is better than if it is the reverse. As affected by milking and birth, the best milk is that which has not been kept too long after milking and which has not been milked immediately after parturition. Of the cheeses which are made from this milk, those made of cow's milk have the most nutriment, but when eaten are discharged with most difficulty; next come those made of sheep's milk,

ovilli, minimi cibi et qui facillime deiciantur caprini. Et etiam est discrimen, utrum casei molles ac recentes sint, an aridi et veteres, cum molles sint magis alibiles, in corpore non resides, veteres et 4 aridi contra. Caseum facere incipiunt a vergiliis vernis exortis ad aestivas vergilias. Mulgent vere ad caseum faciundum mane, aliis temporibus meridianis horis, tametsi propter loca et pabulum disparile non usque quaque idem fit. In lactis duos congios addunt coagulum magnitudine oleae, ut coeat, quod melius leporinum et haedinum quam agninum. Alii pro coagulo addunt de fici ramo lac et acetum, aspargunt item aliis aliquot rebus, quod 5 Graeci appellant alii ὀπόν, alii δάκρυον. Non negarim, inquam, ideo aput divae Ruminae sacellum a pastoribus satam ficum. Ibi enim solent sacrificari lacte pro vino et lactentibus. Mamma enim rumis, ut ante dicebant; a rumi etiam nunc dicuntur subrumi agni, lactantes a lacte. Qui aspargi solent sales, melior fossilis quam marinus.

De tonsura ovium primum animadverto, antequam incipiam facere, num scabiem aut ulcera habeant, ut, si opus est, ante curentur, quam tondeantur.

---

[1] According to the Caesarian calendar, which Varro follows, the Pleiades rose in spring on 10th May. In the farmer's calendar which Columella gives (XI, 2), they are said to "appear fully" on that day, in the morning. He states that they rise in the evening on 10th October. The only explanation of the term "Pleiades in summer" would seem to be that at that time they appear near midnight, which would fix the period from May to mid-July. Columella (XII, 13) advises July for cheese-making. The ancients (Festus and Isidore) derive their name for the constellation, *Vergiliae*, from *ver*, "spring," because of its appearance at that time.

while those made of goat's milk have the least nutriment and are most easily voided. There is also a difference depending on whether the cheeses are soft and fresh or dry and old, as the soft cheeses are more nutritious and less constipating, while the old, dry cheeses are just the opposite. The period for making cheese extends from the rising of the Pleiades in spring until the Pleiades in summer.[1] In spring the milk for cheese making is drawn in the morning, while at other seasons the milking takes place toward midday; but the practice is not entirely uniform because of differences in locality and food. To two congii of milk is added a bit of rennet the size of an olive, to make it coagulate; this is better when made from a hare or a kid than when made from a lamb. Others use, instead of rennet, the milk from the stem of a fig, and vinegar; they also curdle with various other substances—a thing which, in Greek, is sometimes called ὀπός, and sometimes δάκρυον."[2] "I should not be surprised," I remarked, "if that is the reason that a fig tree was planted by shepherds near the shrine of the goddess Rumina; you know at that place sacrifice is offered with milk instead of with wine and sucklings. For people used to call the udder *rumis*, and even to-day we have lambs called *subrumi*[3] from this word, just as they are called *lactantes* (sucklings) from *lac* (milk). Those who sprinkle salt prefer mineral salt to sea salt.

"As to the shearing of sheep, I first am careful to see, before beginning it, whether the sheep have the scab or sores, so that they may be treated if necessary before being sheared. The proper time

[2] ὀπός, the juice of the fig, and δάκρυον (literally "tear") used of the same juice. [3] Cf. II, 1, 20.

Tonsurae tempus inter aequinoctium vernum et solstitium, cum sudare inceperunt oves, a quo sudore 7 recens lana tonsa sucida appellata est. Tonsas recentes eodem die perungunt vino et oleo, non nemo admixta cera alba et adipe suilla; et si ea tecta solet esse, quam habuit pellem intectam, eam intrinsecus eadem re perinungunt et tegunt rursus. Siqua in tonsura plagam accepit, eum locum oblinunt pice liquida. Oves hirtas tondent circiter hordeaceam 8 messem, in aliis locis ante faenisecia. Quidam has bis in anno tondent, ut in Hispania citeriore, ac semenstres faciunt tonsuras; duplicem impendunt operam, quod sic plus putant fieri lanae, quo nomine quidam bis secant prata. Diligentiores tegeticulis subiectis oves tondere solent, nequi flocci intereant. 9 Dies ad eam rem sumuntur sereni, et iis id faciunt fere a quarta ad decimam; cum sole calidiore tonsa, ex sudore eius lana fit mollior et ponderosior et colore meliore. Quam demptam ac conglobatam alii vellera, alii vellimna appellant; ex quo vocabulo animadverti licet prius in lana vulsuram quam tonsuram inventam. Qui etiam nunc vellunt, ante triduo habent ieiunas, quod languidae minus aegre radices lanae retinent.

---

[1] Cf. II, 2, 18.

[2] Since the words *vellus* and *vellimnum* are derived from *vellere*, "pluck."

for shearing is the period from the spring equinox to the solstice, after the sheep have begun to sweat; it is because of this sweat (*sudor*) that freshly shorn wool is called 'juicy' (*sucida*). Freshly clipped sheep are rubbed down on the same day with wine and oil, to which some add a mixture of white chalk and hog lard; and if they have been accustomed to wear a jacket,[1] the skin with which they were covered is greased on the inside with the same mixture and placed on them again. If a sheep has been cut during the shearing, the wound is smeared with soft pitch. Sheep with coarse fleece are shorn about the time of the barley harvest, or at other places before the cutting of the hay. Some shear their sheep twice a year, as is done in Hither Spain, shearing every six months. They undergo the double work on the supposition that more wool is secured by this method—which is the same motive that leads some to mow their meadows twice a year. The more careful farmers spread out cloths and shear the sheep over them to prevent loss of the wool. Calm days are chosen for this work, and on these the shearing is done from about the fourth to the tenth hour. The fleece from a sheep that is clipped when the sun is rather warm is rendered softer by the sweat, as well as heavier and of better colour. When the fleece has been removed and rolled up it is called by some *vellus*, by others *vellimnum*; and it may be seen from these words that in the case of wool, plucking was discovered earlier than shearing.[2] Some people pluck the wool even to-day; and these keep the sheep without food for three days before, as the roots of the wool hold less tightly when the sheep are weak."

10 Omnino tonsores in Italiam primum venisse ex Sicilia dicuntur p. R. c. a. CCCCLIII, ut scriptum in publico Ardeae in litteris extat, eosque adduxisse Publium Titinium Menam. Olim tonsores non fuisse adsignificant antiquorum statuae, quod pleraeque habent capillum et barbam magnam.

11 Suscipit Cossinius: Fructum ut ovis e lana ad vestimentum, sic capra e pilis ministrat ad usum nauticum et ad bellica tormenta et fabrilia vasa. Neque non quaedam nationes harum pellibus sunt vestitae, ut in Gaetulia et in Sardinia. Cuius usum aput anticos quoque Graecos fuisse apparet, quod in tragoediis senes ab hac pelle vocantur diphtheriae et in comoediis qui in rustico opere morantur, ut aput Caecilium in Hypobolimaeo habet adulescens, aput Terentium 12 in Heautontimorumeno senex. Tondentur, quod magnis villis sunt, in magna parte Phrygiae; unde cilicia et cetera eius generis solent fieri. Sed quod primum ea tonsura in Cilicia sit instituta, nomen id Cilicas adiecisse dicunt.

Illi hoc, neque ab hoc quod mutaret Cossinius. Et simul Vituli libertus in urbem veniens ex hortis devertitur ad nos et, Ego ad te missus, inquit, ibam domum rogatum ne diem festum faceres breviorem et mature venires. Itaque discedimus ego et Scrofa in hortos ad Vitulum, Niger Turrani noster, illi partim domum, partim ad Menatem.

---

[1] *i.e.* 300 B.C., as the traditional date of the founding of the city was 753 B.C.

[2] The use of ropes and cloth made from the long hair of the Cilician goat (which we call Angora) for nautical purposes, and for the *catapulta* and the *ballista*, is well known; cf. Virgil, *Georg.*, III, 312. But the "workmen's equipment" has puzzled all readers; among the most plausible suggestions are that their tool-bags were made of this cloth; or that their water-jars were covered with it, in the fashion of the *olla*.

[3] *i.e.* διφθερίας, clad in a leather coat.

In fact, it is claimed that barbers first came to Italy from Sicily 453 years after the founding of the city of Rome[1] (as is recorded still on a public monument at Ardea), and that they were introduced by Publius Titinius Mena. That there were no barbers in early days is evident from the statues of the ancients, many of which have long hair and a large beard."

Cossinius resumed: "As the sheep affords a profit from its wool to be used for clothing, so the goat from her hair is of service for nautical purposes, as well as for military engines and for workmen's equipment.[2] Some barbarous people, too, use their skins for clothing, as, for instance, in Gaetulia and Sardinia. That this usage obtained among the ancient Greeks also is evident from the fact that the old men who appear in the tragedies get their name of *diphtheriae*[3] from the goat skin, and in the comedies those who are engaged in rustic labour, such as the young man in Caecilius's *Hypobolimaeus*, and the old man in Terence's *Heautontimorumenos*. Because they have long hair, goats are clipped over a large part of Phrygia; and it is from this that hair-cloth (*cilicia*) and other fabrics of the kind are made. But it is said that the Cilicians gave the name to it from the fact that this clipping was first practised in Cilicia."

This was their contribution, and Cossinius found nothing to alter in it. At the same time a freedman of Vitulus, on his way to the city from the park, turned aside to us and said: "I was sent to you, and was on my way to your house to ask you not to make the holiday shorter but to come early." And so Scrofa and I set out to Vitulus's place, and the others, my dear Turranius Niger, some for their homes and some to Menates.

BOOK III

LIBER TERTIUS

I. Cum duae vitae traditae sint hominum, rustica et urbana, quidni, Pinni, dubium non est quin hae non solum loco discretae sint, sed etiam tempore diversam originem habeant. Antiquior enim multo rustica, quod fuit tempus, cum rura colerent homines 2 neque urbem haberent. Etenim vetustissimum oppidum cum sit traditum Graecum Boeotiae Thebae, quod rex Ogygos aedificarit, in agro Romano Roma, quam Romulus rex; nam in hoc nunc denique est ut dici possit, non cum Ennius scripsit:

septingenti sunt paulo plus aut minus anni,
augusto augurio postquam inclita condita Roma est.

3 Thebae, quae ante cataclysmon Ogygi conditae dicuntur, eae tamen circiter duo milia annorum et centum sunt. Quod tempus si referas ad illud principium, quo agri coli sunt[1] coepti atque in casis et tuguriis habitabant nec murus et porta quid esset sciebant, immani numero annorum urbanos agricolae 4 praestant. Nec mirum, quod divina natura dedit agros, ars humana aedificavit urbes, cum artes omnes dicantur in Graecia intra mille annorum repertae,

[1] *agri coli sunt* Iucundus: *agricolae sint.*

BOOK III

I. Though there are traditionally two ways in which men live—one in the country, the other in the city—there is clearly no doubt, Pinnius, that these differ not merely in the matter of place but also in the time at which each had its beginning. Country life is much more ancient—I mean the time when people lived on the land and had no cities. For tradition has it that the oldest of all cities is a Greek one, Thebes in Boeotia, founded by King Ogygus; while the oldest on Roman territory is Rome, founded by King Romulus. For we may now say, with regard to this, with more accuracy than when Ennius wrote:

> "Seven hundred years are there, a little more or less,
> Since glorious Rome was founded, with augury august."

Thebes, however, which is said to have been founded before the deluge which takes its name from Ogygus, is some 2,100 years old. If, now, you compare this span of time with that early day when fields were first tilled, and men lived in huts and dugouts, and did not know what a wall or a gate was, farmers antedate city people by an enormous number of years. And no marvel, since it was divine nature which gave us the country, and man's skill that built the cities; since all arts are said to have been discovered in Greece within a thousand years, while there never was a time when there were

agri numquam non fuerint in terris qui coli possint. Neque solum antiquior cultura agri, sed etiam melior. Itaque non sine causa maiores nostri ex urbe in agros redigebant suos cives, quod et in pace a rusticis Romanis alebantur et in bello ab his allevabantur.[1]
5 Nec sine causa terram eandem appellabant matrem et Cererem, et qui eam colerent, piam et utilem agere vitam credebant atque eos solos reliquos esse ex stirpe Saturni regis. Cui consentaneum est, quod initia vocantur potissimum ea quae Cereri fiunt sacra.
6 Nec minus oppidi quoque nomen Thebae indicat antiquiorem esse agrum, quod ab agri genere, non a conditore nomen ei est impositum. Nam lingua prisca et in Graecia Aeolis Boeoti sine afflatu vocant collis tebas, et in Sabinis, quo e Graecia venerunt Pelasgi, etiam nunc ita dicunt, cuius vestigium in agro Sabino via Salaria non longe a Reate miliarius
7 clivus cum appellatur tebae. Agri culturam primo propter paupertatem maxime indiscretam habebant, quod a pastoribus qui erant orti in eodem agro et serebant et pascebant; quae postea creverunt peculia[2] diviserunt, ac factum ut dicerentur alii

[1] *allevabantur* Ellis: *alebantur*.
[2] *peculia* Iucundus: *pecunia*.

---

[1] Properly "the creator." Cf. Servius on *Georg.*, I, 7, *Ceres a creando dicta.* But Varro, *de Ling. Lat.*, V, 64: *quod gerit fruges, Ceres;* and Cicero, *de Nat. Deor.*, II, 67, adopts this etymology.

[2] Properly, "the Sower."

[3] The rites of Ceres are connected with the "beginnings" of civilisation, and so the word *initium* (= initiation) was specially applicable to them. These rites were first established at Eleusis, and in 496 B.C. were introduced at Rome, when Ceres was identified with Demeter. Cf. also Book II, 4, 9.

[4] Aeolis is used here and in III, 12, 6 for the Greek nomina-

not fields on earth that could be tilled. And not only is the tilling of the fields more ancient—it is more noble. It was therefore not without reason that our ancestors tried to entice their citizens back from the city to the country; for in time of peace they were fed by the country Romans, and in time of war aided by them. It was also not without reason that they called the same earth "mother" and "Ceres,"[1] and thought that those who tilled her lived a pious and useful life, and that they were the only survivors of the stock of King Saturnus.[2] And it is in accordance with this that the sacred rites in honour of Ceres are beyond all others called "Initiations."[3] The name of Thebes, too, no less clearly shows that the country is more ancient, in that the name given it comes from a type of land, and not from the name of the founder. For the old language, and the Aeolians[4] of Boeotia in Greece as well, use the word *teba* for hill, leaving out the aspirate; and among the Sabines, a country which was settled by the Pelasgians from Greece, up to this day they use the same word; there is a trace of it in the Sabine country on the Via Salaria, not far from Reate, where a slope of a mile in length is called *tebae*.[5] At first, because of their poverty, people practised agriculture, as a rule, without distinction, the descendants of the shepherds both planting and grazing on the same land; later, as these flocks grew, they made a division, with the result that some were called

tive plural Ἀιολεῖς, as also in *De Ling. Lat.*, V, 102. Cicero, *Flac.* 64, writes it Aeoles.

[5] Most scholars reject the explanation, though Schneider remarks that there is no doubt that many Greek words existed among the Etrurians and other peoples of Italy. We have no other trace of the word. Sir G. Wilkinson derives the Egyptian Thebes from Tapé, "the head."

8 agricolae, alii pastores. Quae ipsa pars duplex est, tametsi ab nullo satis discreta, quod altera est villatica pastio, altera agrestis. Haec nota et nobilis, quod et pecuaria appellatur, et multum homines locupletes ob eam rem aut conductos aut emptos habent saltus; altera villatica, quod humilis videtur, a quibusdam adiecta ad agri culturam, cum esset pastio, neque explicata tota separatim, quod sciam,
9 ab ullo. Itaque cum putarem esse rerum rusticarum, quae constituta sunt fructus causa, tria genera, unum de agri cultura, alterum de re pecuaria, tertium de villaticis pastionibus, tres libros institui, e quis duo scripsi, primum ad Fundaniam uxorem de agri cultura, secundum de pecuaria ad Turranium Nigrum; qui reliquus est tertius de villaticis fructibus, in hoc ad te mitto, quod visus sum debere pro nostra
10 vicinitate et amore scribere potissimum ad te. Cum enim villam haberes opere tectorio et intestino ac pavimentis nobilibus lithostrotis spectandam et parum putasses esse, ni tuis quoque litteris exornati parietes essent, ego quoque, quo ornatior ea esse posset fructu, quod facere possem, haec ad te misi, recordatus de ea re sermones, quos de villa perfecta habuissemus. De quibus exponendis initium capiam hinc.

II. Comitiis aediliciis cum sole caldo ego et Q. Axius

farmers, and others herdsmen. This matter of herding has a twofold division (though no writer has made the distinction clearly), as the feeding around the steading is one thing, and that on the land is another. The latter is well known and highly esteemed, being also called *pecuaria*, and wealthy men frequently have ranches devoted to it, which they have either leased or bought; while the other, that of the steading, as it seems insignificant, has, by some writers, been brought under the head of agriculture, though it is a matter of feeding; and the subject as a whole has not, so far as I know, been treated as a separate topic by anyone. Hence, as I suggested that there are three divisions of rural economy which are instituted for gainful ends—one of agriculture, a second of animal husbandry, and a third of the husbandry of the steading—I fixed on three books, of which I have written two: the first to my wife Fundania, on agriculture, and the second to Turranius Niger, on animal husbandry. The third book, that on the husbandry of the steading, which remains, I am herewith sending to you, thinking that in view of our nearness and our affection it is to you particularly that I should dedicate it. For just as you had a villa noteworthy for its frescoing, inlaid work, and handsome mosaic floors, but thought it was not fine enough until its walls were adorned also by your writings, so I, that it might be farther adorned with fruit, so far as I could make it so, am sending this to you, recalling as I do the conversations which we held on the subject of the complete villa. And in discussing that subject I shall begin as follows.

II. During the election of aediles, Quintus Axius,

senator tribulis suffragium tulissemus et candidato, cui studebamus, vellemus esse praesto, cum domum rediret, Axius mihi, Dum diribentur, inquit, suffragia, vis potius villae publicae utamur umbra, quam privati candidati tabella dimidiata aedificemus nobis? Opinor, inquam, non solum, quod dicitur, "malum consilium consultori est pessimum," sed etiam bonum consilium, qui consulit et qui consulitur, bonum haben-

2 dum. Itaque imus, venimus in villam. Ibi Appium Claudium augurem sedentem invenimus in subselliis, ut consuli, siquid usus poposcisset, esset praesto. Sedebat ad sinistram ei Cornelius Merula consulari familia ortus et Fircellius Pavo Reatinus, ad dextram Minucius Pica et M. Petronius Passer. Ad quem cum accessissemus, Axius Appio subridens, Recipis nos, inquit, in tuum ornithona, ubi sedes inter aves?

3 Ille, Ego vero, inquit, te praesertim, quoius aves hospitales etiam nunc ructor, quas mihi apposuisti paucis ante diebus in Villa Reatina ad lacum Velini eunti de controversiis Interamnatium et Reatinorum.

---

[1] Livy tells us (IV, 22) that it was authorized by the censors in 434 B.C., and that the census was first held there. It was built on the Campus Martius, and remains still exist. The purposes for which it was used are set forth in § 4 of this chapter.

[2] *tabella dimidiata* has never been explained satisfactorily, and the text is probably hopelessly corrupt.

[3] Aulus Gellius (IV, 5) tells a story in illustration of the proverb. The Annales Magni recited that when a statue of Horatius Cocles had been struck by lightning the Etruscan haruspices were consulted. Being hostile to Rome, they purposely gave bad advice, and on their confession were executed. Whereupon this verse (which is a translation of Hesiod, *Works and Days*, I, 266, *ἡ δὲ κακὴ βουλὴ τῷ βουλεύσαντι κακίστη*) was sung in Rome.

[4] A member of this college of priests was at hand on

the senator, a member of my tribe, and I, after casting our ballots, wished, though the sun was hot, to be on hand to escort the candidate whom we were supporting when he returned home. Axius remarked to me: "While the votes are being sorted, shall we enjoy the shade of the Villa Publica,[1] instead of building us one out of the half-plank of our own candidate?"[2] "Well," I replied, "I think that the proverb is correct, 'bad advice is worst for the adviser,'[3] and also that good advice should be considered good both for the adviser and the advised. So we go our way and come to the Villa. There we find Appius Claudius, the augur,[4] sitting on a bench so that he might be on hand for consultation, if need should arise. There were sitting at his left Cornelius Merula ('Blackbird'), member of a consular family, and Fircellius Pavo ('Peacock'), of Reate; and on his right Minucius Pica ('Magpie') and Marcus Petronius Passer ('Sparrow'). When we came up to him, Axius said to Appius, with a smile: "Will you let us come into your aviary, where you are sitting among the birds?" "With pleasure," he replied, "and especially you; I still 'bring up' those hospitable birds which you set before me a few days ago in your villa at Reate, when I was on my way to lake Velinus in the matter of the dispute between the people of Interamna and those of Reate.[5] But,"

public occasions, to give advice on any religious matter which might arise.

[5] Cicero (*ad Atticum*, IV, 15, 5) refers to having stayed with Axius in 54 B.C. when engaged on a dispute between the people of Reate and Interamna. Axius seems to have entertained his guests lavishly, as Appius is still "belching" over the "hospitable" birds which his host furnished. The word *ructor* is used playfully, (*a*) literally, and (*b*) "recall to mind."

Sed non haec, inquit, villa, quam aedificarunt maiores nostri, frugalior ac melior est quam tua illa
4 perpolita in Reatino? Nuncubi hic vides citrum aut aurum? Num minium aut armenium? Num quod emblema aut lithostrotum? Quae illic omnia contra. Et cum haec sit communis universi populi, illa solius tua; haec quo succedant e campo cives et reliqui homines, illa quo equae et asini; praeterea cum ad rem publicam administrandam haec sit utilis, ubi cohortes ad dilectum consuli adductae considant, ubi arma ostendant, ubi censores censu admittant popu-
5 lum. Tua scilicet, inquit Axius, haec in campo Martio extremo utilis et non deliciis sumptuosior quam omnes omnium universae Reatinae? Tua enim oblita tabulis pictis nec minus signis; at mea, vestigium ubi sit nullum Lysippi aut Antiphilu, at crebra sartoris et pastoris. Et cum illa non sit sine fundo magno et eo polito cultura, tua ista neque agrum habeat ullum nec bovem nec equam.
6 Denique quid tua habet simile villae illius, quam tuus avos ac proavos habebat? Nec enim, ut illa, faenisicia vidit arida in tabulato nec vindemiam in cella neque in granario messim. Nam quod extra urbem est aedificium, nihilo magis ideo est villa, quam

---

[1] The *villa* was originally the simple farmstead. This naturally developed into the *villa* described in Book I, Chapters 11–13. But the word was further used of those elaborate establishments to which Varro refers in the Introduction to Book II. The conversation here plays on these various meanings.

[2] The word *emblema* in Latin designates a tessellated pavement of various colours.

[3] As the Villa Publica was on the Campus Martius, it was the natural rendezvous for the mobilization of the army, as well as for the taking of the census, for the elections, and all public occasions which required a large open space.

he added, "isn't this villa, which our ancestors built, simpler and better than that elaborate villa [1] of yours at Reate? Do you see anywhere here citrus wood or gold, or vermilion or azure, or any coloured [2] or mosaic work? At your place everything is just the opposite. Also, while this villa is the common property of the whole population, that one belongs to you alone; this one is for citizens and other people to come to from the Campus, and that one is for mares and asses; and furthermore, this one is serviceable for the transaction of public business—for the cohorts to assemble when summoned by the consul for a levy, for the inspection of arms, for the censors to convoke the people for the census." [3] "Do you really mean," replied Axius, "that this villa of yours on the edge of the Campus Martius is merely serviceable, and isn't more lavish in luxuries than all the villas owned by everybody in the whole of Reate? Why, your villa is plastered with paintings, not to speak of statues; while mine, though there is no trace of Lysippus or Antiphilus,[4] has many a trace of the hoer and the shepherd. Further, while that villa is not without its large farm, and one which has been kept clean by tillage, this one of yours has never a field or ox or mare. In short, what has your villa that is like that villa which your grandfather and great-grandfather had? For it has never, as that one did, seen a cured hay harvest in the loft, or a vintage in the cellar, or a grain-harvest in the bins. For the fact that a building is outside the city no more makes it a villa than the same fact makes villas

[4] Lysippus, the sculptor, and Antiphilus, the painter, were famous contemporaries of Alexander the Great.

eorum aedificia, qui habitant extra portam Flumentanam aut in Aemilianis.

7 Appius subridens, Quoniam ego ignoro, inquit, quid sit villa, velim me doceas, ne labar imprudentia, quod volo emere a M. Seio in Ostiensi villam. Quod si ea aedificia villae non sunt, quae asinum tuum, quem mihi quadraginta milibus emptum ostendebas aput te, non habent, metuo ne pro villa emam 8 in litore Seianas aedes. Quod aedificium hic me Lucius Merula impulit ut cuperem habere, cum diceret nullam se accepisse villam, qua magis delectatus esset, cum apud eum dies aliquot fuisset; nec tamen ibi se vidisse tabulam pictam neque signum aheneum aut marmoreum ullum, nihilo magis torcula vasa vin- 9 demiatoria aut serias olearias aut trapetas. Axius aspicit Merulam et, Quid igitur, inquit, est ista villa, si nec urbana habet ornamenta neque rustica membra? Quoi ille; Num minus villa tua erit ad angulum Velini, quam neque pictor neque tector vidit umquam, quam in Rosia quae est polita opere tectorio eleganter, quam dominus habes communem cum 10 asino? Cum significasset nutu nihilo minus esse villam eam quae esset simplex rustica, quam eam quae esset utrumque, et ea et urbana, et rogasset, quid ex iis rebus colligeret, Quid? inquit, si propter pastiones tuus fundus in Rosia probandus sit, et

---

[1] One of the gates in the Servian Wall, close to the Tiber. The orator Hortensus had a house here.

[2] Another suburb, first mentioned here. It was on the Campus Martius, and seems to have extended to the river.

[3] Storr-Best refers to Aulus Gellius, III, 9, where the story is told of a "certain Seius" who possessed a very beautiful horse, which, however, brought disaster to its successive owners, so that it became proverbial to say of an unlucky man, "He has a Seian horse"; and suggests a play on the proverb.

of the houses of those who live outside the Porta Flumentana[1] or in the Aemiliana.[2]"

To which Appius replied, with a smile: "As I don't know what a villa is, I should like you to enlighten me, so that I shall not go wrong from lack of foresight; since I want to buy a villa from Marcus Seius near Ostia. For if buildings are not villas unless they contain the ass which you showed me at your place, for which you paid 40,000 sesterces, I'm afraid I shall be buying a 'Seian' house[3] instead of a seaside villa. My friend here, Lucius Merula, made me eager to own this house[4] when he told me, after spending several days with Seius, that he had never been entertained in a villa which he liked more; and this in spite of the fact that he saw there no picture or statue of bronze or marble, nor, on the other hand, apparatus for pressing wine, jars for olive oil, or mills." Axius turned to Merula and asked: "How can that be a villa, if it has neither the furnishings of the city nor the appurtenances of the country?" "Why," he replied, "you don't think that place of yours on the bend of the Velinus, which never a painter or fresco-worker has seen, is less a villa than the one in the Rosea which is adorned with all the art of the stucco-worker, and of which you and your ass are joint owners?" When Axius had indicated by a nod that a building which was for farm use only was as much a villa as one that served both purposes, that of farm-house and city residence, and asked what inference he drew from that admission; "Why," he replied, "if your place in the Rosea is to be commended for its pasturage, and is rightly

[4] *i.e.* the house of Marcus Seius.

quod ibi pascitur pecus ac stabulatur, recte villa appellatur, haec quoque simili de causa debet vocari villa, in qua propter pastiones fructus capiuntur 11 magni. Quid enim refert, utrum propter oves, an propter aves fructus capias? Anne dulcior est fructus apud te ex bubulo pecore, unde apes nascuntur, quam ex apibus, quae ad villam Sei in alvariis opus faciunt? Et num pluris tu e villa illic natos verres lanio vendis, quam hinc apros macellario 12 Seius? Qui minus ego, inquit Axius, istas habere possum in Reatina villa? Nisi si apud Seium Siculum fit mel, Corsicum in Reatino; et hic aprum glas cum pascit empticia, facit pinguem, illic gratuita exilem. Appius: Posse ad te fieri, inquit, Seianas pastiones non negavit Merula; ego non esse ipse 13 vidi. Duo enim genera cum sint pastionum, unum agreste, in quo pecuariae sunt, alterum villaticum, in quo sunt gallinae ac columbae et apes et cetera, quae in villa solent pasci, de quibus et Poenus Mago et Cassius Dionysius et alii quaedam separatim ac dispersim in libris reliquerunt, quae Seius legisse videtur et ideo ex iis pastionibus ex una villa maioris fructus capere, quam alii faciunt ex toto fundo. 14 Certe, inquit Merula; nam ibi vidi greges magnos anserum, gallinarum, columbarum, gruum, pavonum,

---

[1] *pastio* is used of feeding bees, birds, etc., and has a wider significance than our word "pasturing."

[2] Cf. Chapter 16, Section 4, of this book.

[3] Honey from Sicily was famous for its excellence (see Chap. 16, 14); that from Corsica was bitter, because the bees fed on wormwood.

called a villa because cattle are fed and stabled there, for a like reason that also should have the name in which a large revenue is derived from pasturing.[1] For if you get a revenue from flocks, what does it matter whether they are flocks of sheep or of birds? Why, is the revenue sweeter on your place from oxen which give birth to bees [2] than it is from the bees which are busy at their task in the hives of Seius's villa? And do you get more from the butcher for the boars born on your place there than Seius does from the market-man for the wild boars from his place?" "Well," replied Axius, "what is there to prevent me from keeping these at my villa at Reate? You don't think that honey is Sicilian if it is produced on Seius's place, and Corsican [3] if it is produced at Reate? And that if mast which has to be bought feeds a boar on his place it makes him fat, while that which is had for nothing on my place makes him thin?" Whereupon Appius remarked: "Merula did not say that you could not have husbandry like Seius's on your place; but I have, with my own eyes, seen that you have not. For there are two kinds of pasturing: one in the fields, which includes cattle-raising, and the other around the farmstead, which includes chickens, pigeons, bees, and the like, which usually feed in the steading; the Carthaginian Mago, Cassius Dionysius, and other writers have left in their books remarks on them, but scattered and unsystematic. These Seius seems to have read, and as a result he gets more revenue from such pasturing out of one villa than others receive from a whole farm." "You are quite right," said Merula; "for I have seen there large flocks of geese, chickens, pigeons, cranes, and peafowl, not to speak

nec non glirium, piscium, aprorum, ceterae venationis. Ex quibus rebus scriba librarius, libertus eius, qui apparuit Varroni et me absente patrono hospitio accipiebat, in annos singulos plus quinquagena milia e villa capere dicebat. Axio admiranti, Certe nosti, inquam, materterae meae fundum, in Sabinis qui est ad quartum vicesimum lapidem via Salaria a

15 Roma. Quidni? inquit, ubi aestate diem meridie dividere soleam, cum eo Reate ex urbe aut, cum inde venio hieme, noctu ponere castra. Atque in hac villa qui est ornithon, ex eo uno quinque milia scio venisse turdorum denariis ternis, ut sexaginta milia ea pars reddiderit eo anno villae, bis tantum quam tuus fundus ducentum iugerum Reate reddit. Quid? sexaginta, inquit Axius, sexaginta, sexaginta?

16 derides. Sexaginta, inquam. Sed ad hunc bolum ut pervenias, opus erit tibi aut epulum aut triumphus alicuius, ut tunc fuit Scipionis Metelli, aut collegiorum cenae, quae nunc innumerabiles excandefaciunt annonam macelli. Reliquis annis omnibus si non[1] hanc expectabis summam, spero, non tibi decoquet ornithon; neque hoc accidit his moribus nisi raro ut decipiaris. Quotus quisque enim est annus, quo non videas epulum aut triumphum aut collegia non epulari? Sed propter luxuriam, inquit, quodam modo epulum cotidianum est intra

17 ianuas Romae. Nonne item L. Abuccius, homo,

[1] *non* added by Goetz.

---

[1] Leading from the Porta Collina into the Sabine country.
[2] There were four sesterces to the denarius.

of numbers of dormice, fish, boars, and other game. His book-keeper, a freedman who waited on Varro and used to entertain me when his patron was away from home, told me that he received, because of such husbandry, more than 50,000 sesterces from the villa every year." When Axius expressed his surprise, I remarked to him: "Doubtless you know my maternal aunt's place in the Sabine country, at the twenty-fourth milestone from Rome on the Via Salaria?"[1] "Of course," he replied; "it is my custom to break the journey there at noon in summer, when I am on my way to Reate from the city, and to camp there at night in winter when I am on my way from there to town." "Well, from the aviary alone which is in that villa, I happen to know that there were sold 5,000 fieldfares, for three denarii apiece, so that that department of the villa in that year brought in sixty thousand sesterces[2]—twice as much as your farm of 200 iugera at Reate brings in." "What? Sixty?" exclaimed Axius, "Sixty? Sixty? You are joking!" "Sixty," I repeated. "But to reach such a haul as that you will need a public banquet or somebody's triumph, such as that of Metellus Scipio at that time, or the club dinners which are now so countless that they make the price of provisions go soaring. If you can't look for this sum in all other years, your aviary, I hope, will not go bankrupt on you; and if fashions continue as they now are, it will happen only rarely that you miss your reckoning. For how rarely is there a year in which you do not see a banquet or a triumph, or when the clubs do not feast?" "Why," said he, "in this time of luxury it may fairly be said that there is a banquet every day within the gates of Rome. Was it not Lucius Abuccius, who is, as

ut scitis, apprime doctus, cuius Luciliano charactere sunt libelli, dicebat in Albano fundum suum pastionibus semper vinci a villa? Agrum enim minus decem milia reddere, villam plus vicena. Idem secundum mare, quo loco vellet, si parasset villam, se supra centum milia e villa recepturum. Age, non M. Cato nuper, cum Luculli accepit tutelam, e piscinis eius quadraginta milibus sestertis vendidit
18 piscis? Axius, Merula mi, inquit, recipe me quaeso discipulum villaticae pastionis. Ille: Quin[1] simulac promiseris minerval, incipiam, inquit. Ego vero non recuso, vel hodie vel ex ista pastione crebro. Appius: Credo simulac primum ex isto villatico pecore mortui erunt anseres aut pavones. Cui ille: Quid enim interest, utrum morticinas editis volucres an pisces, quos nisi mortuos estis numquam? Sed oro te, inquit, induce me in viam disciplinae villaticae pastionis ac vim formamque eius expone.

III. Merula non gravate, Primum, inquit, dominum scientem esse oportet earum rerum, quae in villa circumve eam ali ac pasci possint, ita ut domino sint fructui ac delectationi. Eius disciplinae genera sunt tria: ornithones, leporaria, piscinae. Nunc ornithonas dico omnium alitum, quae intra parietes villae solent
2 pasci. Leporaria te accipere volo non ea quae tritavi nostri dicebant, ubi soli lepores sint, sed

[1] *Quin* Schneider: *qui*.

---

[1] The son of Lucius Lucullus who fought against Mithridates. For the ponds, see pages 444, note 1, and 527.

[2] *minerval*, a satirical word for a fee paid for instruction. Cf. Juvenal, X, 116, *uno parcam colit asse Minervam*, said of a schoolboy.

[3] *morticina* is used of animals which have died a natural death.

you know, an unusually learned man (his writings are quite in the manner of Lucilius), who used to remark likewise that his estate near Alba was always beaten in feeding by his steading? for his land brought in less than 10,000, and his steading more than 20,000 sesterces. He also claimed that if he had got a villa near the sea, where he wanted one, he would take in more than 100,000 from the villa. Come, did not Marcus Cato, when he took over the guardianship of Lucullus[1] recently, sell the fish from his ponds for 40,000 sesterces?" "My dear Merula," said Axius, "take me, I beg, as your pupil in this villa-feeding." "Certainly," he replied; "I will begin as soon as you promise the *minerval*."[2] "That is satisfactory to me; you may have it to-day, or I'll pay it time and again from that feeding." "Humph," replied Appius, "the first time some geese or peacocks out of your flock die!" "Well," retorted Axius, "what does it matter if you eat fowls or fish that have died,[3] seeing that you never eat them unless they are dead? But, I pray you," said he, "lead me into the way of the science of villa-husbandry, and set forth its scope and method."

III. Merula began without hesitation: "In the first place, the owner ought to have so clear an idea of those creatures which can be reared or fed in the villa and around it that they may afford him both profit and pleasure. There are three divisions of this science: the aviary, the hare-warren, and the fish-pond. Under the head of aviary I include enclosures for all fowls which are usually reared within the walls of the villa. Under the head of hare-warrens I wish you to understand, not those which our forefathers called by that name—

omnia saepta, afficta villae quae sunt et habent inclusa animalia, quae pascantur. Similiter piscinas dico eas, quae in aqua dulci aut salsa inclusos habent

3 pisces ad villam. Harum rerum singula genera minimum in binas species dividi possunt: in prima parte ut sint quae terra modo sint contentae, ut sunt pavones turtures turdi; in altera specie sunt quae non sunt contentae terra solum, sed etiam aquam requirunt, ut sunt anseres querquedulae anates. Sic alterum genus illut venaticum duas habet diversas species, unam, in qua est aper caprea lepus; altera item extra villam quae sunt, ut apes cochleae glires.

4 Tertii generis aquatilis item species duae, partim quod habent pisces in aqua dulci, partim quod in marina. De his sex partibus ad ista tria genera item tria genera[1] artificum paranda, aucupes venatores piscatores, aut ab iis emenda quae tuorum servorum diligentia tuearis in fetura ad partus et nata nutricere saginesque, in macellum ut perveniant. Neque non etiam quaedam adsumenda in villam sine retibus aucupis venatoris piscatoris, ut glires cochlias gallinas.

5 Earum rerum cultura instituta prima ea quae in villa habetur; non enim solum augures Romani ad auspicia primum pararunt pullos, sed etiam patres familiae rure. Secunda, quae macerie ad villam

[1] *item tria genera* added by Keil.

---

[1] Cf. Cato, 89, 90.

[2] One of the earliest and most common forms of augury was the feeding of the sacred chickens. If they ate so greedily that parts of the food fell from their beaks, it was called *tripudium solistimum.* Livy, and Cicero, *de Divinatione,* give many anecdotes.

places where there are only hares—but all enclosures which are attached to the villa and keep animals enclosed for feeding. Similarly, by the term fish-pond I mean ponds which keep fish enclosed near the villa, either in fresh or salt water. Each of these divisions may be subdivided into at least two: thus, under the first head, those which are content with the land only, as peafowl, turtle-doves, fieldfares; under the second, those which are not content with the land only, but need water also, as geese, teal, and ducks. In the same way the second head—that of game—contains its two diverse classes, one under which come the boar, the roe, and the hare, and the second, those which are also outside the villa, such as bees, snails, and dormice. There are likewise two divisions of the third class, the aquatic, inasmuch as fish are kept sometimes in fresh water, sometimes in sea-water. For the three classes formed of these six subdivisions must be secured three classes of craftsmen—fowlers, hunters, fishers—or else you must purchase from these those creatures which you are to preserve by the activity of your own servants during the period of gestation and up to the time of birth, and when they are born to rear and fatten so that they may reach the market.[1] And there are, moreover, certain other creatures which are to be brought into the villa without the use of net by fowler or hunter or fisher, such as dormice, snails, and chickens. The rearing of the last named, chickens, was the first to be attempted within the villa; for not only did Roman soothsayers raise chickens first for their auspices,[2] but also the heads of families in the country. Next came the animals which are kept in an enclosure near the villa for hunting, and hard

venationis causa cluduntur et propter alvaria; apes enim subter sugrundas ab initio villatico usae tecto. Tertiae piscinae dulces fieri coeptae et e fluminibus 6 captos recepere ad se pisces. Omnibus tribus his generibus sunt bini gradus; superiores, quos frugalitas antiqua, inferiores, quos luxuria posterior adiecit. Primus enim ille gradus anticus maiorum nostrum erat, in quo essent aviaria duo dumtaxat: in plano cohors, in qua pascebantur gallinae, et earum fructus erat ova et pulli; alter sublimis, in quo erant colum-7 bae in turribus aut summa villa. Contra nunc aviaria sunt nomine mutato, quod vocantur ornithones, quae palatum suave domini paravit, ut tecta maiora habeant, quam tum habebant totas villas, in 8 quibus stabulentur turdi ac pavones. Sic in secunda parti ac leporario pater tuus, Axi, praeterquam lepusculum e venatione vidit numquam. Neque enim erat magnum id saeptum, quod nunc, ut habeant multos apros ac capreas, complura iugera maceriis concludunt. Non tu, inquit mihi, cum emisti fundum Tusculanum a M. Pisone, in leporario 9 apri fuerunt multi? In tertia parti quis habebat piscinam nisi dulcem et in ea dumtaxat squalos ac mugiles pisces? Quis contra nunc minthon [1] non dicit

[1] *minthon* Keil: *mithon* or *rhynton.*

---

[1] Cf. Book II, Introd. § 2.

[2] Both these fishes are unknown elsewhere except as sea-fish. They are frequently referred to by Pliny (*N. H.*, Book IX), who seems to get his information from Aristotle's *History of Animals*.

[3] Keil substitutes *minthon* for the meaningless *mithon* or

by it the bee-hives; for from the first bees took advantage of the roof of the villa under the eaves. Thirdly there began to be built fresh-water ponds, to which were carried fish which had been caught from the streams. Each of these three classes has two stages: the earlier, which the frugality of the ancients observed, and the later, which modern luxury has now added. For instance, first came the ancient stage of our ancestors, in which there were simply two aviaries: the barn-yard on the ground in which the hens fed—and their returns were eggs and chickens—and the other above ground, in which were the pigeons, either in cotes or on the roof of the villa. On the other hand, in these days, the aviaries have changed their name and have become *ornithones*;[1] and those which the dainty palate of the owner has constructed have larger buildings for the sheltering of fieldfares and peafowl than whole villas used to have in those days. So too in the second division, the warren, your father, Axius, never saw any better game from his hunting than a paltry hare. For in his day there was no great preserve, whereas nowadays people enclose many acres within walls, so as to keep numbers of wild boars and roes. When you bought your place near Tusculum from Marcus Piso," he added, turning to me, "were there not many wild boars in the 'hare-warren?' In the third division, who had a fish-pond, except a fresh-water pond, or kept any fish in it except *squali* or *mugiles*?[2] On the other hand what young fop[3] in these days

rhynton of the manuscripts, and quotes for the meaning Philodemus of Gadara, who defines μίνθων as a supercilious fop "who looks down upon everybody and depreciates all whom he meets or hears of even if they be people reputed great," etc.—(Storr-Best). Goetz proposes *malthon*.

sua nihil interesse, utrum iis piscibus stagnum habeat plenum an ranis? Non Philippus, cum ad Ummidium hospitem Casini devertisset et ei e tuo flumine lupum piscem formosum apposuisset atque ille gustasset et expuisset, dixit, "Peream, ni piscem putavi esse"?
10 Sic nostra aetas in quam luxuriam propagavit leporaria, hac piscinas protulit ad mare et in eas pelagios greges piscium revocavit. Non propter has appellati Sergius Orata et Licinius Murena? Quis enim propter nobilitates ignorat piscinas Philippi, Hortensi, Lucullorum? Quare unde velis me incipere, Axi, dic.

IV. Ille, Ego vero, inquit, ut aiunt post principia in castris, id est ab his temporibus quam superioribus, quod ex pavonibus fructus capiuntur maiores quam e gallinis. Atque adeo non dissimulabo, quod volo de ornithone primum, quod lucri fecerunt hoc nomen turdi. Sexaginta enim milia Fircelina excande me fecerunt[1] cupiditate.

2 Merula, Duo genera sunt, inquit, ornithonis: unum delectationis causa, ut Varro hic fecit noster sub Casino, quod amatores invenit multos; alterum fructus causa, quo genere macellarii et in urbe quidam habent loca clausa et rure, maxime conducta in Sabinis, quod ibi propter agri naturam frequentes

[1] i.e. *me excandefecerunt.*

---

[1] He returns to the subject in Chapter 18 of this book, and especially §§ 5 ff. For these fish-ponds built out into the sea cf. Hor., *Odes*, II, 15, 3: latius | extenta visentur Lucrino | stagna lacu (cf. Marquardt, *Privatleben der Römer*, pp. 433 ff.).

[2] *i.e.* "behind the front rank (of *principes*)," but Varro seems to use it in the sense of "after the beginning," not going too far back.

[3] Cf. III, 2, 14–15.

will not tell you that he would as soon have his pond full of frogs as of such fish as these? You remember that Philippus once, when he had turned aside to visit his friend Ummidius at Casinum, was served with a fine pike from your river; he tasted it, spat it out, and exclaimed: 'I'll be hanged if I didn't think it was fish!' So our generation, with the same extravagance with which it extended the boundaries of its warrens, has thrust its fish-ponds to the sea, and has brought into them whole schools of deep-sea fish.[1] Was it not from these that Sergius Orata (Goldfish) and Licinius Murena ('Lamprey') got their names? And, indeed, who does not know, on account of their fame, the fish-ponds of Philippus, Hortensius, and the Luculli? So, then, where do you wish me to begin, Axius?"

IV. "Personally," he replied, "if I may use a military figure, I should like you to begin *post principia*,[2] that is, with the present rather than the former times, as larger returns are had from peafowl than from chickens. And what is more, I will make no secret of the fact that I want to hear first about the *ornithon*, because those fieldfares have made the word mean 'gain'; for those sixty thousand sesterces of Fircellia have set me on fire with greed."[3]

"There are," resumed Merula, "two kinds of *ornithon*; one merely for pleasure, such as our friend Varro has built near Casinum, which has found many admirers, and the other for profit. Of the latter class are the enclosures which those who supply fowl for the market keep, some in the city, others in the country; especially the leased enclosures in the Sabine district, as, because of the nature of the country, large flocks of fieldfares are found there.

3 apparent turdi. Ex iis tertii generis voluit esse Lucullus coniunctum aviarium, quod fecit in Tusculano, ut in eodem tecto ornithonis inclusum triclinium haberet, ubi delicate cenitaret et alios videret in mazonomo positos coctos, alios volitare circum fenestras captos. Quod inutile invenerunt. Nam non tantum in eo oculos delectant intra fenestras aves volitantes, quantum offendit quod alienus odor opplet nares.

V. Sed quod te malle arbitror, Axi, dicam de hoc quod fructus causa faciunt, unde, non ubi, sumuntur pingues turdi. Igitur testudo, aut[1] peristylum tectum tegulis aut rete, fit magna, in qua milia aliquot 2 turdorum ac merularum includere possint, quidam cum eo adiciant praeterea aves alias quoque, quae pingues veneunt care, ut miliariae ac coturnices. In hoc tectum aquam venire oportet per fistulam et eam potius per canales angustas serpere, quae facile extergeri possint (si enim late ibi diffusa aqua, et inquinatur facilius et bibitur inutilius), et ex eis caduca quae abundat per fistulam exire, ne luto aves laborent. 3 Ostium habere humile et angustum et potissimum eius generis, quod cocliam appellant, ut solet esse in cavea, in qua tauri pugnare solent; fenestras raras, per quas non videantur extrinsecus arbores aut aves,

[1] *aut* Schneider; all other editors *ut*.

---

[1] Varro has already commented on Greek names in common use. Horace uses the same word (*Sat.*, II, 8, 86).

[2] Columella (VIII, 3, 8) advises that in the chicken house the water and food be kept in covered lead troughs which contain openings to admit the heads of the fowl.

[3] The word is used only here. Schneider interprets it as a trap-door, *cataracta*; but Pollack, in Pauly-Wissowa, cites Procopius, *de Bell. Pers.*, I, 24, who states that a door in the Hippodrome at Constantinople was called κοχλίας because of

Lucullus claimed that the aviary which he built on his place near Tusculum, formed by a combination of these two, constituted a third class. Under the same roof he had an aviary and a dining-room, where he could dine luxuriously, and see some birds lying cooked on the dish [1] and others fluttering around the windows of their prison. But they found it unserviceable; for in it the birds fluttering around the windows do not give pleasure to the eyes to the same extent that the disagreeable odour which fills the nostrils gives offence.

V. "I shall, however, as I suppose you prefer, Axius, discuss the aviary which is built for profit—the place from which fat fieldfares are taken, and not the place where they are taken. Well, there is built a large domed building, or a peristyle covered with tiles or netting, in which several thousand fieldfares and blackbirds can be enclosed; though some breeders add besides other birds which, when fattened, bring a high price, such as ortolans and quails. Into this building water should be conducted through a pipe and allowed to spread preferably through narrow channels which can easily be cleaned (for if the water spreads there in pools, it more easily becomes foul and is not good for drinking), and the superfluous drip-water from these should run out through a pipe, so that the birds may not be troubled by mud.[2] It should have a low, narrow door, and preferably of the kind which they call *coclia*,[3] such as usually are seen in the pit where bullfights are held. The windows should be few, and so arranged that trees and birds outside cannot be seen; for the sight of these, and

its winding course (ἀπὸ τῆς καθόδου κυκλοτεροῦς οὔσης). The word properly means a snail-shell with spirals.

quod earum aspectus ac desiderium marcescere facit volucres inclusas. Tantum locum luminis habere oportet, ut aves videre possint, ubi assidant, ubi cibus, ubi aqua sit. Tectorio tacta esse levi circum ostia ac fenestras, nequa intrare mus aliave quae
4 bestia possit. Circum huius aedifici parietes intrinsecus multos esse palos, ubi aves assidere possint, praeterea perticis inclinatis ex humo ad parietem et in eis traversis gradatim modicis intervallis perticis adnexis ad speciem cancellorum scenicorum ac theatri. Deorsum in terram esse aquam, quam bibere possint, cibatui offas positas. Eae maxime glomerantur ex ficis et farre mixto. Diebus viginti antequam tollere vult turdos, largius dat cibum, quod plus ponit et farre subtiliore incipit alere. In hoc tecto caveas, quae caveae tabulata habeant aliquot
5 ad perticarum[1] supplementum. Contra hic aviarius[2] quae mortuae ibi sunt aves, ut domino numerum reddat, solet ibidem servare. Cum opus sunt, ex hoc aviario ut sumantur idoneae, excludantur in minusculum aviarium, quod est coniunctum cum maiore ostio, lumine illustriore, quod seclusorium appellant. Ibi cum eum numerum habet
6 exclusum, quem sumere vult, omnes occidit. Hoc ideo in secluso clam, ne reliqui, si videant, despondeant animum atque alieno tempore venditoris moriantur. Non ut advenae volucres pullos faciunt,

[1] *ad perticarum* Schneider: *adportat.*

[2] *hic aviarius* Storr-Best: *hoc* (*hic* Politian) *aviarium*, MSS. and most editors. Lacuna indicated by Keil and Goetz before *quae* suspected by Iucundus, whose conjectured reading (*est aliud minus, in quo*) was accepted by early editors. Storr-Best rightly defends the reading without lacuna.

---

[1] *Cancellus* means, in general, any sort of screen on windows or doors. Here, as in Ovid, *Am.*, III, 2, 64, it refers to tiers of seats guarded by such grilles.

the longing for them, makes the imprisoned birds grow thin. It should have only enough openings for light to enable the birds to see where to perch, and where the food and water is. It should be faced around the doors and windows with smooth plaster, so that no mice or other vermin can enter anywhere. Around the walls of this building on the inside there should be a number of poles for the birds to perch on; and, in addition, rods sloping from ground to wall, with transverse rods fastened to them in steps at moderate intervals, after the fashion of the balustrades of the theatre or the arena.[1] At the bottom, on the ground, there should be water for them to drink, and here should be placed cakes for their food. These are usually made by kneading a mixture of figs and spelt. Twenty days before the breeder desires to remove fieldfares, he feeds them more liberally, giving larger quantities and beginning to feed them on spelt ground finer. In this building there should be recesses, equipped with several shelves, as a supplement to the perches; it is here, facing the perches, that the caretaker usually keeps on hand the birds which have died in the place, so as to render account to his master. When it becomes necessary to remove from this aviary birds which are fit for market, they should be taken out and put into a smaller aviary, called the *seclusorium* (coop), which is connected by a door with the larger aviary and better lighted. When he has the number which he desires to take shut up here, he kills them all. The reason for doing this privately in a separate room is to prevent the others, if they should see it, from moping and dying at a time which would be inopportune for the seller. Fieldfares do not rear their young here and there as do the other migratory

in agro ciconiae, in tecto hirundines, sic aut hic aut illic turdi, qui cum sint nomine mares, re vera feminae quoque sunt. Neque id non secutum ut esset in merulis, quae nomine feminino mares quoque sunt.

7 Praeterea volucres cum partim advenae sint, ut hirundines et grues, partim vernaculae, ut gallinae ac columbae, de illo genere sunt turdi adventicio ac quotannis in Italiam trans mare advolant circiter aequinoctium autumnale et eodem revolant ad aequinoctium vernum, et alio tempore turtures ac coturnices immani numero. Hoc ita fieri apparet in insulis propinquis Pontiis, Palmariae, Pandateriae. Ibi enim in prima volatura cum veniunt, morantur dies paucos requiescendi causa itemque faciunt, cum ex Italia trans mare remeant.

8 Appius Axio, Si quinque milia hoc coieceris, inquit, et erit epulum ac triumphus, sexaginta milia quae vis statim in fenus des licebit multum. Tum mihi, tu dic illut alterum genus ornithonis, qui animi causa constitutus a te sub Casino fertur, in quo diceris longe vicisse non modo archetypon inventoris nostri ornithotrophion M. Laeni Strabonis, qui Brundisii hospes noster primus in peristylo habuit exhedra conclusas aves, quas pasceret obiecto rete, sed etiam

9 in Tusculano magna aedificia Luculli. Quoi ego: Cum habeam sub oppido Casino flumen, quod per villam fluat, liquidum et altum marginibus lapideis, latum pedes quinquaginta septem, et e villa in villam pontibus transeatur, longum pedes DCCCCL derectum ab insula, quae est in imo fluvio, ubi con-

---

[1] This strange addition may be part of Varro's queer grammarian humour.

birds, storks in the field, swallows under the roof [and though their name (*turdi*) is masculine, there are in fact females too; nor is the case otherwise as regards blackbirds (*merulae*)—though they have a feminine name, there are also males].[1] Again, birds being partly migratory, as swallows and cranes, and partly indigenous, as hens and doves, fieldfares belong to the former class, the migratory, and fly yearly across the sea into Italy about the time of the autumnal equinox, and back again whence they came about the spring equinox, as do turtle-doves and quail at another season in vast numbers. The proof of this is seen in the near-by islands of Pontiae, Palmaria, and Pandateria; for when they arrive in these at the first migration, they remain there for a few days to rest, and do the same when they leave Italy for their return across the sea.

"If you put 5,000 birds into this aviary," said Appius to Axius, "and there comes a banquet and a triumph, you may at once put out at high interest that 60,000 sesterces which you want." Then, turning to me, he said: "Do you now describe that other kind of aviary which I am told you built for your amusement near Casinum, in the construction of which you are reputed to have far surpassed not only the archetype built by its inventor, our friend Marcus Laenius Strabo, our host at Brundisium, who was the first to keep birds penned up in a recess in his peristyle, feeding them through a net covering, but also Lucullus' huge buildings on his place at Tusculum." I replied: "I own, near the town of Casinum, a stream which runs through my villa, clear and deep, with a stone facing, 57 feet wide, and requiring bridges for passage from one side of the villa to the other; it is 950 feet in a straight line from the island in the lowest part of

fluit altera amnis, ad summum flumen, ubi est
10 museum, circum huius ripas ambulatio sub dio pedes
lata denos, ab hac est in agrum versus ornithonis
locus ex duabus partibus dextra et sinistra maceriis
altis conclusus. Inter quas locus qui est ornithonis
deformatus ad tabulae litterariae speciem cum
capitulo, forma qua est quadrata, patet in latitudinem
pedes XLVIII, in longitudinem pedes LXXII; qua ad
11 capitulum rutundum est, pedes XXVII. Ad haec, ita
ut in margine quasi infimo tabulae descripta sit,
ambulatio, ab ornithone † plumula, in qua media sunt
caveae, qua introitus in aream est. In limine, in
lateribus dextra et sinistra porticus sunt primoribus
columnis lapideis, pro mediis arbusculis humilibus
ordinatae, cum a summa macerie ad epistylum tecta
porticus sit rete cannabina et ab epistylo ad stylobaten.
Hae sunt avibus omnigenus oppletae, quibus
cibus ministratur per retem et aqua rivolo tenui
12 affluit. Secundum stylobatis interiorem partem
dextra et sinistra ad summam aream quadratam e
medio diversae duae non latae oblongae sunt piscinae
ad porticus versus. Inter eas piscinas tantummodo

---

[1] The comparison with the *tabula litteraria* or school-boy's "slate," clarifies the description. As the *tabula* was provided with a loop or ring at the top with which to carry it (cf. Horace, *Sat.* I, 6, 74, *Epist.* I, 1, 56), so the quadrangle was topped off with a projection rounded on the upper end, the *capitulum*. The circular building, *tholos*, referred to below, seems to have been erected in the rounded upper end of this *capitulum*.

Various attempts have been made to reconstruct the aviary described in this paragraph. The reader may be referred to an important contribution (with ground-plan, translation of the passage, and commentary) by A. W. Van Buren and R. M. Kennedy, "Varro's Aviary at Casinum," *Journal of Roman Studies*, IX, 59–66. See also plates in editions of

the stream, where another stream runs into it, to the upper part of the stream, where the Museum is situated. Along the banks of this stream there runs an uncovered walk 10 feet broad; off this walk and facing the open country is the place in which the aviary stands, shut in on two sides, right and left, by high walls. Between these lies the site of the aviary, shaped in the form of a writing-tablet with a top-piece,[1] the quadrangular part being 48 feet in width and 72 feet in length, while at the rounded top-piece it is 27 feet. Facing this, as it were a space marked off on the lower margin of the tablet, is an uncovered walk with a *plumula*[2] extending from the aviary, in the middle of which are cages; and here is the entrance to the courtyard. At the entrance, on the right side and the left, are colonnades, arranged with stone columns in the outside rows and, instead of columns in the middle, with dwarf trees; while from the top of the wall to the architrave the colonnade is covered with a net of hemp, which also continues from the architrave to the base. These colonnades are filled with all manner of birds, to which food is supplied through the netting, while water flows to them in a tiny rivulet. Along the inner side of the base of the columns, on the right side and on the left, and extending from the middle to the upper end of the open quadrangle, are two oblong fish-basins, not very wide, facing the colonnades. Between these basins is merely a path

Gesner and Schneider, and frontispiece in translation of Storr-Best.

[2] *Plumula* (lit. "little wing") is generally regarded by editors as quite unintelligible and corrupt. Van Buren and Kennedy (*op. cit.*, p. 64) make out a good case for their translation "façade."

accessus semita in tholum, qui est ultra rutundus columnatus, ut est in aede Catuli, si pro parietibus feceris columnas. Extra eas columnas est silva manu sata grandibus arboribus,[1] ut infima perluceat, 13 tota saepta maceriis altis. Intra tholi columnas exteriores lapideas et totidem interiores ex abiete tenues locus est pedes quinque latus. Inter columnas exteriores pro pariete reticuli e nervis sunt, ut prospici in silvam possit et quae ibi sunt videri neque avis ea transire. Intra interiores columnas pro pariete rete aviarium est obiectum. Inter has et exteriores gradatim substructum ut theatridion avium, mutuli crebri in omnibus columnis impositi, sedilia avium. 14 Intra retem aves sunt omnigenus, maxime cantrices, ut lusciniolae ac merulae, quibus aqua ministratur per canaliculum, cibus obicitur sub retem. Subter columnarum stylobaten est lapis a falere pedem et dodrantem alta; ipsum falere ad duo pedes altum a stagno, latum ad quinque, ut in culcitas et columellas convivae pedibus circumire possint. Infimo intra falere est stagnum cum margine pedali et insula in medio parva. Circum falere et navalia sunt 15 excavata anatium stabula. In insula est columella, in qua intus axis, qui pro mensa sustinet rotam radia-

[1] *arboribus* Keil: *arboribus tecta.*

---

[1] *i.e.* small shelf-like projections suggesting in appearance the mutules of Doric architecture.

[2] The word *falere* occurs only in this passage, and we must conjecture its meaning; cf. *fala,* " platform." It seems to be a platform serving as the *lectus* or couch at the repast.

[3] Keil holds that previous editors have erred in thinking that Varro entertained his guests in this building, and that the word " guests " is a playful reference to the birds; and he reminds us of Lucullus's disappointing experience in holding banquets in an aviary (Chap. 4, Sec. 3). But certainly the

giving access to the *tholos*, which is a round domed building outside the quadrangle, faced with columns, such as is seen in the hall of Catulus, if you put columns instead of walls. Outside these columns is a wood planted by hand with large trees, so that the light enters only at the lower part, and the whole is enclosed with high walls. Between the outer columns of the rotunda, which are of stone, and the equal number of slender inner columns, which are of fir, is a space five feet wide. Between the exterior columns, instead of a wall there is netting of gut, so that there is a view into the wood and the objects in it, while not a bird can get out into it. In the spaces between the interior columns the aviary is enclosed with a net instead of a wall. Between these and the exterior columns there is built up step by step a sort of little bird-theatre, with brackets [1] fastened at frequent intervals to all the columns as bird-seats. Within the nettings are all manner of birds, chiefly songsters, such as nightingales and blackbirds, to which water is supplied by means of a small trench, while food is passed to them under the netting. Below the base of the columns is stone-work rising a foot and nine inches above the platform; [2] the platform itself rises about two feet above a pond, and is about five feet wide, so that the guests [3] can walk in among the benches and the small columns. At the foot of the platform inside, is the pond, with a border a foot wide, and a little island in the middle. Along the platform also docks [4] have been hollowed out as shelters for ducks. On the island is a small column, and on the inside of it is a post, which holds up, instead of a table, a wheel

arrangements named below seem better suited to people than to birds.

[4] *i.e.* miniature ship-sheds.

tam, ita ut ad extremum, ubi orbile solet esse, arcuata[1] tabula cavata sit ut tympanum in latitudinem duo pedes et semipedem, in altitudinem palmum. Haec ab uno puero, qui ministrat, ita vertitur, ut omnia una ponantur et ad bibendum et ad edendum

16 et admoveantur ad omnes convivas. Ex suggesto faleris, ubi solent esse peripetasmata, prodeunt anates in stagnum ac nant, e quo rivus pervenit in duas, quas dixi, piscinas, ac pisciculi ultro ac citro commetant, cum et aqua calida et frigida ex orbi ligneo mensaque, quam dixi in primis radiis esse, epitoniis versis ad

17 unum quemque factum sit ut fluat convivam. Intrinsecus sub tholo stella lucifer interdiu, noctu hesperus, ita circumeunt ad infimum hemisphaerium ac moventur, ut indicent, quot sint horae. In eodem hemisphaerio medio circum cardinem est orbis ventorum octo, ut Athenis in horologio, quod fecit Cyrrestes; ibique eminens radius a cardine ad orbem ita movetur, ut eum tangat ventum, qui flet, ut intus scire possis.

18 Cum haec loqueremur, clamor fit in campo. Nos athletae comitiorum cum id fieri non miraremur propter studia suffragatorum et tamen scire vellemus,

[1] *arcuata* Keil: *acuitum*.

---

[1] The *peripetasmata* were the richly embroidered coverlets which were spread over the couches and hung down the sides to the floor; but in this case the guests would see, instead of the usual side-hangings, the open side of the platform with its duck-shelters facing the pond.

[2] *Epitonium*, signifying originally "key," is used in the sense of "cock" by Vitruvius and by Seneca.

[3] This is the water-clock, popularly called the "Tower of the Winds," built by Andronicus of Cyrrhus, in the first century B.C., and still to be seen. Each of its eight sides corresponded to one of the eight winds and held a picture of

with spokes, in such fashion that on the outer rim, where the felloe usually stands, there is a curved board with raised edges like a tambourine, two and a half feet in width and a palm in height. This is revolved by a single manservant in such a way that everything to drink and eat is placed on it at once and moved around to all the guests. From the side of the platform, on which there are usually coverlets,[1] the ducks come out into the pond and swim about; from this pond a stream runs into the two fish-basins which I have described, and the minnows dart back and forth, while it is so arranged that cold and warm water flows for each guest from the wooden wheel and the table which, as I have said, is at the ends of the spokes, by the turning of cocks.[2] Inside, under the dome of the rotunda, the morning-star by day and the evening-star at night circle around near the lower part of the hemisphere, and move in such a manner as to show what the hour is. In the middle of the same hemisphere, running around the axis, is a compass of the eight winds, as in the horologium at Athens, which was built by the Cyrrestrian;[3] and there a pointer, projecting from the axis, runs about the compass in such a way that it touches the wind which is blowing, so that you can tell on the inside which it is."

While we were thus conversing, a shouting arose in the Campus. We old hands at politics were not surprised at this occurrence, as we knew how excited an election crowd could become, but still we wanted to know what it meant; thereupon Pantuleius

that wind. The water-clock was so arranged that it marked the hour, as Varro here describes, and the vane on the roof directed the pointer to the figure of the wind then blowing.

quid esset, venit ad nos Pantuleius Parra, narrat ad tabulam, cum diriberent, quendam deprensum tesserulas coicientem in loculum, eum ad consulem tractum a fautoribus competitorum. Pavo surgit, quod eius candidati custos dicebatur deprensus.

VI. Axius, De pavonibus, inquit, libere licet dicas, quoniam discessit Fircellius, qui, secus siquid diceres de iis, gentilitatis causa fortasse an tecum duceret serram. Quoi Merula, De pavonibus nostra memoria, inquit, greges haberi coepti et venire magno. Ex iis M. Aufidius Lurco supra sexagena milia nummum in anno dicitur capere. Ii aliquanto pauciores esse debent mares quam feminae, si ad fructum spectes; si 2 ad delectationem, contra; formosior enim mas. Pascendi greges agrestes. Transmarini esse dicuntur in insulis, Sami in luco Iunonis, item in Planasia insula M. Pisonis. Hi ad greges constituendos parantur bona aetate et bona forma. Huic enim natura formae e volucribus dedit palmam. Ad admissuram haec minores bimae non idoneae nec iam maiores 3 natu. Pascuntur omne genus obiecto frumento, maxime hordeo. Itaque Seius iis dat in menses singulos hordei singulos modios, ita ut in fetura det uberius, antequam salire incipiant. In[1] has a procuratore ternos pullos exigit eosque, cum creverunt,

[1] *In* added by Keil.

---

[1] *Parra* is probably the barn owl, but in any case a bird of evil omen (Horace, *Odes*, III, 27, 1) as Parra, with his evil tidings, is here.

[2] Fircellius Pavo ("Peacock"), whose full name was given, III, 2, 2.

[3] Literally, "pull a saw"; from the alternate pulling and yielding of the sawyers.

[4] A small island, now Pianosa, about 20 miles due south of Elba.

Parra[1] comes to us, and tells us that a man had been caught, while they were sorting the ballots in the office, in the act of casting ballots into the ballot-box; and that he had been dragged off to the consul by the supporters of the other candidates. Pavo arose, as it was the watcher for his candidate who was reported to have been arrested.

VI. "You may speak freely about peafowl," said Axius, "since Fircellius[2] has gone; if you should say anything out of the way about them, he would perhaps have a bone to pick[3] with you for the credit of the family." To whom Merula said: "As to peafowl, it is within our memory that flocks of them began to be kept and sold at a high price. From them Marcus Aufidius Lurco is said to receive an income of more than 60,000 sesterces a year. There should be somewhat fewer males than females if you have an eye to the financial returns; but the opposite if you look at the pleasure, for the male is handsomer. They should be pastured in flocks in the fields. Across the water they are said to be reared in the islands—on Samos, in the grove of Juno, and likewise in Marcus Piso's island of Planasia.[4] For the forming of a flock they are to be secured when they are young and of good appearance; for nature has awarded the palm of beauty to this fowl over all winged things. The hens are not suited for breeding under two years, and are no longer suited when they get rather old. They eat any kind of grain placed before them, and especially barley; so Seius issues a modius of barley a month per head, with the exception that he feeds more freely during the breeding season, before they begin to tread. He requires of his breeder three

quinquagenis denariis vendit, ut nulla avis hunc
4 assequatur fructum. Praeterea ova emit ac supponit gallinis, ex quibus excusos pullos refert in testudinem eam, in qua pavones habet. Quod tectum pro multitudine pavonum fieri debet et habere cubilia discreta, tectorio levata, quo neque serpens
5 neque bestia accedere ulla possit; praeterea habere locum ante se, quo pastum exeant diebus apricis. Utrumque locum purum esse volunt hae volucres. Itaque pastorem earum cum vatillo circumire oportet ac stercus tollere ac conservare, quod et ad agri culturam idoneum est et ad substramen pullorum.
6 Primus hos Q. Hortensius augurali aditiali cena posuisse dicitur, quod potius factum tum luxuriosi quam severi boni viri laudabant. Quem cito secuti multi extulerunt eorum pretia, ita ut ova eorum denariis veneant quinis, ipsi facile quinquagenis, grex centenarius facile quadragena milia sestertia ut reddat, ut quidem Abuccius aiebat, si in singulos ternos exigeret pullos, perfici sexagena posse.

VII. Interea venit apparitor Appi a consule et augures ait citari. Ille foras exit e villa. At in villam intro involant columbae, de quibus Merula Axio: Si umquam peristerotrophion constituisses, has tuas esse putares, quamvis ferae essent. Duo enim genera earum in peristerotrophio esse solent:

---

[1] Columella, VIII, 11, 3, gives a detailed description of such a building.

[2] Modern authorities are unable to give any reason for this latter statement.

[3] This is Cicero's contemporary, the famous orator, Quintus Hortensius Hortalus. His love of luxury is often mentioned.

[4] The word is formed from περιστερός, "pigeon," and τρέφειν, "rear." In Sec. 2 below the author uses *peristeron* also, from περιστερών.

chicks for each hen, and these, when they are grown, he sells for fifty denarii each, so that no other fowl brings in so high a revenue. He buys eggs, too, and places them under hens, and the chicks which are hatched from these he places in that domed building[1] in which he keeps his peafowl. This building should be made of a size proportioned to the number of peafowl, and should have separate sleeping quarters, coated with smooth plaster, so that no serpent or animal can get in; it should also have an open place in front of it, to which they may go out to feed on sunny days. These birds require that both places be clean; and so their keeper should go around with a shovel and pick up the droppings and keep them, as they are useful for fertilizer and as litter for chicks.[2] It is said that Quintus Hortensius[3] was the first to serve these fowl; it was on the occasion of his inauguration as aedile, and the innovation was praised at that time rather by the luxurious than by those who were strict and virtuous. As his example was quickly followed by many, the price has risen to such a point that the eggs sell for five denarii each, the birds themselves sell readily for 50 each, and a flock of 100 easily brings 40,000 sesterces—in fact, Abuccius used to say that if one required three chicks to every hen, the total might amount to 60,000.

VII. Meanwhile Appius's bailiff comes with a message from the consul that the augurs are summoned, and he leaves the villa. But pigeons fly into the villa, and Merula, pointing to them, remarks to Axius: "If you had ever built a dove-cote[4] you might think these were your doves, wild though they are. For in a dove-cote there are usually two species of these: one the wild, or as some call them, the

unum agreste, ut alii dicunt, saxatile, quod habetur in turribus ac columinibus villae, a quo appellatae columbae, quae propter timorem naturalem summa loca in tectis captant; quo fit ut agrestes maxime sequantur turres, in quas ex agro evolant suapte
2 sponte ac remeant. Alterum genus columbarum est clementius, quod cibo domestico contentum intra limina ianuae solet pasci. Hoc genus maxime est colore albo, illut alterum agreste sine albo, vario. Ex iis duabus stirpibus fit miscellum tertium genus fructus causa, atque incedunt in locum unum, quod alii vocant peristerona, alii peristerotrophion, in quo
3 uno saepe vel quinque milia sunt inclusae. Peristeron fit ut testudo magna, camara tectus, uno ostio angusto, fenestris punicanis aut latioribus reticulatis utrimque, ut locus omnis sit illustris, neve quae serpens aliutve quid animal maleficum introire queat. Intrinsecus quam levissimo marmorato toti parietes ac camarae oblinuntur et extrinsecus circum fenestras, ne mus aut lacerta qua adrepere ad columbaria
4 possit. Nihil enim timidius columba. Singulis paribus columbaria fiunt rutunda in ordinem crebra, ordines quam plurimi possunt a terra usque ad camaram. Columbaria singula esse oportet ut os habeat, quo modo introire et exire possit, intus ternarum palmarum ex omnibus partibus. Sub ordines singulos tabulae fictae ut sint bipalmes, quo utantur
5 vestibulo ac prodeant. Aquam esse oportet quae

---

[1] We do not know what a "Punic window" was. Cato in 14, 2 refers to such a window, and in 18, 9 he speaks of a "Punic joint," the character of which is also unknown. There seems to be here none of the slighting tone found in Cicero's allusion to *lectuli Punicani* in *Pro Murena*, 75. The larger windows were guarded by lattice-work without and within.

rock-pigeon, which lives in turrets and gable-ends (*columina*) of the farmhouse—whence the name *columbae*—and these, because of their natural shyness, hunt for the highest peak of the roof; hence the wild pigeons chiefly hunt for the turrets, flying into them from the fields and back again, as the fancy takes them. The other species of pigeon is gentler, and being content with the food from the house usually feeds around the doorstep. This species is generally white, while the other, the wild, has no white, but is variously coloured. From these two stocks is bred for profit a third hybrid species; these are put in a place called by some *peristeron* and by others *peristerotrophion*, and often a single one of these will contain as many as 5,000. The *peristeron* is built in the form of a large building, with a vaulted roof; it has one narrow door and windows of the Punic style,[1] or wider ones with double lattice-work, so that the whole interior is light, but so that no snake or other noxious creature can get in. The whole of the walls and chambers in the interior is covered with the smoothest possible plaster made of marble dust, and the exterior is also plastered around the windows, so that no mouse or lizard can crawl into the pigeon nests; for nothing is more timid than a pigeon. Round nests are constructed for each pair, side by side in a row, and as many rows as possible are run from the floor up to the vaulted roof. Each nest should be so constructed as to have an opening large enough to allow only entrance and exit, and on the interior should be three palms in all directions. Under each row there should be fixed a board two palms wide, to serve as an entrance and walk-way. Provision should be made for water to flow in, so that

influat, unde et bibere et ubi lavari possint. Permundae enim sunt hae volucres. Itaque pastorem columbarum quotquot mensibus crebro oportet everrere; est enim quod eum inquinat locum appositum ad agri culturam, ita ut hoc optimum esse scripserint aliquot. Siquae columba quid offenderit, ut medeatur; siquae perierit, ut efferatur; siqui pulli 6 idonei sunt ad vendendum, promat. Item quae fetae sunt, certum locum ut disclusum ab aliis rete habeat, quo transferantur, e quo foras ex peristerone evolare[1] possint matres. Quod faciunt duabus de causis: una, si fastidiunt aut inclusae consenescunt, quod libero aere, cum exierint in agros, redintegrentur; altera de causa propter inlicium. Ipsae enim propter pullos, quos habent, utique redeunt, nisi a corvo occisae aut ab accipitre inter-7 ceptae. Quos columbarii interficere solent duabus virgis viscatis defictis in terra inter se curvatis, cum inter eas posuerint obligatum animal, quod petere soleant accipitres, qui ita decipiuntur, cum se obleverunt visco. Columbas redire solere ad locum licet animadvertere, quod multi in theatro e sinu missas faciunt, atque ad locum redeunt, quae nisi 8 reverterentur, non emitterentur. Cibus apponitur circum parietes in canalibus, quas extrinsecus per fistulas supplent. Delectantur milio, tritico, hordeo, piso, fasiolis, ervo. Item fere haec, in turribus ac summis villis qui habent agrestes columbas, quaad

[1] *evolare* Ursinus: *evocare.*

they may have a place to drink and bathe, for these birds are extremely cleanly. So the pigeon-keeper should sweep them out frequently every month; for the droppings which make the place filthy are so well suited for fertilizing that several writers have stated that it is the best kind. He should see to it that any pigeon which has been hurt be treated, and that any dead one be removed, and should remove the squabs which are fit for market. He should also have a place shut off by a net from the rest, to which the brooding birds may be transferred, and from which the mother-birds may be able to fly away from the pigeon-house. This they do for two reasons: first, if they lose their appetite or grow sickly from confinement, as they are refreshed by the open air when they fly over the fields, or secondly for a decoy; for they will themselves return in any case, because of the young they have, unless they are killed by a crow or cut off by a hawk. These birds the pigeon-keepers make a practice of killing by planting two limed twigs in the ground, leaning toward each other, after placing between them, with its legs tied, some animal which hawks are in the habit of chasing; and they are caught in this way, when they have smeared themselves with the lime. You may see that doves do return to a place, from the fact that many people let them loose from their bosoms in the theatre and they return to their homes; and if they did not come back they would not be turned loose. Food is furnished them in troughs running around the walls, which are filled from the outside through pipes. Their favourite foods are millet, wheat, barley, peas, kidney-beans, and vetch. Those who have wild pigeons in turrets and in the tops of their villas should

possunt, imitandum. In peristeronas aetate bona parandum, neque pullos neque vetulas, totidem 9 mares quot feminas. Nihil columbis fecundius. Itaque diebus quadragenis concipit et parit et incubat et educat. Et hoc fere totum annum faciunt; tantummodo intervallum faciunt a bruma ad aequinoctium vernum. Pulli nascuntur bini, qui simulac creverunt et habent robor, cum matribus pariunt. Qui solent saginare pullos columbinos, quo pluris vendant, secludunt eos, cum iam pluma sunt tecti. Deinde manducato candido farciunt pane; hieme hoc bis, aestate ter, mane meridie vesperi; 10 hieme demunt cibum medium. Qui iam pinnas incipiunt habere, relincunt in nido inlisis cruribus et matribus, uberius ut cibo uti possint, obiciunt. Eo enim totum diem se et pullos pascunt. Qui ita educantur, celerius pinguiores fiunt quam alii, et candidae fiunt parentes eorum. Romae, si sunt formosi, bono colore, integri, boni seminis, paria singula volgo veneunt ducenis nummis nec non eximia singulis milibus nummum. Quas nuper cum mercator tanti emere vellet a L. Axio, equite Romano, minoris quadringentis denariis daturum negavit. 11 Axius, Si possem emere, inquit, peristerona factum, quem ad modum, in aedibus cum habere vellem, emi fictilia columbaria, iam issem emptum et misissem [1] ad

[1] *et misissem* Iucundus : *emissem.*

imitate these methods so far as they can. Those which are placed in the pigeon-house should be of a proper age, neither squabs nor old birds; and there should be an equal number of cocks and hens. Nothing is more prolific than the pigeon; thus, within a period of forty days it conceives, lays, hatches, and brings off its young. And they continue this, too, through practically the entire year, leaving an interval only between the winter solstice and the vernal equinox. Two chicks are born each time, and as soon as they have grown and have their strength they breed along with their mothers. Those who practise the fattening of squabs to increase their selling price, shut them up as soon as they are covered with down; then they stuff them with white bread which has been chewed, twice a day in winter and three times in summer—morning, noon, and evening; in winter they omit the noon feeding. When they begin to have feathers they are left in the nest, with their legs broken, and are left to their mothers so that they can eat the food more freely; for they feed themselves and their young on it all day long. Birds which are reared in this way fatten more quickly than others, and their parents become white. At Rome, if the birds are handsome, of good colour, sound, and of good breed, single pairs sell usually for 200 sesterces; but unusually fine ones sometimes for 1,000 sesterces. When a trader wanted recently to buy such birds at this price from Lucius Axius, a Roman knight, he said he would not sell for less than 400 denarii." Axius remarked: "If I could buy a ready-made pigeon-house, as I bought an earthenware dove-cote when I wanted one in my town house, I should already have gone

villam. Quasi vero, inquit Pica, non in urbe quoque sint multi. An tibe[1] columbaria qui in tegulis habent, non videntur habere peristeronas, cum aliquot supra centum milium sestertium habeant instrumentum? E quis alicuius totum emas censeo, et antequam aedificas rure, magnum condiscas hic in urbe cotidie lucrum assem semissem condere in loculos. Tu, Merula, sic perge deinceps.

VIII. Ille, Turturibus item, inquit, locum constituendum proinde magnum, ac multitudinem alere velis; eumque item ut de columbis dictum est, ut habeat ostium ac fenestras et aquam puram ac pari-
2 etes camaras munitas tectorio; sed pro columbariis in pariete mutulos aut palos in ordinem, supra quos tegeticulae cannabinae sint impositae. Infimum ordinem oportet abesse a terra non minus tres pedes, inter reliquos dodrantes, a summo ad camaram semipedem, aeque latum ac mutulus a pariete extare
3 potest, in quibus dies noctesque pascuntur. Cibatui quod sit, obiciunt triticum siccum, in centenos vicenos turtures fere semodium, cottidie everrentes eorum stabula, a stercore ne offendantur, quod item servatur ad agrum colendum. Ad saginandum adpositissimum tempus circiter messem. Etenim matres eorum tum optimae sunt, cum pulli plurimi gignuntur, qui ad farturam meliores. Itaque eorum fructus id temporis maxime consistit.

[1] *multi an tibe* (*antibe*) *columbaria* MSS. and later editions: *multi columbaria q. i. t. h. An tibi non videntur* old editions following Iucundus.

---

[1] Perhaps an instance of Varro's humour. They are constantly poking fun at Axius's greed, and Pica ironically advises him to learn from the city man a thing or two about "big" profits before setting up in the business in the country. Another interpretation, with no humour intended, comes from

to buy it and have sent it to the farm-house." "Just as if," replied Pica, "there weren't many of them in the city, also. Or doesn't it seem to you that people who have dove-cotes on their roof-tiles possess pigeon-houses, inasmuch as some of them have equipment worth more than 100,000 sesterces? I suggest that you buy the complete outfit from one of these, and before you build in the country learn here in the city to put in your purse every day the big profit of a penny or two.[1] But go ahead with your subject, Merula."

VIII. "For turtle-doves, also," he resumed, "a place should be built of a size proportioned to the number you wish to raise; and this, too, as was remarked of pigeons, so that it has a door and windows, clear water, walls and cupola covered with plaster. But instead of nests set in the walls it should have brackets or poles in a row, and over these there should be placed small mats of hemp. The bottom row should be not less than three feet from the ground, between the other rows there should be a space of nine inches, with a half-foot interval between the top and the cupola; and the row should be as wide as the bracket can stand out from the wall, as they feed on the brackets day and night. As to food, dry wheat is given them, about a half-modius for 120 turtle-doves, and their quarters are swept out every day so that they may not suffer harm from the dung—and this is also kept for fertilizing the ground. The most suitable time for fattening is about harvest, for at that time their mothers are at their best, when most chicks are being born, these latter being better for fattening; and hence the income from them is greatest at this time."

[1] Iucundus's conjecture, *ex asse semissem*: i.e. a gain of one-half *as* from every *as* invested, or a profit of fifty per cent.

IX. Axius, Ego quae requiro farturae membra, de gallinis dic sodes, Merula: tum de reliquis siquid idoneum fuerit ratiocinari, licebit. Igitur sunt gallinae quae vocantur generum trium: villaticae 2 et rusticae et Africanae. Gallinae villaticae sunt, quas deinceps rure habent in villis. De his qui ornithoboscion instituere vult, id est adhibita scientia ac cura ut capiant magnos fructus, ut factitaverunt Deliaci, haec quinque maxime animadvertant oportet: de emptione, cuius modi et quam multas parent; de fetura, quem ad modum admittant et pariant; de ovis, quem ad modum incubent et excudant; de pullis, quem ad modum et a quibus educentur; hisce appendix adicitur pars quinta, quem ad 3 modum saginentur. Ex quis tribus generibus proprio nomine vocantur feminae quae sunt villaticae gallinae, mares galli, capi semimares, qui sunt castrati. Gallos castrant, ut sint capi, candenti ferro inurentes ad infima crura, usque dum rumpatur, et quod extat 4 ulcus, oblinunt figlina creta. Qui spectat ut ornithoboscion perfectum habeat, scilicet genera ei tria paranda, maxime villaticas gallinas. E quis in parando eligat oportet fecundas, plerumque rubicunda pluma, nigris pinnis, imparibus digitis, magnis capitibus, crista erecta, amplas; hae enim ad par-

---

[1] Merula resumes. [2] See note 1, page 478.

[3] Guinea-fowl. [4] ὀρνιθοβόσκιον, breeding-place for birds.

[5] The outstanding success of the Delians as poultrymen is attested by Columella (VIII, 2, 4). Pliny (*N.H.*, X, 139) states that they were the first to fatten poultry for market.

[6] This strange operation is more clearly described by Columella, VIII, 2, 3, "And yet they experience this (lack of desire) not through the loss of their generative organs, but by having their spurs burned off with a red-hot iron." See also Pliny, *N.H.*, X, 50.

IX. "I wish, Merula," said Axius, "you would tell us of the division of fattening in which I am interested—that of chickens; then if there is anything in the other topics that is worth taking into account we may do so." "Well,[1] under the term poultry are included three kinds of fowl: the barn-yard, the wild,[2] and the African.[3] Barn-yard fowls are the species which are kept continuously in farmsteads. One who wants to set up a poultry-farm [4] of these—that is, wants to gain a large profit by the exercise of knowledge and care, as the Delians [5] generally have done—should observe especially the following five points: purchase, including the breed and number to secure; breeding, including the manner of mating and laying; eggs, including the manner of sitting and hatching; chicks, including the manner of rearing and the birds by which they are reared; and to these is added, as an appendix, the fifth topic—the method of fattening. Of the three species, the proper name for the female of the barn-yard fowl is hen, for the male is cock, while that of the half-males, which have been castrated, is capon. Cocks are castrated, to make them capons, by burning with a red-hot iron at the lowest part of the leg until it bursts;[6] and the sore which results is smeared with potter's clay. One who intends to have a complete poultry-farm should, of course, procure all three species, but chiefly the barn-yard fowls. In buying these he should choose hens which are prolific, usually of a reddish plumage, with black wing feathers, toes of uneven length, large heads, upright crest, full-bodied, as these are better fitted for laying. Cocks should be amorous; and this is judged from their being muscular, with

5 tiones sunt aptiores. Gallos salaces qui animadvertunt,[1] si sunt lacertosi, rubenti crista, rostro brevi pleno acuto, oculis ravis aut nigris, palea rubra subalbicanti, collo vario aut aureolo, feminibus pilosis, cruribus brevibus, unguibus longis, caudis magnis, frequentibus pinnis; item qui elati sunt ac vociferant saepe, in certamine pertinaces et qui animalia quae nocent gallinis non modo non per-
6 timescant, sed etiam pro gallinis propugnent. Nec tamen sequendum in seminio legendo Tanagricos et Melicos et Chalcidicos, qui sine dubio sunt pulchri et ad proeliandum inter se maxime idonei, sed ad partus sunt steriliores. Si ducentos alere velis, locus saeptus adtribuendus, in quo duae caveae coniunctae magnae constituendae, quae spectent ad exorientem versus, utraeque in longitudinem circiter decem pedum, latitudine dimidio minores, altitudine paulo humiliores: in utraque fenestra lata tripedalis,[2] et eae pede altiores e viminibus factae raris, ita ut lumen praebeant multum, neque per eas quicquam ire intro
7 possit, quae nocere solent gallinis. Inter duas ostium sit, qua gallinarius, curator earum, ire possit. In caveis crebrae perticae traiectae sint, ut omnes sustinere possint gallinas. Contra singulas perticas

[1] The text is confused. Ursinus reads *quod animadvertunt*, from "an old codex"; Gesner proposes *animadvertuntur*. Both these, and Schneider, would understand *eligat oportet* from the preceding sentence.

[2] *in utraque . . . tripedalis* Keil: *utraque . . . tripedali*.

---

[1] For the spelling "Median" vs. "Melian," see Sec. 19 of this chapter, with note 3.

[2] The hen-house described by Columella in greater detail

comb reddish, beak short, wide, and sharp, eyes yellowish or black, wattles red with a trace of white, neck particoloured or golden, thighs feathered, lower leg short, claws long, tail large, feathers thick; also by their stretching and crowing often, being stubborn in a fight—those which not only do not fear animals which attack the hens but even fight for the hens. In choosing a strain, however, it is not well to go after the Tanagrian, Median,[1] or Chalcidian; these are undoubtedly handsome birds and very well fitted for fighting one another, but they are rather poor for laying. If you wish to raise 200 you should assign them an enclosed place, and on it construct two large connecting hen-houses, facing eastward, each about ten feet in length, one-half smaller in width, and a little less in height.[2] In each of these there should be a window three feet wide and one foot higher; these should be made of withes so spaced as to allow plenty of light to enter, and yet to keep from passing through them any of the things which usually injure fowls. Between the two houses there should be a door through which their keeper, the *gallinarius*, can enter. In the houses should be run a number of perches sufficient to hold all the hens. Facing the several perches separate nests should be

(VIII, 3) corresponds, in general, to this, except in the matter of the room for the caretaker, which here is called "large," while in Columella's it is but seven feet in every dimension. Moreover, in Columella it contains only a fireplace, the smoke from which "is very salutary for hens." Columella states that this room connects the two hen-houses, but we can only infer that this is true in Varro. It appears that the caretaker's room was surrounded by nests, but the word *plena* (Sec. 7) seems very odd. Editors therefore suspect the text at this point; see critical note, page 474.

in pariete exclusa sint cubilia earum. Ante sit, ut dixi, vestibulum saeptum, in quo diurno tempore esse possint atque in pulvere volutari. Praeterea sit cella grandis, in qua curator habitet, ita ut in parietibus circum omnia plena[1] sint cubilia gallinarum aut exculpta aut adficta firmiter. Motus enim, cum 8 incubat, nocet. In cubilibus, cum parturient, acus substernendum; cum pepererunt, tollere substramen et recens aliut subicere, quod pulices et cetera nasci solent, quae gallinam conquiescere non patiuntur; ob quam rem ova aut inaequabiliter maturescunt aut consenescunt. Quae velis incubet, negant plus XXV oportere ova incubare, quamvis propter fecun- 9 ditatem pepererit plura, optimum esse partum ab aequinoctio verno ad autumnale. Itaque quae ante aut post nata sunt et etiam prima eo tempore, non supponenda; et ea quae subicias, potius vetulis quam pullitris, et quae rostra aut ungues non habeant acutos, quae debent potius in concipiendo occupatae esse quam incubando. Adpositissimae ad partum 10 sunt anniculae aut bimae. Si ova gallinis pavonina subicias, cum iam decem dies fovere coepit, tum denique gallinacia subicere, ut una excudat. Gallinaciis enim pullis bis deni dies opus sunt, pavoninis ter noveni. Eas includere oportet, ut diem et noctem incubent, praeterquam mane et vespere, dum 11 cibus ac potio is detur. Curator oportet circumeat

[1] The text is suspected. Earlier editors substituted *posita* for *plena* ("all around are placed nests for hens").

---

[1] Unless the speaker refers to the "enclosed place" in Sec. 6, this is a slip.

built for them in the wall. In front of it, as I said,[1] should be an enclosed yard, in which they may run during the daytime and dust themselves. In addition there should be a large room for the caretaker to live in, so built that the surrounding walls may be entirely filled with hens' nests, either built in the walls or firmly attached; for movement in harmful to a sitting hen. In their nests at laying-time chaff should be spread under them; and when they have laid their eggs, the bedding should be removed and other fresh bedding spread, as in old bedding lice and other vermin generally breed, and these keep the hen from resting quietly, the result being that the eggs either develop unevenly or become stale. If you wish the hen to cover the eggs, it is claimed that a sitting should number not more than 25, even if the hen has been so prolific as to lay more, and that the laying is best from the vernal to the autumnal equinox. So eggs which are laid before or after that period, and even the first laid within the period, should not be set; and the eggs which you set should be put under old hens (and such hens should not have sharp beaks or claws) rather than under pullets, as the latter ought to be busy at laying rather than at sitting. They are best fitted for laying when one year or two years old. If you are putting peafowl eggs under a hen, you should put the hen's eggs under her only at the beginning of the tenth day of sitting, so that she will hatch them together; for chicks require twice ten days, and peafowl chicks thrice nine. The hens should be shut up, so that they may sit day and night, except at the times, morning and evening, when food and drink are being given them. The caretaker should go around at

diebus interpositis aliquot ac vertere ova, ut aequabiliter concalefiant. Ova plena sint atque utilia necne, animadverti aiunt posse, si demiseris in aquam, quod inane natet, plenum desidit. Qui ut hoc intellegant concutiant, errare, quod vitales venas confundant in iis. Idem aiunt, cum ad lumen sustu-

12 leris, quod perluceat, id esse inane. Qui haec volunt diutius servare, perfricant sale minuto aut muria tres aut quattuor horas eaque abluta condunt in furfures aut acus. In supponendo ova observant ut sint numero imparia. Ova, quae incubantur, habeantne semen pulli, curator quadriduo post quam incubari coepit intellegere potest. Si contra lumen tenuit et purum unius modi esse animadvertit, putant

13 eiciendum et aliud subiciundum. Excusos pullos subducendum ex singulis nidis et subiciendum ei quae habeat paucos; ab eaque, si reliqua sint ova pauciora, tollenda et subicienda aliis, quae nondum excuderunt et minus habent triginta pullos. Hoc enim gregem maiorem non faciendum. Obiciendum pullis diebus XV primis mane subiecto pulvere, ne rostris noceat terra dura, polentam mixtam cum nasturti semine et aqua aliquanto ante factam intritam, ne tum denique in eorum corpore turgescat;

---

[1] Pliny (*N.H.*, X, 148, 151) repeats this, and says that all eggs have, in the centre of the yolk, a sort of blood-drop, which some consider to be the heart of the embryo chick. Aristotle, however, believed (*Hist. Anim.*, VI, 3) that this drop, from which developed the body of the chick, was in the white of the egg, and that the yolk supplied nourishment to the embryo.

[2] Columella, whose directions in this matter are, as often, somewhat fuller and clearer than Varro's, recommends (VIII, 5, 7) that two or three newly hatched broods, up to a maximum of thirty chicks, be put under one hen on the very day of their hatching; but he advises (VIII, 5, 15) that they be kept in

intervals of several days and turn the eggs so that they will warm evenly. It is said that you can tell whether eggs are full and fertile or not if you drop them into water, as the empty egg floats, while the full one sinks. Those who shake an egg to find this out make a mistake, as they break up the vital veins in them.[1] The same authorities state that when you hold it up to the light, the one that the light shines through is infertile. Those who wish to keep eggs a considerable time rub them down thoroughly with fine salt or brine for three or four hours, and when this is washed off pack them in bran or chaff. In setting eggs, care is taken that the number be uneven. The caretaker can find out four days after the sitting begins whether the incubating eggs contain the embryo of a chick. If he holds one against a light and observes it to be uniformly clear, the belief is that it should be thrown out and another substituted. The chicks, when hatched, should be taken from the several nests and placed under a hen which has few chicks; and if a few eggs are left they should be taken away from this hen and put under others which have not yet hatched and those which have fewer than 30 chicks;[2] for the batch must not exceed this number. During the first fifteen days there should be fed to the chicks in the morning, on a bed of dust, so that the hard earth may not injure their beaks, a mixture of barley-meal and cress-seed which has been worked up some time before with water, so that when it is eaten it may not swell up in their crops; and they must be kept away from water.

the nests with their mothers for that day rather than removed one by one as they are hatched. The unhatched eggs, he says (*ibid.*), must be removed after twenty-one days of incubation.

14 aqua prohibendum. Qua de clunibus coeperint habere pinnas, e capite, e collo eorum crebro eligendi pedes; saepe enim propter eos consenescunt. Circum caveas eorum incendendum cornum cervinum, ne quae serpens accedat, quarum bestiarum ex odore solent interire. Prodigendae in solem et in stercilinum, ut volutare possint, quod ita alibi-
15 liores fiunt; neque pullos, sed omne ornithoboscion cum aestate, tum utique cum tempestas sit mollis atque apricum; intento supra rete, quod prohibeat eas extra saepta evolare et in eas involare extrinsecus accipitrem aut quid aliut; evitantem caldorem et frigus, quod utrumque iis adversum. Cum iam pinnas habebunt, consuefaciundum ut unam aut duas sectentur gallinas, ceterae ut potius ad pariendum
16 sint expeditae, quam in nutricatu occupatae. Incubare oportet incipere secundum novam lunam, quod fere quae ante, pleraque non succedunt. Diebus fere viginti excudunt. De quibus villaticis quoniam vel nimium dictum, brevitate reliqua compensabo.

Gallinae rusticae sunt in urbe rarae nec fere nisi mansuetae in cavea videntur Romae, similes facie non his gallinis villaticis nostris, sed Africanis.
17 Aspectu ac facie incontaminatae in ornatibus publicis solent poni cum psittacis ac merulis albis, item aliis id genus rebus inusitatis. Neque fere in villis ova ac pullos faciunt, sed in silvis. Ab his gallinis dicitur

---

[1] Schneider and Keil both think these are Italian partridges. Durand de la Malle thinks they are domestic fowl which have reverted. Still others think they are our heath-fowl.

[2] Guinea fowl.

When they begin to grow feathers from the rump, the lice must be picked from their heads and necks often, for they frequently waste away because of these. Around their houses stag horns should be burned, to keep snakes from coming in; for the smell of these animals is usually fatal to them. They should be driven out into the sunshine and on to the dung-hill so they can flutter about, as in that way they grow healthier—not only the chicks but the whole poultry yard, both in summer and whenever the air is mild and it is sunny, with a net spread above them to keep them from flying outside the enclosure, and to keep hawks and the like from flying into it from outside; avoiding heat and cold, each of which is harmful to them. As soon as they have their wing-feathers they should be trained to follow one or two hens, so that the others may be free for laying rather than busied with the rearing of young. They should begin to sit after the new moon, for the sittings which begin before that time usually do not turn out well. They are hatched in about twenty days. As really too much has been said about these barnyard fowls, I shall make up for it by brevity in speaking of the rest.

"Wild hens[1] are found rarely in town and are hardly seen in Rome, except the tamed ones in cages. In appearance they are not like these barnyard fowls of ours, but rather like the African fowl.[2] Birds whose appearance and shape show that they are of unmixed breed are usually displayed in public ceremonies, along with parrots, white blackbirds, and other unusual things of that sort. Usually they do not produce eggs and chicks in farmsteads, but in the forests. It is from these fowls that the island

insula Gallinaria appellata, quae est in mari Tusco secundum Italiam contra montes Liguscos, Intimilium, Album Ingaunum; alii ab his villaticis
18 invectis a nautis, ibi feris factis procreatis. Gallinae Africanae sunt grandes, variae, gibberae, quas meleagridas appellant Graeci. Haec novissimae in triclinium cenantium[1] introierunt e culina propter
19 fastidium hominum. Veneunt propter penuriam magno. De tribus generibus gallinae saginantur maxime villaticae. Eas includunt in locum tepidum et angustum et tenebricosum, quod motus earum et lux pinguitudinis vindicta, ad hanc rem electis maximis gallinis, nec continuo his, quas Melicas appellant falso, quod antiqui, ut Thetim Thelim dicebant, sic Medicam Melicam vocabant. Hae primo dicebantur, quae ex Medica propter magni-
20 tudinem erant allatae quaeque ex iis generatae, postea propter similitudinem amplae omnes. Ex iis evulsis ex alis pinnis et e cauda farciunt turundis hordeaceis partim admixtis farina lolleacia aut semine lini ex aqua dulci. Bis die cibum dant, observantes ex quibusdam signis ut prior sit concoctus, antequam secundum dent. Dato cibo, quom perpurgarunt caput, nequos habeat pedes, rursus eas concludunt. Hoc faciunt usque ad dies
21 XXV; tunc denique pingues fiunt. Quidam et triticeo pane intrito in aquam, mixto vino bono et

[1] *cenantium* Keil: *genanium.*

---

[1] The modern name of Album Ingaunum is Albenga; that of Intimilium is Vintimiglia; and that of Gallinaria is Isola d'Albegna.

[2] Greek μελεαγρίς, from which our name for the turkey family, Meleagridae.

[3] But Columella (VIII, 2, 4) says this was a mistake made, not by the ancients, but by "the ignorant rabble."

Gallinaria, in the Tuscan Sea off the coast of Italy opposite the Ligurian mountains, Intimilium, and Album Ingaunum, is said to have got its name;[1] others hold that they are the descendants of those barnyard fowls which were carried there by sailors and became wild. The African hens are large, speckled, with rounded back, and the Greeks call them 'meleagrides.'[2] These are the latest fowls to come from the kitchen to the dining-room because of the pampered tastes of people. On account of their scarcity they fetch a high price. Of the three species, it is chiefly the barnyard fowls which are fattened. These are shut into a warm, narrow, darkened place, because movement on their part and light free them from the slavery of fat. For this purpose the largest hens are chosen, but not necessarily those which are mistakenly called "Melic"; for the ancients[3] said "Melic" for "Medic," just as they said "Thelis" for "Thetis." Those were called so originally which, because of their size, were imported from Media, and the descendants of these; but later on all large hens got the name on account of their likeness. On these hens the feathers are pulled from wings and tail, and they are fattened on pellets of barley-meal, sometimes mixed with darnel flour, or with flax seed soaked in fresh water. They are fed twice a day, and are watched to see, from certain symptoms, that the last food taken has been digested before more is given. When they have eaten, and their heads have been cleaned to prevent their having lice, they are again shut up. This is continued as long as twenty-five days, and at this time they finally become fat. Some breeders fatten them also on wheat bread softened in water mixed

odorato, farciunt, ita ut diebus XX pingues reddant ac teneras. Si in farciendo nimio cibo fastidiunt, remittendum in datione pro portione, ac decem primis processit, in posterioribus ut deminuat eadem ratione, ut vicesimus dies et primus sint pares.[1] Eodem modo palumbos farciunt ac reddunt pingues.

X. Transi, inquit Axius, nunc in illud genus, quod non est ulla villa ac terra contentum, sed requirit piscinas, quod vos philograeci vocatis amphibium. In quibus ubi anseres aluntur, nomine alieno chenoboscion appellatis. Horum greges Scipio Metellus et M. Seius habent magnos aliquot. Merula, Seius, inquit, ita greges comparavit anserum, ut hos quinque gradus observaret, quos in gallinis dixi. Hi sunt de genere, de fetura, de ovis, de pullis, de sagina.

2 Primum iubebat servum in legendo observare ut essent ampli et albi, quod plerumque pullos similes sui faciunt. Est enim alterum genus varium, quod ferum vocatur, nec cum iis libenter congregantur,

3 nec aeque fit mansuetum. Anseribus ad[2] admittendum tempus est aptissimum a bruma, ad pariendum et incubandum a Kalendis Februariis vel Martiis[3] usque ad solstitium. Saliunt fere in aqua, iniguntur in flumen aut piscinam. Singulae non plus quam ter in anno pariunt. Singulis, ubi pariant, faciendum haras quadratas circum binos pedes et semipedem; eas substernendum palea. Notandum

[1] *sint pares* Schneider: *sit pari.*
[2] *ad* supplied by Keil, from Cod. Sang. of Columella, VIII, 14, 4.
[3] *a . . . Martiis* supplied by Keil from Columella, VIII, 14, 4.

---

[1] χηνοβοσκεῖον, "a place for feeding geese."
[2] Quintus Caecilius Metellus Pius Scipio, consul with Pompey for part of the year 52 B.C.
[3] Columella's goose-pens (VIII, 14, 1) were walled enclosures, nine feet high, with three-foot coops built into the walls.

with a sound, fragrant wine, which results in making them fat and tender within twenty days. If, in the course of the fattening, they lose their appetites from too much food, the amount fed should be lessened, diminishing in the last ten days in the same proportion as it increased in the first ten, so that the twentieth day will be equal to the first. The same method is followed in fattening wood-pigeons and making them plump."

X. "Pass on now," said Axius, "to that kind of fowl which is not content with any farmstead and land, but wants ponds—the kind you Greek-lovers call amphibious. The place where geese are reared you call by the foreign name of *chenoboscion*.[1] Scipio Metellus[2] and Marcus Seius have several large flocks of geese." "Seius," continued Merula, "in making provision for his flocks of geese, observed the five steps which I have described in the case of chickens, and which had to do with strain, mating, eggs, chicks, and fattening. His first injunction to his servant was to see in choosing them that they were full-bodied and white, as usually they have goslings like themselves. For there is another species, mottled, which is called 'wild,' and these do not like to flock with the others, and are not tamed so easily. The most suitable time for mating, in the case of geese, is after the winter solstice, for laying and sitting from the first of February or March up to the summer solstice. As they usually mate in the water, they are driven into a stream or a pond. Individuals do not lay more than three times in a year, and when they do, square coops should be built for each, about two and a half feet on each side,[3] and these should be carpeted with straw. Their eggs should be distinguished by some

earum ova aliquo signo, quod aliena non excudunt. Ad incubandum supponunt plerumque novem aut undecim, qui hoc minus, quinque, qui hoc plus, XV. Incubat tempestatibus dies triginta, tepidioribus 4 XXV. Cum excudit, quinque diebus primis patiuntur esse cum matre. Deinde cotidie, serenum cum est, producunt in prata, item piscinas aut paludes, iisque faciunt haras supra terram aut suptus, in quas non inducunt plus vicenos pullos, easque cellas provident ne habeant in solo umorem et ut molle habeant substramen e palea aliave qua re, neve qua eo accedere possint mustelae aliaeve quae bestiae 5 noceant. Anseres pascunt in umidis locis et pabulum serunt, quod aliquem ferat fructum, seruntque his herbam, quae vocatur seris,[1] quod ea aqua tacta, etiam cum est arida, fit viridis. Folia eius decerpentes dant, ne, si eo inegerint, ubi nascitur, aut obterendo perdant aut ipsi cruditate pereant; voraces enim sunt natura. Quo temperandum iis, qui propter cupiditatem saepe in pascendo, si radicem prenderunt, quam educere velint e terra, abrumpunt collum; perimbecillum enim id, ut caput molle. Si haec herba non est, dandum hordeum aut frumentum aliut. Cum est tempus farraginis, dandum, ut in 6 seri dixi. Cum incubant, hordeum iis intritum in aqua apponendum. Pullis primum biduo polenta aut hordeum apponitur, tribus proximis nasturtium

[1] *seris* Iucundus : *heris.*

---

[1] Columella (VIII, 14, 2) makes clear what Varro evidently means: "A swampy, but grassy place is set apart for them, and other forage plants are grown, such as vetch, trefoil, fenugreek; but especially a species of endive which the Greeks call *σέρις*."

mark, as they do not hatch the eggs of another. Usually nine eggs or eleven form a sitting; if fewer are set, five, if more, fifteen. In cold weather they sit thirty days, in warmer weather twenty-five. When they hatch they are allowed to stay with the mother for the first five days; then they are driven out daily, when the weather is good, into meadows, and also into ponds or swamps. Coops are made for them above ground or under it, and not more than twenty goslings are placed in each; and care is taken that these quarters do not have moisture in the ground, and that they do have a soft cushion of straw or some other material, and that weasels cannot get in, or any other harmful beasts. Geese feed in damp places; so a food is sowed which will bring in a profit,[1] and also there is sowed for them an herb which is called *seris*, because this, even when it is dry, if touched by water becomes green. The leaves of this are plucked and fed to them, for if they are driven into the place where it is growing they either ruin it by their trampling or die from over-eating; for they are naturally ravenous. For this reason you must restrain them, for, as often happens in their feeding because of their greed, if they catch hold of a root which they want to pull out of the ground, they break their necks; for the neck is exceedingly weak, just as the head is soft. If there is none of this herb, they should be fed on barley or other grain. When the season for mixed forage comes, this should be fed as I said in regard to *seris*. While the geese are sitting they should be fed on barley soaked in water. The goslings are fed first on barley-meal or barley for two days, and for the next three on green cress cut fine,

viride consectum minutatim ex aqua in vas aliquod. Cum autem sunt inclusi in haras aut speluncas, ut dixi, viceni, obiciunt iis polentam hordeaceam aut farraginem herbamve teneram aliquam concisam.

7 Ad saginandum eligunt pullos circiter sesquimensem[1] qui sunt nati; eos includunt in saginario ibique polentam et pollinem aqua madefacta dant cibum, ita ut ter die saturent. Secundum cibum large ut bibant faciunt potestatem. Sic curati circiter duobus mensibus fiunt pingues. Quotienscumque sumpserunt, locus solet purgari, quod amant locum purum neque ipsi ullum, ubi fuerunt, relincunt purum.

XI. Qui autem volunt greges anatium habere ac constituere nessotrophion, primum locum, quoi est facultas, eligere oportet palustrem, quod eo maxime delectantur; si id non, potissimum ibi, ubi sit naturalis aut lacus aut stagnum aut manu facta

2 piscina, quo gradatim descendere possint. Saeptum altum esse oportet, ubi versentur, ad pedes quindecim, ut vidistis ad villam Sei, quod uno ostio claudatur. Circum totum parietem intrinsecus crepido lata, in qua secundum parietem sint tecta cubilia, ante ea vestibulum earum exaequatum tectorio opere testaceo. In eo perpetua canalis, in quam et cibus imponitur iis et immittitur aqua; sic

3 enim cibum capiunt. Omnes parietes tectorio levigantur, ne faeles aliave quae bestia introire ad nocendum possit, idque saeptum totum rete grandibus maculis integitur, ne eo involare aquila possit neve

[1] *sesquimensem* Keil: *sesquimense* AB, *sexquimense* P.

---

[1] In Section 4.

[2] νεσσοτροφεῖον, "a place where ducks are reared"; with Varro's description may be compared Columella, VIII, 15.

soaked in water and turned into a vessel. But after they are shut into the coops or the underground nests, twenty to the nest, as I have said,[1] they are fed on ground barley or mixed forage or tender grass cut fine. For fattening, goslings are chosen which are about one and one-half months old; these are enclosed in the fattening pen, and there they are fed on a food consisting of barley-meal and flour dampened with water, being surfeited three times a day. After eating, they are allowed the opportunity of drinking as much as they want. When they are treated in this way they become fat in about two months. After every feeding the place is cleaned out; for they like a clean place, and yet never leave any place clean where they have been.

XI. "One who wishes to keep flocks of ducks and build a duck-farm[2] should choose, first, if he has the opportunity, a place which is swampy, for they like this best of all; if this is not available, a place preferably where there is a natural pond or pool or an artificial pond, to which they can go down by steps. There should be an enclosure in which they can move about, some fifteen feet high, as you saw at Seius's place, closed by one entrance. Around the entire wall on the inside should run a wide ledge, along which, next to the wall, are the covered resting places, and in front of them their vestibule levelled with plastered brickwork. In this is a continuous trough, in which food is placed for them and water is admitted; for in this way they take their food. All the walls are smoothed with plaster, so that no weasel or other beast can get in to harm them; and the entire enclosure is covered with a wide-meshed net, so that an eagle cannot fly in or

evolare anas. Pabulum iis datur triticum, hordeum, vinacei, non numquam etiam ex aqua cammari et quaedam eius modi aquatilia. Quae in eo saepto erunt piscinae, in eas aquam large influere oportet, ut semper recens sit.

4 Sunt item non dissimilia alia genera, ut querquedulae, phalarides, sic perdices, quae, ut Archelaus scribit, voce maris audita concipiunt. Quae, ut superiores, neque propter fecunditatem neque propter suavitatem saginantur et sic pascendo fiunt pingues. Quod ad villaticarum pastionum primum actum pertinere sum ratus, dixi.

XII. Interea redit Appius, et percontati nos ab illo et ille a nobis, quid esset dictum ac factum. Appius, Sequitur, inquit, actus secundi generis adficticius ad villam qui solet esse, ac nomine antico a parte quadam leporarium appellatum. Nam neque solum lepores in eo includuntur silva, ut olim in iugero agelli aut duobus, sed etiam cervi aut capreae in iugeribus multis. Quintus Fulvius Lippinus dicitur habere in Tarquiniensi saepta iugera quadraginta, in quo sunt inclusa non solum ea quae dixi, sed etiam oves ferae, etiam hoc maius hic in Statoniensi et 2 quidam in locis aliis; in Gallia vero transalpina

---

[1] The same story is told by Aristotle, *Hist. Anim.*, V, 2: "With partridges, by the way, if the female gets to leeward of the male, she becomes thereby impregnated. And often when they happen to be in heat she is affected in this wise by the voice of the male, or by his breathing down on her as he flies overhead." He repeats this story several times, as do other writers, both Greek and Latin. The myth, probably Egyptian, is, like the similar one of the vulture, referred to frequently by the Fathers. A very similar story is told by Varro, II, 1, 19.

the ducks fly out. For food they are given wheat, barley, grape-skins, and sometimes water-crabs and certain aquatic food of that sort. Any ponds in the enclosure should have a large inflow of water, so that it may always be fresh.

"There are also other species not unlike them, such as the teal, coot, and partridge, which, as Archelaus writes, conceive when they hear the voice of the male.[1] These are not stuffed as are those above mentioned, either to increase their fecundity or to improve their flavour, but they become fat by merely feeding them as described. I have finished telling what seems to belong to the first act of the husbandry of the steading."[2]

XII. Meanwhile Appius returns, and we are asked by him and he by us what has been said and done. Appius continues: "There follows the second act, which is usually an appendage to the villa and retains its old name of hare-warren because of one part of it—for not only are hares enclosed in it in woods, as used to be the case on an acre or two of land, but also stags and roes on many acres. It is reported that Quintus Fulvius Lippinus[3] has a preserve in the vicinity of Tarquinii of forty *iugera*, in which are enclosed, not only the animals I have named, but also wild sheep; and an even larger one near Statonia, and some in other places; while in Transalpine

[2] Martial's description of the villa of his friend Faustinus at Baiae (III, 58), which might well serve as a poetical preface to this work, mentions most of the animals and birds named by Varro.

[3] Known only from this passage, and two passages in Pliny (VIII, 211; IX, 173) which give the same facts. Tarquinii, now Corneto Tarquinia, was a very ancient and important city in Etruria. Statonia was also in southern Etruria.

T. Pompeius tantum saeptum venationis, ut circiter ∞ ∞ ∞ ∞ passum locum inclusum habeat. Praeterea in eodem consaepto fere habere solent cocliaria atque alvaria atque etiam dolia, ubi habeant conclusos glires. Sed horum omnium custodia, incrementum et pastio aperta, praeterquam de apibus.

3 Quis enim ignorat saepta e maceriis ita esse oportere in leporario, ut tectorio tacta sint et sint alta? Alterum ne faelis aut maelis aliave quae bestia introire possit, alterum ne lupus transilire; ibique esse latebras, ubi lepores interdiu delitiscant in virgultis atque herbis, et arbores patulis ramis, quae

4 aquilae impediant conatus. Quis item nescit, paucos si lepores, mares ac feminas, intromiserit, brevi tempore fore ut impleatur? Tanta fecunditas huius quadripedis. Quattuor modo enim intromisit in leporarium, brevi solet repleri. Etenim saepe, cum habent catulos recentes, alios in ventre habere reperiuntur. Itaque de iis Archelaus scribit, annorum quot sit qui velit scire, inspicere oportere foramina

5 naturae, quod sine dubio alius alio habet plura. Hos quoque nuper institutum ut saginarent plerumque, cum exceptos e leporario condant in caveis et loco clauso faciant pingues. Quorum ergo tria genera fere sunt: unum Italicum hoc nostrum pedibus

---

[1] See Chapter 14.

[2] Aristotle makes this statement several times, *e.g.*, *Hist. Anim.*, V, 9: "The greater part of wild animals bring forth once and once only in the year, except in the case of animals like the hare, where the female can become superfoetally impregnated." Many other authors repeat it, and marvellous tales are told of the fecundity of the hare. Aristotle explains this (*Hist. Anim.*, III. 1): "The females of horned non-ambidental animals are furnished with cotyledons in the womb when they are pregnant, and such is the case, among ambidentals, with the hare, the mouse, and the bat."

Gaul, Titus Pompeius has a hunting preserve so large that he keeps a tract of about four square miles enclosed. In addition to this, in the same enclosure are usually kept places for snails[1] and bee-hives, and also casks in which dormice are kept confined. But the care, increase, and feeding of all these, except the bees, is evident. For everybody knows that walled enclosures in warrens ought to be covered with plaster and ought to be high—in the one case to make it impossible for a weasel or a badger or other animal to enter, and in the other to keep a wolf from leaping over; and they should have coverts in which the hares may hide in the day-time under the brush and grass, and trees with spreading branches to hinder the swooping of an eagle. Who also does not know that if he puts in a few hares, male and female, in a short time the place will be filled? Such is the fecundity of this animal. For place only four in a warren and it is usually filled in a short time; for often, while they have a young litter they are found to have others in the womb.[2] And so Archelaus writes of them that one who wishes to know how old they are should examine the natural openings, for undoubtedly one has more than another.[3] There is a recent general practice of fattening these, too, by taking them from the warren and shutting them up in hutches and fattening them in an enclosed space. There are, then, some three species of these: one, this Italian species of ours, with short fore-legs and long hind

[3] Pliny (*N.H.*, VIII, 218): "Archelaus is our authority for the statement that the hare has as many years as it has in its body openings for excrement. Certainly a varying number is found."

primis humilibus, posterioribus altis, superiore parte pulla, ventre albo, auribus longis. Qui lepus dicitur, cum praegnas sit, tamen concipere. In Gallia Transalpina et Macedonia fiunt permagni, in Hispania 6 et in Italia mediocres. Alterius generis est, quod in Gallia nascitur ad Alpis, qui hoc fere mutant, quod toti candidi sunt; ii raro perferuntur Romam. Tertii generis est, quod in Hispania nascitur, similis nostro lepori ex quadam parte, sed humile, quem cuniculum appellant. L. Aelius putabat ab eo dictum leporem a celeritudine, quod levipes esset. Ego arbitror a Graeco vocabulo antico, quod eum Aeolis λέποριν appellabant. Cuniculi dicti ab eo, quod sub terra cuniculos ipsi facere solent, ubi 7 lateant in agris. Horum omnium tria genera, si possis, in leporario habere oportet. Duo quidem utique te habere puto, quod in Hispania annis ita fuisti multis, ut inde te cuniculos persecutos credam.

XIII. Apros quidem posse haberi in leporario nec magno negotio ibi et captivos et cicuris, qui ibi nati sint, pingues solere fieri scis, inquit, Axi. Nam quem fundum in Tusculano emit hic Varro a M. Pupio Pisone, vidisti ad bucinam inflatam certo tempore apros et capreas convenire ad pabulum, cum ex superiore loco e palaestra apris effunderetur glas, 2 capreis vicia aut quid aliut. Ego vero, inquit ille, apud Q. Hortensium cum in agro Laurenti essem, ibi

---

[1] Cf. Section 4.

[2] *e palaestra* is obscure. We have references to *palaestrae* attached to private houses, and I have assumed that it was an open space, a platform or terrace, on which the guests stood to watch the feeding. This seems to be in harmony with the statement immediately below.

legs, the upper part of the body dark, belly white, and ears long. This hare is said to conceive even while it is pregnant.[1] In Transalpine Gaul and Macedonia they grow very large; in Spain and in Italy they are medium-sized. Belonging to the second species is the hare which is born in Gaul near the Alps, which usually differs in the fact that it is entirely white; these are not often brought to Rome. To the third species belongs the one which is native to Spain—like our hare in some respects, but with short legs—which is called cony. Lucius Aelius thought that the hare received its name *lepus* because of its swiftness, being *levipes*, nimble-foot. My own opinion is that it comes from an old Greek word, as the Aeolians called it λέπορις. The conies are so named from the fact that they have a way of making in the fields tunnels (*cuniculos*) in which to hide. You should have all these three species in your warren if you can. You surely have two species anyway, I suppose, as you were in Spain for so many years that I imagine the conies followed you all the way from there.

XIII. "You know, Axius," Appius continued, "that boars can be kept in the warren with no great trouble; and that both those that have been caught and the tame ones which are born there commonly grow fat in them. For on the place that our friend Varro here bought from Marcus Pupius Piso near Tusculum, you saw wild boars and roes gather for food at the blowing of a horn at a regular time, when mast was thrown from a platform[2] above to the boars, and vetch or the like to the roes." "Why," said he, "I saw it carried out more in the Thracian fashion at Quintus Hortensius's place near Laurentum when

istuc magis θρᾳκικῶς[1] fieri vidi. Nam silva erat, ut dicebat, supra quinquaginta iugerum maceria saepta, quod non leporarium, sed therotrophium appellabat. Ibi erat locus excelsus, ubi tricilinio posito cena-
3 bamus, quo Orphea vocari iussit. Qui cum eo venisset cum stola et cithara cantare esset iussus, bucina inflavit, ut tanta circumfluxerit nos cervorum aprorum et ceterarum quadripedum multitudo, ut non minus formosum mihi visum sit spectaculum, quam in Circo Maximo aedilium sine Africanis bestiis cum fiunt venationes.

XIV. Axius, Tuas partes, inquit, sublevavit Appius, O Merula noster. Quod ad venationem pertinet, breviter secundus trasactus est actus, nec de cochleis ac gliribus quaero, quod relicum est; neque enim magnum molimentum esse potest. Non istuc tam simplex est, inquit Appius, quam tu putas, O Axi noster. Nam et idoneus sub dio sumendus locus cochleariis, quem circum totum aqua claudas, ne, quas ibi posueris ad partum, non liberos earum, sed
2 ipsas quaeras. Aqua, inquam, finiendae, ne fugitivarius sit parandus. Locus is melior, quem et non coquit sol et tangit ros. Qui si naturalis non est, ut fere non sunt in aprico loco, neque habeas in opaco ubi facias, ut sunt sub rupibus ac montibus, quorum

[1] Keil's emendation for *magis tragicos* of the first edition. The MSS. have *magistracicos* and *magistraicos*.

---

[1] θηροτροφεῖον, "a place for rearing game."

[2] Panthers. Pliny tells us (VIII, 64) that there was an old decree of the Senate which forbade panthers to be brought into Italy; and that Gnaeus Aufidius, tribune of the plebs (probably in 170 B.C.), had the people pass a law permitting their importation for use in the Circus.

[3] Cf. III, 3, 1; III, 12, 1.

[4] Snails, a favourite article of the Roman diet, are still

I was there. For there was a forest which covered, he said, more than fifty *iugera*; it was enclosed with a wall and he called it, not a warren, but a game-preserve.[1] In it was a high spot where was spread the table at which we were dining, to which he bade Orpheus be called. When he appeared with his robe and harp, and was bidden to sing, he blew a horn; whereupon there poured around us such a crowd of stags, boars, and other animals that it seemed to me to be no less attractive a sight than when the hunts of the aediles take place in the Circus Maximus without the African beasts.[2]"

XIV. "Appius has lightened your task, my dear Merula," said Axius. "So far as game is concerned, the second act[3] has been completed briefly; and I do not ask for the rest of it—snails[4] and dormice—as that cannot be a matter of great effort." "The thing is not so simple as you think, my dear Axius," replied Appius. "You must take a place fitted for snails, in the open, and enclose it entirely with water; for if you do not, when you put them to breed it will not be their young which you have to search for, but the old snails. They have to be shut in, I repeat, with water, so that you need not get a runaway-catcher. The best place is one which the sun does not parch, and where the dew falls. If there is no such natural place—and there usually is not in sunny ground—and you have no place where you can build one in the shade, as at the foot of a cliff or a mountain with a

commonly eaten in Italy and France, where they are bred in snail gardens (*escargotières*). Pliny (*N.H.*, IX, 173) ascribes to Fulvius Lippinus (cf. 12, 1, above) the establishment of vivaries for snails in the vicinity of Tarquinii, shortly before the Civil War, and the segregation of the various species named by Varro in this chapter.

adluant radices lacus ac fluvii, manu facere oportet roscidum. Qui fit, si adduxeris fistula et in eam mammillas imposueris tenues, quae eructent aquam, ita ut in aliquem lapidem incidat ac late dissipetur.
3 Parvus iis cibus opus est, et is sine ministratore, et hunc, dum serpit, non solum in area reperit, sed etiam, si rivus non prohibet, parietes stantes invenit. Denique ipsae et ruminantes ad propolam [1] vitam diu producunt, cum ad eam rem pauca laurea folia intericiant et aspergant furfures non multos. Itaque cocus has vivas an mortuas coquat, plerumque nescit.
4 Genera cochlearum sunt plura, ut minutae albulae, quae afferuntur e Reatino, et maximae, quae de Illyrico apportantur, et mediocres, quae ex Africa afferuntur; non quo non in his regionibus quibusdam locis ac magnitudinibus sint disperiles; nam et valde amplae sunt ex Africa, quae vocantur solitannae, ita ut in eas LXXX quadrantes coici possint, et sic in aliis regionibus eaedem inter se collatae minores ac
5 maiores. Hae in fetura pariunt innumerabilia. Earum semen minutum ac testa molli diuturnitate obdurescit. Magnis insulis in areis factis magnum bolum deferunt aeris. Has quoque saginare solent ita, ut ollam cum foraminibus incrustent sapa et farri, ubi pascantur, quae foramina habeat, ut intrare aer possit; vivax enim haec natura.

XV. Glirarium autem dissimili ratione habetur, quod non aqua, sed maceria locus saepitur; tota levi

[1] Numerous emendations have been proposed for the manuscript *ex gruminantes ad propalam*. The reading here followed is that conjectured by Keil (*et ruminantes*) and Scaliger (*propolam*).

---

[1] See critical note above.

[2] About two and one-half gallons. This statement is repeated by Pliny (*N.H.*, IX, 56); and he adds, rather significantly, that his authority is Marcus Varro!

pool or stream at the bottom, you should make an artificially dewy one. This can be done if you will run a pipe and attach to it small teats to squirt out the water in such a way that it will strike a stone and be scattered widely in a mist. They need little food, and require no one to feed them; they get their food, not only in the open while crawling around, but even discover any upright walls, if the stream does not prevent. In fact, even at the dealer's they keep alive for a long time by chewing the cud,[1] a few laurel leaves being thrown them for the purpose, sprinkled with a little bran. Hence the cook usually doesn't know whether they are alive or dead when he is cooking them. There are several varieties of snails, such as the small whites, which come from Reate, the large-sized, which are brought from Illyricum, and the medium-sized, which come from Africa. Not that they do not vary in these regions in distribution and size; thus, very large ones do come from Africa—the so-called *solitannae*—so large that 80 quadrantes[2] can be put into their shells; and so in other countries the same species are relatively larger or smaller. They produce innumerable young; these are very small and with a soft shell, but it hardens with time. If you build large islands in the yards, they will bring in a large haul of money. Snails, too, are often fattened as follows: a jar for them to feed in, containing holes, is lined with must and spelt—it should contain holes in order to allow the air to enter, for the snail is naturally hardy.

XV. "The place for dormice is built on a different plan, as the ground is surrounded not by water but by a wall, which is covered on the inside with smooth

lapide aut tectorio intrinsecus incrustatur, ne ex ea erepere possit. In eo arbusculas esse oportet, quae ferant glandem. Quae cum fructum non ferunt, intra maceriem iacere oportet glandem et castaneam, 2 unde saturi fiant. Facere iis cavos oportet laxiores, ubi pullos parere possint; aquam esse tenuem, quod ea non utuntur multum et aridum locum quaerunt. Hae saginantur in doliis, quae etiam in villis habent multi, quae figuli faciunt multo aliter atque alia, quod in lateribus eorum semitas faciunt et cavum, ubi cibum constituant. In hoc dolium addunt glandem aut nuces iuglandes aut castaneam. Quibus in tenebris cum operculum impositum [1] est in doleis, fiunt pingues.

XVI. Appius, Igitur relinquitur, inquit, de pastione villatica tertius actus de piscinis. Quid tertius? inquit Axius. An quia tu solitus es in adulescentia tua domi mulsum non bibere propter parsimoniam, nos mel neclegemus? Appius nobis, Verum dicit, 2 inquit. Nam cum pauper cum duobus fratribus et duabus sororibus essem relictus, quarum alteram sine dote dedi Lucullo, a quo hereditate me cessa primum et primus mulsum domi meae bibere coepi ipse, cum interea nihilo minus paene cotidie in convivio omnibus 3 daretur [2] mulsum. Praeterea meum erat, non tuum, eas novisse volucres, quibus plurimum natura ingeni atque artis tribuit. Itaque eas melius me nosse quam te ut scias, de incredibili earum arte naturali

[1] *operculum impositum* Keil: *cumularim positum.*
[2] *daretur* Keil, following Ursinus; *darem.*

---

[1] Cf. Chapter 14.
[2] The statement of Appius seems to be made in order to emphasize his poverty; but Schneider remarks that it is silly.
[3] A pun on the name Appius, from *apis.*

stone or plaster over the whole surface, so that they cannot creep out of it. In this place there should be small nut-bearing trees; when they are not bearing, acorns and chestnuts should be thrown inside the walls for them to glut themselves with. They should have rather roomy caves built for them in which they can bring forth their young; and the supply of water should be small, as they do not use much of it, but prefer a dry place. They are fattened in jars, which many people keep even inside the villa. The potters make these jars in a very different form from other jars, as they run channels along the sides and make a hollow for holding the food. In such a jar acorns, walnuts, or chestnuts are placed; and when a cover is placed over the jars they grow fat in the dark."

XVI. "Well," remarked Appius, "the third act[1] of the husbandry of the steading is left—fishponds." "Why third?" inquired Axius. "Or, just because you were accustomed in your youth not to drink honey-wine at home for the sake of thrift, are we to overlook honey?" "It is the truth he is telling," Appius said to us. "For I was left in straitened circumstances, together with two brothers and two sisters, and gave one of them to Lucullus without a dowry; it was only after he relinquished a legacy in my favour that I, for the very first time, began to drink honey-wine at home myself, though meantime mead was none the less commonly served at banquets almost daily to all guests.[2] And furthermore, it was my right[3] and not yours to know these winged creatures, to whom nature has given so much talent and art. And so, that you may realize that I know bees better than you do, hear of the incredible art that nature has given

audi. Merula, ut cetera fecit, historicos quae sequi melitturgoe soleant demonstrabit.

4 Primum apes nascuntur partim ex apibus, partim ex bubulo corpore putrefacto. Itaque Archelaus in epigrammate ait eas esse

βοὸς φθιμένης πεπλανημένα τέκνα,

idem

ἵππων μὲν σφῆκες γενεά, μόσχων δὲ μέλισσαι.

Apes non sunt solitaria natura, ut aquilae, sed ut homines. Quod si in hoc faciunt etiam graculi, at non idem, quod hic societas operis et aedificiorum, quod illic non est, hic ratio atque ars, ab his opus facere discunt, ab his aedificare, ab his cibaria con-
5 dere. Tria enim harum: cibus, domus, opus, neque idem quod cera cibus, nec quod mel, nec quod domus. Non in favo sex angulis cella, totidem quot habet ipsa pedes? Quod geometrae hexagonon fieri in orbi rutundo ostendunt, ut plurimum loci includatur. Foris pascuntur, intus opus faciunt, quod dulcissimum quod est, et deis et hominibus est acceptum, quod favus venit in altaria et mel ad principia convivi et
6 in secundam mensam administratur. Haec ut hominum civitates, quod hic est et rex et imperium et societas. Secuntur omnia pura. Itaque nulla harum adsidit in loco inquinato aut eo qui male oleat, neque etiam in eo qui bona olet unguenta. Itaque iis

---

[1] Cf. II, 5, 5.

[2] The Roman dinner usually consisted of three parts: (1) the *gustus* or *promulsis*, containing chiefly the *hors d'oeuvres*, especially eggs (cf. I, 2, 11), whence came the expression *ab ovo usque ad mala*; (2) the dinner proper; then, after an offering to the gods, (3) the *mensa secunda* or dessert. *Mulsum*,

them. Our well-versed Merula, as he has done in other cases, will tell you of the practice followed by bee-keepers.

"In the first place, bees are produced partly from bees, and partly from the rotted carcass of a bullock.[1] And so Archelaus, in an epigram, says that they are 'the roaming children of a dead cow'; and the same writer says: 'While wasps spring from horses, bees come from calves.' Bees are not of a solitary nature, as eagles are, but are like human beings. Even if jackdaws in this respect are the same, still it is not the same case; for in one there is a fellowship in toil and in building which does not obtain in the other; in the one case there is reason and skill —it is from these that men learn to toil, to build, to store up food. They have three tasks: food, dwelling, toil; and the food is not the same as the wax, nor the honey, nor the dwelling. Does not the chamber in the comb have six angles, the same number as the bee has feet? The geometricians prove that this hexagon inscribed in a circular figure encloses the greatest amount of space. They forage abroad, and within the hive they produce a substance which, because it is the sweetest of all, is acceptable to gods and men alike; for the comb comes to the altar and the honey is served at the beginning of the feast and for the second table.[2] Their commonwealth is like the states of men, for here are king, government, and fellowship. They seek only the pure; and hence no bee alights on a place which is befouled or one which has an evil odour, or even one which smells of sweet perfume. So one who

which was wine sweetened with honey, appeared both in the *promulsis* and with the dessert.

unctus qui accessit, pungunt, non, ut muscae, ligurriunt, quod nemo has videt, ut illas, in carne aut sanguine aut adipe. Ideo modo considunt in eis quo-
7 rum sapor dulcis. Minime malefica, quod nullius opus vellicans facit deterius, neque ignava, ut non, qui eius conetur disturbare, resistat; neque tamen nescia suae imbecillitatis. Quae cum causa Musarum esse dicuntur volucres, quod et, si quando displicatae sunt, cymbalis et plausibus numero redducunt in locum unum; et ut his dis Helicona atque Olympon adtribuerunt homines, sic his floridos et incultos
8 natura adtribuit montes. Regem suum secuntur, quocumque it, et fessum sublevant, et si nequit volare, succollant, quod eum servare volunt. Neque ipsae sunt inficientes nec non oderunt inertes. Itaque insectantes ab se eiciunt fucos, quod hi neque adiuvant et mel consumunt, quos vocificantes plures persecuntur etiam paucae. Extra ostium alvi opturant omnia, qua venit inter favos spiritus, quam
9 erithacen appellant Graeci. Omnes ut in exercitu vivunt atque alternis dormiunt et opus faciunt pariter et ut colonias mittunt, iique duces conficiunt quaedam ad vocem ut imitatione tubae. Tum id faciunt, cum inter se signa pacis ac belli habeant. Sed, O Merula, Axius noster ne, dum haec audit physica, macescat,

---

[1] We do not know the author; but Aristophanes (*Eccles.* 974) calls a girl μέλιττα Μούσης, "honey-bee of the Muse." The reason given here is their response to the music of the cymbals, or the rhythmical beating of the hands.

[2] Even Aristotle did not know that the queen bee was the common mother of the hive. This discovery, made by Swammerdam in the seventeenth century, is the beginning of the modern knowledge of the subject.

approaches them smelling of perfume they sting, and do not, as flies do, lick him; and one never sees bees, as he does flies, on flesh or blood or fat—so truly do they alight only on objects which have a sweet savour. The bee is not in the least harmful, as it injures no man's work by pulling it apart; yet it is not so cowardly as not to fight anyone who attempts to break up its own work; but still it is well aware of its own weakness. They are with good reason called[1] 'the winged attendants of the Muses,' because if at any time they are scattered they are quickly brought into one place by the beating of cymbals or the clapping of hands; and as man has assigned to those divinities Helicon and Olympus, so nature has assigned to the bees the flowering untilled mountains. They follow their own king[2] wherever he goes, assist him when weary, and if he is unable to fly they bear him upon their backs, in their eagerness to save him. They are themselves not idle, and detest the lazy; and so they attack and drive out from them the drones, as these give no help and eat the honey, and even a few bees chase larger numbers of drones in spite of their cries. On the outside of the entrance to the hive they seal up the apertures through which the air comes between the combs with a substance which the Greeks call *erithace*.[3] They all live as if in an army, sleeping and working regularly in turn, and send out as it were colonies, and their leaders give certain orders with the voice, as it were in imitation of the trumpet, as happens when they have signals of peace and war with one another. But, my dear Merula, that our friend Axius may not waste away while hearing this

[3] See Section 23.

quod de fructu nihil dixi, nunc cursu lampada tibe trado.

10 Merula, De fructu, inquit, hoc dico, quod fortasse an tibi satis sit, Axi, in quo auctorem habeo non solum Seium, qui alvaria sua locata habet quotannis quinis milibus pondo mellis, sed etiam hunc Varronem nostrum, quem audivi dicentem duo milites se habuisse in Hispania fratres Veianios ex agro Falisco locupletis, quibus cum a patre relicta esset parva villa et agellus non sane maior iugero uno, hos circum villam totam alvaria fecisse et hortum habuisse ac relicum thymo et cytiso opsevisse et apiastro, quod alii meliphyllon, alii melissophyllon, quidam melittae-

11 nam appellant. Hos numquam minus, ut peraeque ducerent, dena milia sestertia [1] ex melle recipere esse solitos, cum dicerent velle expectare, ut suo potius tempore mercatorem admitterent, quam celerius alieno. Dic igitur, inquit, ubi et cuius modi me facere oporteat alvarium, ut magnos capiam fructus.

12 Ille, melittonas ita facere oportet, quos alii melitrophia appellant, eandem rem quidam mellaria. Primum secundum villam potissimum, ubi non resonent imagines (hic enim sonus harum fugae existimatur esse protelum [2]), esse oportet aere temperato, neque aestate fervido neque hieme non

[1] Ursinus would change to *sestertium*.

[2] *protelum* Scaliger: *procerum*. Scaliger's conjecture is here followed as the most plausible emendation that has been suggested for the impossible *procerum*.

---

[1] A metaphor frequently used. It is taken from the torch-race at Athens, in which a lighted torch was handed on from one runner to another; it resembled the modern relay race.

[2] This use of the distributive numeral with *milia sestertia* is paralleled in Chapter 6, 6, and Chapter 17, 3. For the apparently adjectival use of *sestertia* with *milia*, see also II, 1, 14, and Columella, III, 3, 8–10.

essay on natural history, in which I have made no mention of gain, I hand over to you the torch in the race."[1]

Whereupon Merula: "As to the gain I have this to say, which will perchance be enough for you, Axius, and I have as my authorities not only Seius, who has his apiaries let out for an annual rental of 5,000 pounds of honey, but also our friend Varro here. I have heard the latter tell the story that he had two soldiers under him in Spain, brothers named Veianius, from the district near Falerii. They were well-off, because, though their father had left only a small villa and a bit of land certainly not larger than one *iugerum*, they had built an apiary entirely around the villa, and kept a garden; and all the rest of the land had been planted in thyme, snail-clover, and balm—a plant which some call honey-leaf, others bee-leaf, and some call bee-herb. These men never received less than 10,000 sesterces[2] from their honey, on a conservative estimate, as they said they preferred to wait until they could bring in the buyer at the time they wanted rather than to rush into market at an unfavourable time." "Tell me, then," said he, "where I ought to build an apiary and of what sort, so as to get a large profit." "The following," said Merula, "is the proper method for building apiaries, which are variously called *melitrophia* and *mellaria:*[3] first, they should be situated preferably near the villa, but where echoes do not resound (for this sound is thought to be a signal for flight in their case); where the air is temperate, not too hot in summer, and not without sun in winter;

[3] *Melitton* is the Greek μελιττών or μελισσών, "a bee-house"; *melitrophion* is μελιτροφεῖον (a shortened form of μελιττοτροφεῖον), "a place for raising bees"; and *mellarium* is the Latin equivalent.

aprico, ut spectet potissimum ad hibernos ortus, qui prope se loca habeat ea, ubi pabulum sit frequens et 13 aqua pura. Si pabulum naturale non est, ea oportet dominum serere, quae maxime secuntur apes. Ea sunt rosa, serpyllon, apiastrum, papaver, faba, lens, pisum, ocimum, cyperum, medice, maxime cytisum, quod minus valentibus utilissimum est. Etenim ab aequinoctio verno florere incipit et permanet ad 14 alterum aequinoctium. Sed ut hoc aptissimum ad sanitatem apium, sic ad mellificium thymum. Propter hoc Siculum mel fert palmam, quod ibi thymum bonum frequens est. Itaque quidam thymum contundunt in pila et diluunt in aqua tepida; eo conspergunt omnia seminaria consita apium 15 causa. Quod ad locum pertinet, hoc genus potissimum eligendum iuxta villam, non quo non in villae porticu quoque quidam, quo tutius esset, alvarium collocarint. Ubi sint, alii faciunt ex viminibus rutundas, alii e ligno ac corticibus, alii ex arbore cava, alii fictiles, alii etiam ex ferulis quadratas longas pedes circiter ternos, latas pedem, sed ita, ubi parum sunt quae compleant, ut eas conangustent, in vasto loco inani ne despondeant animum. Haec omnia vocant a mellis alimonio alvos, quas ideo videntur

---

[1] "Alfalfa was one of the stand-bys of ancient agriculture. According to Pliny, it was introduced into Italy from Greece, whither it had been brought from Asia during the Persian wars, and so derived its Greek and Roman name *Medica*. As Cato does not mention it with the other legumes he used, it is probable that the Romans had not yet adopted it in Cato's day, but by the time of Varro and Virgil it was well established in Italy. In Columella's day it was already a feature of the agriculture of Andalusia, and there the Moors, who loved plants, kept it alive, as it were a Vestal fire, while it died out of Italy

that it preferably face the winter sunrise, and have near by a place which has a good supply of food and clear water. If there is no natural food, the owner should sow crops which are most attractive to bees. Such crops are: the rose, wild thyme, balm, poppy, bean, lentil, pea, clover, rush, alfalfa,[1] and especially snail-clover, which is extremely wholesome for them when they are ailing. It begins flowering at the vernal equinox and continues until the second equinox. But while this is most beneficial to the health of bees, thyme is best suited to honey-making; and the reason that Sicilian honey bears off the palm is that good thyme is common there. For this reason some bruise thyme in a mortar and soak it in lukewarm water, and with this sprinkle all the plots planted for the bees. So far as the situation is concerned, one should preferably be chosen close to the villa—and some people place the apiary actually in the portico of the villa, so that it may be better protected. Some build round hives of withes for the bees to stay in, others of wood and bark, others of a hollow tree, others build of earthenware, and still others fashion them of fennel stalks,[2] building them square, about three feet long and one foot deep, but making them narrower when there are not enough bees to fill them, so that they will not lose heart in a large empty space. All such hives are called *alvi*, 'bellies,' because of the nourishment (*alimonium*), honey, which they contain; and it seems that the reason they are made

during the Dark Ages; from Spain it spread again all over southern Europe, and with America it was a fair exchange for tobacco."—Fairfax Harrison.

[2] For the merits of various types of hives, see Sec. 17, below, and cf. Columella, IX, 6.

medias facere angustissimas, ut figuram imitentur
16 earum. Vitiles fimo bubulo oblinunt intus et extra, ne asperitate absterreantur, easque alvos ita collocant in mutulis parietis, ut ne agitentur neve inter se contingant, cum in ordinem sint positae. Sic intervallo interposito alterum et tertium ordinem infra faciunt et aiunt potius hinc demi oportere, quam addi quartum. Media alvo, qua introeant apes, faciunt
17 foramina parva dextra ac sinistra. Ad extremam,[1] qua mellarii favum eximere possint, opercula imponunt. Alvi optimae fiunt corticeae, deterrimae fictiles, quod et frigore hieme et aestate calore vehementissime haec commoventur. Verno tempore et aestivo fere ter in mense mellarius inspicere debet fumigans leniter [2] eas et ab spurcitiis purgare alvum
18 et vermiculos eicere. Praeterea ut animadvertat ne reguli plures existant; inutiles enim fiunt propter seditiones. Et quidam dicunt, tria genera cum sint ducum in apibus, niger ruber varius, ut Menecrates scribit, duo, niger et varius, qui ita melior, ut expediat mellario, cum duo sint in eadem alvo, interficere nigrum, cum sit cum altero rege, esse seditiosum et corrumpere alvom, quod fuget aut cum multitudine
19 fugetur. De reliquis apibus optima est parva varia rutunda. Fur qui vocabitur, ab aliis fucus, est ater

[1] *extremam* Keil: *extrema*. [2] *leniter* Keil: *leviter*.

---

[1] But *earum* may refer to *alvos*, and so mean "the shape of the belly."

[2] *ad extremam* naturally means "at the end"; but since Pliny says (*N.H.*, XXI, 80) that the best cover was *a tergo*, and that it was a sliding cover, the interpretation here given seems probable.

[3] Varro is quoting (and misquoting) Aristotle, who says (*Hist. Anim.*, IX, 40) that there are two varieties, not three: the black, and the red and striped.

with a very narrow middle is that they may imitate the shape of the bees.[1] Those that are made of withes are smeared, inside and out, with cow-dung, so that the bees may not be driven off by any roughness; and these hives are so placed on brackets attached to the walls that they will not be shaken nor touch one another when they are arranged in a row. In this method, a second and a third row are placed below it at an interval, and it is said that it is better to reduce the number than to add a fourth. At the middle of the hive small openings are made on the right and left, by which the bees may enter; and on the back,[2] covers are placed through which the keepers can remove the comb. The best hives are those made of bark, and the worst those made of earthenware, because the latter are most severely affected by cold in winter and by heat in summer. During the spring and summer the bee-keeper should examine them about thrice a month, smoking them lightly, and clear the hive of filth and sweep out vermin. He should further see to it that several chiefs do not arise, for they become nuisances because of their dissensions. Some authorities state also that, as there are three kinds of leaders among bees—the black, the red, and the striped[3]—or, as Menecrates[4] states, two—the black and the striped—the latter is so much better that it is good practice for the keeper, when both occur in the same hive, to kill the black; for when he is with the other king he is mutinous and ruins the hive, because he either drives him out or is driven out and takes the swarm with him. Of ordinary bees, the best is the small round striped one. The one called by some the

[4] Cited in I, 1, 9, as a poet of Ephesus; Pliny also names him, XI, 17.

et lato ventre. Vespa, quae similitudinem habet apis, neque socia est operis et nocere solet morsu, quam apes a se secernunt. Hae differunt inter se, quod ferae et cicures sunt. Nunc feras dico, quae in silvestribus locis pascitant, cicures, quae in cultis. Silvestres minores sunt magnitudine et pilosae, sed opifices magis.

In emendo emptorem videre oportet, valeant an 20 sint aegrae. Sanitatis signa, si sunt frequentes in examine et si nitidae et si opus quod faciunt est aequibile ac leve. Minus valentium signa, si sunt pilosae et horridae, ut pulverulentae, nisi opificii eas urget tempus; tum enim propter laborem asper- 21 antur ac marcescunt. Si transferendae sunt in alium locum, id facere diligenter oportet et tempora, quibus id potissimum facias, animadvertendum et loca, quo transferas, idonea providendum: tempora, ut verno potius quam hiberno, quod hieme difficulter consuescunt quo translatae manere, itaque fugiunt plerumque. Si e bono loco transtuleris eo, ubi idonea pabulatio non sit, fugitivae fiunt. Nec, si ex alvo in alvum in eodem loco traicias, neglegenter 22 faciendum, sed et in quam transiturae sint apes, ea apiastro perfricanda, quod inlicium hoc illis, et favi melliti intus ponendi a faucibus non longe, ne, cum[1] animadverterint aut[2] inopiam esse[3] . . .[4] habuisse[5] dicit.[6]

[1] *ne cum*] *nec cum* Scaliger. [2] *aut*] *haud* Scaliger, *cibi* Ursinus.
[3] *esse*] *escae* early editors. [4] Lacuna suspected by Gesner.
[5] *habuisse*] *abivisse* Scaliger.
[6] *habuisse dicit. Is ait*] *h. dicatur* (*dicantur*) *aut* early editors, *abeant. Menecrates ait* Ursinus.

---

[1] *Locus desperatus.* The text followed is that of Keil, accepted by Goetz, as being best attested by the MSS.; though Scaliger's emendation has palaeographical possibilities at least. If a lacuna be assumed with Gesner (cf. Zahlfeldt, "Quaest. Crit. in Varr. *Rer. Rust.*" p. 24), we have no hint in other

thief, and by others the drone, is black, with a broad belly. The wasp, though it has the appearance of a bee, is not a partner in its work, and frequently injures it by its sting, and so the bees keep it away. Bees differ from one another in being wild or tame; by wild, I mean those which feed in wooded places, and by tame those which feed in cultivated ground. The former are smaller in size, and hairy, but are better workers.

"In purchasing, the buyer should see whether they are well or sick. The signs of health are their being thick in the swarm, sleek, and building uniformly smooth comb. When they are not so well, the signs are that they are hairy and shaggy, as if dusted over —unless it is the working season which is pressing them; for at this time, because of the work, they get rough and thin. If they are to be transferred to another place, it should be done carefully, and the proper time should be observed for doing it, and a suitable place be provided to which to move them. As to the time, it should be in spring rather than in winter, as in winter it is difficult for them to form the habit of staying where they have been moved, and so they generally fly away. If you move them from a good situation to one where there is no suitable pasturage, they become runaways. And even if you move them from one hive into another at the same place, the operation should not be carried out carelessly, but the hive into which the bees are going should be smeared with balm, which has a strong attraction for them, and combs full of honey should be placed inside not far from the entrance, for fear that, when they notice either a lack of food. . .[1] He

writers as to what is lost. Editors are generally agreed that Varro is quoting from Menecrates.

Is ait, cum sint apes morbidae propter primoris vernos pastus, qui ex floribus nucis graecae et cornus fiunt,
23 coeliacas fieri atque urina pota reficiendas. Propolim vocant, e quo faciunt ad foramen introitus protectum ante alvum maxime aestate. Quam rem etiam nomine eodem medici utuntur in emplastris, propter quam rem etiam carius in sacra via quam mel venit. Erithacen vocant, quo favos extremos inter se conglutinant, quod est aliut melle et propoli; itaque in hoc vim esse illiciendi. Quocirca examen ubi volunt considere, eum ramum aliamve quam rem
24 oblinunt hoc admixto apiastro. Favus est, quem fingunt multicavatum e cera, cum singula cava sena latera habeant, quot singulis pedes dedit natura. Neque quae afferunt ad quattuor res faciendas, propolim, erithacen, favum, mel, ex iisdem omnibus rebus carpere dicunt. Simplex, quod e malo punico et asparago cibum carpant solum, ex olea arbore
25 ceram, e fico mel, sed non bonum. Duplex ministerium praeberi, ut e faba, apiastro, cucurbita, brassica ceram et cibum; nec non aliter duplex quod

---

[1] "A red, resinous, odorous substance having some resemblance to wax and smelling like storax. It is collected by bees from the viscid buds of various trees, and used to stop the holes and crevices in their hives to prevent the entrance of cold air, to strengthen the cells, etc. Also called *bee-glue*."—*Century Dict.*

Varro puns on the literal meaning of the word πρὸ-πολις, "before the city," by using the term *protectum*.

[2] "Bee-bread." "The pollen of flowers settles on the hairs with which their body is covered, whence it is collected into pellets by a brush on their second pair of legs, and deposited in a hollow in the third pair. It is called bee-bread and is the food of the larvae or young. The adult bees feed on honey."—*Century Dict.*

[3] Aristotle, *Hist. Anim.*, IX, 40: "They have also another

says that when bees are sickly, because of their feeding in the early spring on the blossoms of the almond and the cornel, it is diarrhœa that affects them, and they are cured by drinking urine. *Propolis*[1] is the name given to a substance with which they build a *protectum* ('gable') over the entrance opening in front of the hive, especially in summer. This substance is used, and under the same name, by physicians in making poultices, and for this reason it brings even a higher price than honey on the Via Sacra. *Erithace*[2] is the name given a substance with which they fasten together the ends of the comb (it is a different substance than either honey or *propolis*) and it is in it that the force of the attraction[3] lies. So they smear with this substance, mixed with balm, the bough or other object on which they want the swarm to settle. The comb is the structure which they fashion in a series of cells of wax, each separate cell having six sides, the same number as that of the feet given to each bee by nature. It is said that they do not gather wholly from the same sources the materials which they bring in for making the four substances, *propolis*, *erithace*, comb, and honey. Sometimes what they gather is of one kind, since from the pomegranate and the asparagus they gather only food, from the olive tree wax, from the fig honey, but of a poor quality. Sometimes a double service is rendered,[4] as both wax and food from the bean, the balm, the gourd, and the cabbage; and similarly a double service of food and honey from the apple and wild pear, and still another double service in com-

food which is called bee-bread; this is scarcer than honey, and has a sweet fig-like taste."

[4] *i.e.* by the flower furnishing the "material."

fit e malo et piris silvestribus, cibum et mel; item aliter duplex quod e papavere, ceram et mel. Triplex ministerium quoque fieri, ut ex nuce Graeca et e lapsano cibum, mel, ceram. Item ex aliis floribus ita carpere, ut alia ad singulas res sumant, alia ad plures,
26 nec non etiam aliut discrimen sequantur in carptura aut eas sequatur, ut in melle, quod ex alia re faciant liquidum mel, ut e siserae flore, ex alia contra spissum, ut e rore marino; sic ex alia re, ut e fico mel
27 insuave, e cytiso bonum, e thymo optimum. Cibi pars quod potio et ea iis aqua liquida, unde bibant esse oportet, eamque propinquam, quae praeterfluat aut in aliquem lacum influat, ita ut ne altitudine escendat duo aut tres digitos; in qua aqua iaceant testae aut lapilli, ita ut extent paulum, ubi adsidere et bibere possint. In quo diligenter habenda cura ut aqua sit pura, quod ad mellificium bonum vehe-
28 menter prodest. Quod non omnis tempestas ad pastum prodire longius patitur, praeparandus his cibus, ne tum melle cogantur solo vivere aut relinquere exinanitas alvos. Igitur ficorum pinguium circiter decem pondo decoquont in aquae congiis sex, quas coctas in offas prope apponunt. Alii aquam mulsam in vasculis prope ut sit curant, in quae addunt lanam puram, per quam sugant, uno tempore

---

[1] *i.e.* while the bees seem to make this further distinction, as it is done unconsciously the distinction really governs the bees.

[2] Supposed to be the skirret, *Sium sisarum*, L.

bination, since they get wax and honey from the poppy. A threefold service, too, is rendered, as food, honey, and wax from the almond and the charlock. From other blossoms they gather in such a way that they take some materials for just one of the substances, other materials for more than one; they also follow another principle of selection in their gathering (or rather the principle follows the bees[1]); as in the case of honey, they make watery honey from one flower, for instance the sisera,[2] thick honey from another, for instance from rosemary; and so from still another they make an insipid honey, as from the fig, good honey from snail-clover, and the best honey from thyme. As drink is a component of food, and as this, in the case of bees, is clear water, they should have a place from which to drink, and this close by; it should flow past their hives, or run into a pool in such a way that it will not rise higher than two or three fingers, and in this water there should lie tiles or small stones in such a way that they project a little from the water, so that the bees can settle on them and drink. In this matter great care should be taken to keep the water pure, as this is an extremely important point in making good honey. As it is not every kind of weather that allows them to go far afield for feeding, food should be provided for them, so that they will not have to live on the honey alone at such times, or leave the hives when it is exhausted. So about ten pounds of ripe figs are boiled in six congii of water, and after they are boiled they are rolled into lumps and placed near the hives. Other apiarists have water sweetened with honey placed near the hives in vessels, and drop clean pieces of wool into it through which they can suck, for the

ne potu nimium impleantur aut ne incidant in aquam. Singula vasa ponunt ad alvos, haec supplentur. Alii uvam passam et ficum cum pisierunt, affundunt sapam atque ex eo factas offas apponunt ibi, quo foras hieme in pabulum procedere tamen possint.

29 Cum examen exiturum est, quod fieri solet, cum adnatae prospere sunt multae ac progeniem ut coloniam emittere volunt, ut olim crebro Sabini factitaverunt propter multitudinem liberorum, huius quod duo solent praeire signa, scitur: unum, quod superioribus diebus, maxime vespertinis, multae ante

30 foramen ut uvae aliae ex aliis pendent conglobatae; alterum, quod, cum iam evolaturae sunt aut etiam inceperunt, consonant vehementer, proinde ut milites faciunt, cum castra movent. Quae primum exierunt, in conspectu volitant reliquas, quae nondum congregatae sunt, respectantes, dum conveniant. A mellario cum id fecisse sunt animadversae, iaciundo in eas pulvere et circumtinniendo aere perterritae,

31 quo volunt perducere, non longe inde oblinunt erithace atque apiastro ceterisque rebus, quibus delectantur. Ubi consederunt, afferunt alvum eisdem inliciis litam intus et prope apposita fumo leni circumdato cogunt eas intrare. Quae in novam coloniam cum introierunt, permanent adeo libenter, ut etiam si proximam posueris illam alvum, unde exierunt, tamen novo domicilio potius sint contentae.

---

[1] But the word *adnatae*, "cousins," is playfully used, and reminds us of Cicero, *de Off.*, 53, where he describes the propagation of the family by the addition of brothers and cousins, and adds: "When these can no longer be contained in one home, they go out into other homes as into colonies."

[2] Cf. Section 23 above.

double purpose of keeping them from surfeiting themselves with the drink and from falling into the water. A vessel is placed near each hive and is kept filled. Others pound raisins and figs together, soak them in boiled wine, and put pellets made of this mixture in a place where they can come out to feed even in winter.

"The time when the bees are ready to swarm, which generally occurs when the well hatched new brood[1] is over large and they wish to send out their young as it were a colony (just as the Sabines used to do frequently on account of the number of their children), you may know from two signs which usually precede it: first, that on preceding days, and especially in the evenings, numbers of them hang to one another in front of the entrance, massed like a bunch of grapes; and secondly, that when they are getting ready to fly out or even have begun the flight, they make a loud humming sound exactly as soldiers do when they are breaking camp. Those which have gone out first fly around in sight, looking back for the others, which have not yet gathered, to swarm. When the keeper observes that they have acted so, he frightens them by throwing dust on them and by beating brass around them; and the place to which he wishes to carry them, and which is not far away, is smeared with bee-bread and balm and other things by which they are attracted.[2] When they have settled, a hive, smeared on the inside with the same enticing substances, is brought up and placed near by; and then by means of a light smoke blown around them they are induced to enter. When they have moved into the new colony, they remain so willingly that even if you place near by the hive from which they came, still they are content rather with their new home.

32 Quod ad pastiones pertinere sum ratus quoniam dixi, nunc iam, quoius causa adhibetur ea cura, de fructu dicam. Eximendorum favorum signum sumunt ex ipsis † uiris alvos habeat nem congerminarit †[1] coniecturam capiunt, si intus faciunt bombum et, cum intro eunt ac foras, trepidant et si, opercula alvorum cum remoris, favorum foramina obducta videntur
33 membranis, cum sint repleti melle. In eximendo quidam dicunt oportere ita ut novem partes tollere, decumam relinquere; quod si omne eximas, fore ut discedant. Alii hoc plus relincunt, quam dixi. Ut in aratis qui faciunt restibiles segetes, plus tollunt frumenti ex intervallis, sic in alvis, si non quotannis eximas aut non aeque multum, et magis his assiduas
34 habeas apes et magis fructuosas. Eximendorum favorum primum putant esse tempus vergiliarum exortu, secundum aestate acta, antequam totus exoriatur arcturus, tertium post vergiliarum occasum, et ita, si fecunda sit alvos, ut ne plus tertia pars eximatur mellis, reliquum ut hiemationi relinquatur; sin alvus non sit fertilis, ne quid eximatur. Exemptio cum est maior, neque universam neque palam facere oportet, ne deficiant animum. Favi qui eximuntur, siqua pars nihil habet aut habet incunatum, cultello
35 praesicatur. Providendum ne infirmiores a valentioribus opprimantur, eo enim minuitur fructus; itaque imbecilliores secretas subiciunt sub alterum

[1] This is perhaps the most corrupt passage in the whole work, and editors have struggled with it in vain.

---

[1] Columella (IX, 15, 4) merely says that when you observe frequent battles between the drones and the bees, you should open the hives and see whether the comb is half-filled or full and covered with a membrane.

[2] These dates are, respectively, 10th May, early September, and early November.

" As I have given my views on the subject of feeding, I shall now speak of the thing on account of which all this care is exercised—the profit. The signal for removing the comb is given by the following occurrences . . . if the bees make a humming noise inside, if they flutter when going in and out, and if, when you remove the covers of the hives, the openings of the combs are seen to be covered with a membrane, the combs being filled with honey.[1] Some authorities hold that in taking off honey nine-tenths should be removed and one-tenth left; for if you take all, the bees will quit the hive. Others leave more than the amount stated. Just as in tilling, those who let the ground lie fallow reap more grain from the interrupted harvests, so in the matter of hives if you do not take off honey every year, or not the same amount, you will by this method have bees which are busier and more profitable. It is thought that the first season for removing the comb is at the time of the rising of the Pleiades, the second at the end of summer, before Arcturus is wholly above the horizon, and the third after the setting of the Pleiades.[2] But in this case, if the hive is well filled no more than one-third of the honey should be removed, the remainder being left for the wintering; but if the hive is not well filled no honey should be taken out. When the amount removed is large, it should not all be taken at one time or openly, for fear the bees may lose heart. If some of the comb removed contains no honey or honey that is dirty, it should be cut off with a knife. Care should be taken that the weaker bees be not imposed upon by the stronger, for in this case their output is lessened; and so the weaker are separated and

regem. Quae crebrius inter se pugnabunt, aspargi eas oportet aqua mulsa. Quo facto non modo desistunt pugna, sed etiam conferciunt se lingentes, eo magis, si mulso sunt asparsae, quo propter odorem avidius applicant se atque obstupescunt potantes.
36 Si ex alvo minus frequentes evadunt ac subsidit aliqua pars, subfumigandum et prope apponendum bene olentium herbarum maxime apiastrum et
37 thymum. Providendum vehementer ne propter aestum aut propter frigus dispereant. Si quando subito imbri in pastu sunt oppressae aut frigore subito, antequam ipsae providerint id fore, quod accidit raro ut decipiantur, et imbris guttis uberibus offensae iacent prostratae, ut efflictae, colligendum eas in vas aliquod et reponendum in tecto loco ac tepido, proximo die[1] quam maxime tempestate bona cinere facto e ficulneis lignis infriandum paulo plus caldo quam tepidiore. Deinde concutiendum leviter ipso
38 vaso, ut manu non tangas, et ponendae in sole. Quae enim sic concaluerunt, restituunt se ac revivescunt, ut solet similiter fieri in muscis aqua necatis. Hoc faciendum secundum alvos, ut reconciliatae ad suum quaeque opus et domicilium redeant.

XVII. Interea redit ad nos Pavo et, Si vultis, inquit, ancoras tollere, latis tabulis sortitio fit tribuum, ac coepti sunt a praecone recini, quem quaeque tribus

[1] *proximo die* Keil: *promum e die.*

---

[1] To decide, in case two candidates received an equal number of votes, which should be aedile. Cicero, *Pro Planco,* 22, says: "For our ancestors would never have set up the lottery for aediles if they had not seen that a case could occur in which the candidates received an equal number of votes."

placed under another king. Those which often fight one another should be sprinkled with honey-water. When this is done they not only stop fighting, but swarm over one another, licking the water; and even more so if they are sprinkled with mead, in which case the odour causes them to attach themselves more greedily, and they drink until they are stupefied. If they leave the hive in smaller numbers and a part of the swarm remains idle, light smoke should be applied, and there should be placed near by some sweet-smelling herbs, especially balm and thyme. The greatest possible care should be taken to prevent them from dying from heat or from cold. If at any time they are knocked down by a sudden rain while harvesting, or overtaken by a sudden chill before they have foreseen that this would happen (though it is rarely that they are caught napping), and if, struck by the heavy rain-drops, they lie prostrate as if dead, they should be collected into a vessel and placed under cover in a warm spot; the next day, when the weather is at its best, they should be dusted with ashes made of fig wood, and heated a little more than warm. Then they should be shaken together gently in the vessel, without being touched with the hand, and placed in the sun. Bees which have been warmed in this way recover and revive, just as happens when flies which have been killed by water are treated in the same way. This should be carried out near the hives, so that those which have been revived may return each to his own work and home."

XVII. Meantime Pavo returns to us and says: "If you wish to weigh anchor, the ballots have been cast and the casting of lots for the tribes is going on;[1] and the herald has begun to announce who has been

fecerit aedilem. Appius confestim surgit, ut ibidem candidato suo gratularetur ac discederet in hortos. Merula, Tertium actum de pastionibus villaticis postea, inquit, tibi reddam, Axi. Consurgentibus illis, Axius mihi respectantibus nobis, quod et candidatum nostrum venturum sciebamus, Non laboro,

2 inquit, hoc loco discessisse Merulam. Reliqua enim fere mihi sunt nota, quod, cum piscinarum genera sint duo, dulcium et salsarum, alterum apud plebem et non sine fructu, ubi Lymphae aquam piscibus nostris villaticis ministrant; illae autem maritimae piscinae nobilium, quibus Neptunus ut aquam et piscis ministrat, magis ad oculos pertinent, quam ad vesicam, et potius marsippium domini exinaniunt, quam implent. Primum enim aedificantur magno, secundo implentur magno, tertio aluntur magno.

3 Hirrus circum piscinas suas ex aedificiis duodena milia sestertia capiebat. Eam omnem mercedem escis, quas dabat piscibus, consumebat. Non mirum; uno tempore enim memini hunc Caesari duo milia murenarum mutua dedisse in pondus et propter piscium multitudinem quadragies sestertio villam venisse. Quae nostra piscina mediterranea ac plebeia recte dicitur dulcis et illa amara; quis

---

[1] Book II ends in the same way. *Horti* has several meanings, among them " country house "; and it seems probable that in this instance Appius's candidate was so sure of his election that he had arranged a dinner-party to celebrate his victory.

[2] Cf. III, 3, 1.

[3] For the meaning, see Section 3 below.

[4] Latin writers, and especially poets, write Lympha or Nympha, and use it as synonymous with water. The contrast here is, of course, between the fresh water of the ordinary pond and the sea-water of the more elaborate pond.

[5] He is mentioned also II, 1, 2.

elected aedile by each tribe." Appius arose hurriedly, so as to congratulate his candidate at once and then go on to his home.[1] And Merula remarked: "I'll give you the third act[2] of the husbandry of the steading later, Axius." As they were rising, and we were looking back, because we knew that our candidate was coming also, Axius remarked to me: "I am not sorry that Merula left at this point, for the rest is pretty well known to me. There are two kinds of fish-ponds, the fresh and the salt. The one is open to common folk,[3] and not unprofitable, where the Nymphs[4] furnish the water for our domestic fish; the ponds of the nobility, however, filled with sea-water, for which only Neptune can furnish the fish as well as the water, appeal to the eye more than to the purse, and exhaust the pouch of the owner rather than fill it. For in the first place they are built at great cost, in the second place they are stocked at great cost, and in the third place they are kept up at great cost. Hirrus[5] used to take in 12,000 sesterces from the buildings around his fish-ponds; but he spent all that income for the food which he gave his fish. No wonder; for I remember that he lent to Caesar on one occasion 2,000 lampreys by weight;[6] and that on account of the great number of fish his villa sold for 4,000,000 sesterces. Our inland pond, which is for the common folk, is properly called 'sweet,' and the other 'bitter';[7] for who

[6] Pliny (IX, 171) relates that the loan was on the occasion of one of Caesar's triumphs as dictator; but he says it was 6000.

[7] Varro indulges in much punning, on "sweet" and "bitter," and on "mullets" and "mules." Thus, *dulcis* means both "fresh" and "delightful," while *amarus* has both the meanings of "bitter." In Section 6 the *mulli*, mullets, are contrasted with his *muli*, mules.

enim nostrum non una contentus est hac piscina? Quis contra maritumas non ex piscinis singulis coniunctas habet pluris? Nam ut Pausias et ceteri pictores eiusdem generis loculatas magnas habent arculas, ubi discolores sint cerae, sic hi loculatas habent piscinas, ubi dispares disclusos habeant pisces, quos, proinde ut sacri sint ac sanctiores quam illi in Lydia, quos sacrificanti tibi, Varro, ad tibicinem gregatim venisse dicebas ad extremum litus atque aram, quod eos capere auderet nemo, cum eodem tempore insulas Lydorum ibi χορευούσας vidisses, sic hos piscis nemo cocus in ius vocare audet. Quintus Hortensius, familiaris noster, cum piscinas haberet magna pecunia aedificatas ad Baulos, ita saepe cum eo ad villam fui, ut illum sciam semper in cenam pisces Puteolos mittere emptum solitum. Neque satis erat eum non pasci e piscinis, nisi etiam ipse eos pasceret ultro ac maiorem curam sibi haberet, ne eius esurirent mulli, quam ego habeo, ne mei in Rosea esuriant asini, et quidem utraque re, et cibo et potione, cum non paulo sumptuosius, quam ego, ministraret victum. Ego enim uno servulo, hordeo non multo, aqua domestica meos multinummos alo

(Section numbers in the margin: 4 at "iunctas habet pluris"; 5 at "sic hos piscis"; 6 at "pisces Puteolos".)

---

[1] Pliny (*N.H.*, XXXV, 123) tells us he was a native of Sicyon, a contemporary of Apelles, and the discoverer of the art of foreshortening.

[2] Pliny devotes a chapter (II, 95) to such islands and says: "In the Nymphaeus, certain small islands called the 'Dancers,' because when choruses are sung they move in tune." Varro was a *legatus* to Pompey in Lydia in 67 or 66 B.C.

[3] The pun is common in Latin writers, as *ius* means both "sauce" and "justice." Hence, *vocare in ius*, "to call to *ius*, sauce or justice."

[4] He has already been mentioned, III, 6, 6, and III, 13, 2. Many other anecdotes are told of him: that he watered favourite trees with wine, and on one occasion asked Cicero to

of us is not content with one such pond? Who, on the other hand, who starts with one of the sea-water ponds doesn't go on to a row of them? For just as Pausias[1] and the other painters of the same school have large boxes with compartments for keeping their pigments of different colours, so these people have ponds with compartments for keeping the varieties of fish separate, as if they were holy and more inviolate than those in Lydia about which, Varro, you used to say that while you were sacrificing, they would come up in schools, at the sound of a flute, to the edge of the shore and the altar, because no one dared catch them (the same time as that at which you saw the 'dancing islands' of the Lydians);[2] just so no cook dares 'haul these fish over the coals.'[3] Though our friend Quintus Hortensius[4] had ponds built at great expense near Bauli, I was at his villa often enough to know that it was his custom always to send to Puteoli to buy fish for dinner.[5] And it was not enough for him not to feed from his ponds—nay, he must feed his fish with his own hands; and he actually took more pains to keep his mullets from getting hungry than I do to keep my mules at Rosea from getting hungry, and indeed he furnished them nourishment in the way of both food and drink much more generously than I do in caring for my donkeys. For I keep my very valuable asses with the help of a single stable-boy, a bit of barley, and water

exchange places with him in speaking, that he might go home to water with his own hands a plane tree he had planted; that he wept over the death of a favourite *muraena*, etc. Cicero has frequent sneers at those whom he calls *piscinarios*; as, *e.g.*, *Ad Att.*, II, 1, 7: "who think they can touch the sky with their finger, if they have barbed mullets in their ponds which are tame enough to come when called."

[5] This would be three or four miles.

asinos; Hortensius primum qui ministrarent piscatores habebat complures, et ei pisciculos minutos aggerebant frequenter, qui a maioribus absumerentur. 7 Praeterea salsamentorum in eas piscinas emptum coiciebat, cum mare turbaret ac per tempestatem macellum piscinarum obsonium praeberet neque everriculo in litus educere possent vivam saginam, plebeiae cenae piscis. Celerius voluntate Hortensi ex equili educeres redarias, ut tibi haberes, mulas, 8 quam e piscina barbatum mullum. Atque, ille inquit, non minor cura erat eius de aegrotis piscibus, quam de minus valentibus servis. Itaque minus laborabat ne servos aeger aquam frigidam, quam ut recentem biberent sui pisces. Etenim hac incuria laborare aiebat M. Lucullum ac piscinas eius despiciebat, quod aestuaria idonea non haberet, ac reside aqua in locis pestilentibus habitarent pisces eius; 9 contra ad Neapolim L. Lucullum, posteaquam perfodisset montem ac maritumum flumen immisisset in piscinas, qui reciproce fluerent ipsae, Neptuno non cedere de piscatu. Factum esse enim ut amicos pisces suos videatur propter aestus eduxisse in loca frigidiora, ut Apuli solent pecuarii facere, qui per calles in montes Sabinos pecus ducunt. In Baiano autem aedificans tanta ardebat cura, ut architecto permiserit vel ut suam pecuniam consumeret, dum-

---

[1] The two were brothers, and are often mentioned together. The ponds mentioned here were near Bauli, which lay between Baiae and Misenum, and remains may still be seen. Vedius Pollio, Vergil, Hortensius, and many others had villas in this neighbourhood.

[2] Cf. II, 1, 16, and II, 2, 9.

from the place; while Hortensius in the first place kept an army of fishermen to supply food, and they were continually heaping up minnows for the larger fish to eat. Besides, he used to buy salted fish and throw them into the ponds when the sea was disturbed and on account of bad weather this source of supply of the ponds failed to furnish food, and the live food—the fish which supplies the people with supper—could not be brought ashore with the net." "You could more easily get Hortensius's consent to take the carriage mules (*mulas*) from his travelling-carriage and keep them for your own," said I, "than take a barbed mullet (*mullum*) from his pond." "And," he continued, "he was no less disturbed over his sick fish than he was over his ailing slaves. And so he was less careful to see that a sick slave did not drink cold water than that his fish should have fresh water to drink. In fact he used to say that Marcus Lucullus[1] suffered from carelessness in this respect, and he looked down on his ponds because they did not have suitable tidal-basins, and so, as the water became stagnant, his fish lived in unwholesome quarters; while, on the other hand, after Lucius Lucullus[1] had cut through a mountain near Naples and let a stream of sea-water into his ponds, so that they ebbed and flowed, he had no need to yield to Neptune himself in the matter of fishing—for he seemed, because of the hot weather, to have led his beloved fish into cooler places, just as the Apulian shepherds are wont to do when they lead their flocks along the cattle-trails into the Sabine hills.[2] But while he was building near Baiae he became so enthusiastic that he allowed the architect to spend money as if it were his own, provided he would

modo perduceret specus e piscinis in mare obiecta mole, qua aestus bis cotidie ab exorta luna ad proximam novam introire ac redire rursus in mare posset ac refrigerare piscinas.

10 Nos haec. At strepitus ab dextra et cum lata candidatus noster designatus aedilis in villam. Cui nos occedimus et gratulati in Capitolium persequimur. Ille inde endo suam domum, nos nostram, o Pinni noster, sermone de pastione villatica summatim hoc, quem exposui, habito.

---

[1] The whole statement is surprising, as it is well known that the Mediterranean is practically tideless. But Columella, in the corresponding passage (VIII, 17), uses language which also seems to assume a tide. Livy, in his description of the capture of New Carthage (XXVI, 45, 8), says that the swamp, because the water was carried out naturally by the tide and also by a brisk wind blowing in the same direction, was at

run a tunnel from his ponds into the sea and throw up a mole, so that the tide might run into the pond and back to the sea twice a day from the beginning of the moon until the next new moon, and cool off the ponds."[1]

So far we. Then a noise on the right, and our candidate, as aedile-elect, came into the villa wearing the broad stripe.[2] We approach and congratulate him and escort him to the Capitoline. Thence, he to his home, we to ours, my dear Pinnius, after having had the conversation on the husbandry of the villa, the substance of which I have given you.

some places waist-deep and at others knee-deep. The United States Hydrographic Office reports that the mean tidal range at Naples is 0·8 of a foot; the spring tide range (twice a month) is 1·0 foot.

[2] *i.e.* the *toga praetexta*, which had a broad border of purple. As aedile, he had a right to wear the official robe.

GLOSSARY OF TERMS

MEASURES of capacity, weights, and monetary units are translated into approximate U.S. and British equivalents, in most cases slightly smaller than the former and a trifle larger than the latter.

Acetabulum: = 1½ *cyathi*, ⅛ pint, liquid or dry measure.
Amphora: = 2 *urnae*, 6 gallons, liquid measure.
Concha: a small, shell-shaped vessel.
Congius: = 6 *sextarii*, 3 quarts, liquid measure.
Cotyla: a small vessel with capacity of 1 *hemina*, ½ pint.
Culleus: = 20 *amphorae*, 120 gallons, liquid measure.
Cyathus: = 1/12 pint, liquid or dry measure.
Denarius: = 4 *sestertii*, about 16 cents or 8*d*.
Eugeneum: cf. Gk. "well-born." A kind of grape said by Pliny (*N.H.*, XIV. 4) to have originated in Sicily.
Greek block: an ancient mechanical lifting device of pulleys and ropes, resembling our "block and tackle" mounted on uprights and worked by ropes and windlasses. The contrivance is described by Vitruvius (*De Architectura*, X, 2–5) alone of classical writers, though mentioned elsewhere. See also Hugo Blümner, *Technologie und Terminologie der Gewerbe und Künste bei Griechen und Romern*, Vol. III, p. 112, with drawings.
Hemina: = 2 *quartarii*, ½ pint, liquid or dry measure.
Iugerum: an area of 28,800 square feet, approximately ⅝ acre.
Iugum vinarium: apparently in Cato, 11, 2, a yoke from which wine-buckets were suspended for transportation; perhaps elsewhere some sort of framework for the support of young vines. Such a *iugum* is often mentioned in connection with vine-growing, and is described by Columella, IV, 17, 19, 22; cf. Varro, I, 8, 1.
Libella: a silver coin, worth about 1/10 denarius.
Mina: a Greek weight = 100 Attic drachmas, slightly less than 1 pound.
Modius: = 1 peck, dry measure.
Orcite: a kind of olive, whose Greek name (cf. ὄρχις) suggests its shape—oblong.
Posea: a kind of olive, valued for its oil.
Quadrantal: = 1 *amphora*.
Sestertius: = 2½ *asses*, about 4 cents or 2*d*.
Sextarius: = 2 *heminae*, 1 pint, liquid or dry measure.
Triobolus: as a weight = ½ drachma.
Urna: = 4 *congii*, 3 gallons, liquid measure.
Victoriatus: a silver coin, worth about ½ *denarius*, 8 cents or 4*d*., in the time of Varro.

INDEX TO CATO

The references are to pages in the English translation.

INDEX TO CATO

INDEX TO VARRO

The references are to pages in the English translation

INDEX TO VARRO

INDEX TO VARRO

D

E

INDEX TO VARRO

INDEX TO VARRO

T

INDEX TO VARRO